# Bivariate Integer-Valued Time Series Models

This book proposes some novel models based on the autoregressive and moving average structures under various distributional assumptions of the innovation series for analysing non-stationary bivariate time series of counts.

Time series of count responses are recorded for different correlated variables which may be marginally dispersed relative to their means, may exhibit different levels of dispersion and may be commonly influenced by one or more dynamic explanatory variables. Analysis of such type of bivariate time series data is quite challenging and the challenge mounts up further if these time series are non-stationary. This book proposes some bivariate models that allow for different levels of dispersion as well as non-stationarity. Specifically, BINAR(1) and BINMA(1) models under Poisson, NB and COM-Poisson innovations are constructed and tested. Another important contribution of this book is in developing a novel estimation procedure for estimating the parameters of the proposed BINAR(1) and BINMA(1) models. Hence, a new estimation approach based on the GQL is proposed. Monte-Carlo simulations are implemented to assess the performance of the GQL. In some simple cases of stationarity, we also compare the GQL with the other estimation techniques such as CMLE and FGLS.

This book is a useful resource for undergraduate students, postgraduate students, researchers and academics in the field of time series models.

# Bivariate Integer-Valued Time Series Models

## Bivariate Models

Sunecher Yuvraj, Mamode Khan Naushad
and Vandna Jowaheer

CRC Press
Taylor & Francis Group
Boca Raton London New York

CRC Press is an imprint of the
Taylor & Francis Group, an **informa** business
A CHAPMAN & HALL BOOK

First edition published 2025
by CRC Press
2385 NW Executive Center Drive, Suite 320, Boca Raton FL 33431

and by CRC Press
4 Park Square, Milton Park, Abingdon, Oxon, OX14 4RN

*CRC Press is an imprint of Taylor & Francis Group, LLC*



ISBN: 978-1-032-98767-5 (hbk)
ISBN: 978-1-032-98765-1 (pbk)
ISBN: 978-1-003-60043-5 (ebk)

DOI: 10.1201/9781003600435

Typeset in CMR10 font
by KnowledgeWorks Global Ltd.

# Contents

# Preface

This book proposes some novel models based on the autoregressive and moving average structures under various distributional assumptions of the innovation series for analyzing non-stationary bivariate time series of counts.

In various studies, time series of count responses are recorded for different correlated variables which may be marginally dispersed relative to their means and different series may exhibit different levels of dispersion. In addition, such series may be commonly influenced by one or more dynamic explanatory variables. Analysis of such type of bivariate time series data is quite challenging and the challenge mounts up further if these time series are non-stationary. The existing bivariate time series models are not suitable to analyze such type of series since these models have been developed under time-independent covariates and can only model count responses with same level of dispersion. Thus, a major limitation of these models is that they cannot cater for series that exhibit over- or under-dispersion mutually at different levels and fail to model non-stationary moment conditions, which in fact limit their applications in many practical real-life situations.

To overcome these limitations, this book proposes bivariate models that allow for different levels of dispersion as well as non-stationarity. Specifically, first-order bivariate integer-valued autoregressive (BINAR(1)) and first-order bivariate integer-valued moving average (BINMA(1)) models under Poisson, Negative Binomial (NB) and Conway-Maxwell-Poisson (COM-Poisson) innovations are constructed and tested. Models are constructed in two ways: first, assuming that the inter-relation between the series is induced by the innovation series only which ultimately yields a constrained BINAR(1) or BINMA(1) model under each distributional assumption and, second, in an unconstrained set-up, where the cross-correlation is influenced by relating the current series observation with previous-lagged observation of the other series and vice-versa. Non-stationary nature of the series is taken in account under each model.

Another important contribution of this book is in developing a novel estimation procedure for estimating the parameters of the proposed BINAR(1) and BINMA(1) models. The Conditional Maximum Likelihood Estimation (CMLE), Feasible Generalized Least Squares (FGLS) and Method of Moments (MoM) are some of the popular techniques used to estimate the model parameters in this domain where CMLE, as expected, yields the most efficient estimates. Nevertheless, the CMLE is computationally intensive to implement and requires numerical approximations. Moreover, in the complex unconstrained BINAR(1) and BINMA(1) non-stationary processes, it is difficult to identify the marginal distributions of the counting series which limits the application of likelihood-based methods. Hence, a new estimation approach based on the Generalized Quasi-Likelihood (GQL) is proposed. This approach requires only the correct specification of the mean score and the robust auto-covariance structure. Monte-Carlo simulations are implemented to assess the performance of the GQL. In some simple cases of stationarity, we also compare the GQL with the other estimation techniques such as CMLE and FGLS.

# Acknowledgments

There are few people who have been instrumental in the development of this book. We extend our deepest gratitude to all for their moral and technical support.

Our thanks are due to Professor Dimitris Karlis and Professor Miroslav Ristic for their valuable suggestions.

This work is also part of the Post-Doctoral Fellowship of Dr Yuvraj Sunecher titled "The Family of Bivariate Integer-Valued Autoregressive Models" with SaLLY-Statistical Learning Laboratory and the UFBA-CNPq research group Statistics and Data Science, University of Bahia, Brazil. We are thankful to Professor Paulo Jorge Canas Rodrigues.

# Authors

**Sunecher Yuvraj** is a Senior Lecturer of Statistics and Finance in the Department of Accounting, Finance and Economics at the School of Business, Management and Finance at the University of Technology Mauritius. He earned a PhD in statistics at the University of Mauritius in 2018.

Dr. Yuvraj's research interests span modeling time series of counts, statistical modeling, computational statistics and longitudinal data analysis. He has published extensively in recognized international journals published by Springer, Taylor & Francis, Wiley, DeGruyter and MDPI. Dr. Yuvraj has contributed significantly to the literature by developing novel univariate and multivariate integer-valued auto-regressive and integer-valued moving average time series models of different orders, with applications in various socioeconomic issues. He is also involved in MPhil/PhD supervision of research students and has served as Editor and Reviewer for many journals in the field of computational modeling, financial modeling and finance. Dr. Yuvraj has successfully completed numerous research projects locally with societally impactful research. He has actively contributed to the dissemination of knowledge to the research community, industry and policymakers. Dr. Yuvraj has provided his services to the Ministry of Gender Equality and Family Welfare as Consultant for the preparation of a ten-year strategic plan for children.

Dr. Yuvraj is the Vice-President of the Society of Statistics and Data Analytics in Mauritius and a Member of the International Statistical Institute, South African Statistical Association and the Australian Econometric Society. He has also served as Visiting Professor at several institutions, including Savitrabai Phule University of Pune, India; Ibn Zohr University of Agadir, Morocco; and Chhatrapati Shahu Institute of Business Education and Research (CSIBER) of Kohlapur, India.

**Naushad Mamode Khan** is an Associate Professor in Statistics in the Department of Economics and Statistics at the Faculty of Social Sciences and Humanities at the University of Mauritius. He is also currently the Head of the Department of Economics and Statistics at the University of Mauritius and also the President of the Society of Statistics and Data Analytics in Mauritius. He earned a PhD in statistics at the University of Mauritius in 2010, and his thesis focused on the development of statistical models for longitudinal count data.

Dr. Khan's research areas include statistical modeling, distribution theory, longitudinal count data analysis, computational statistics, spatial count data modeling and integer-valued time series modeling. His most popular research works in recent years encompasses the significant development in uni- and multivariate integer-valued auto-regressive time series models and its applications in various disciplines.

Dr. Khan was elected as the academic editor of *Plos One* and has been actively publishing in renowned rated statistics journals by Taylor & Francis, Wiley, DeGruyter, Springer, Plos One, MDPI and Elsevier. He has several successfully completed local and international PhD and post-doc supervisions mainly in the field of time series modeling and has also offered his services as a PhD examiner at the University of Pretoria, South Africa.

Dr. Khan has served as a Visiting Professor at several institutions, including the University of Shahrood, Iran; INSA Universite de Lyon, France; Savitrabai University of Pune, India; and the University of Agadir, Morocco. He was also nominated from 2015 to 2018 to the Statistics Board as the representative of the Vice-Chancellor of the University of Mauritius at Statistics Mauritius, under the aegis of the Ministry of Finance and Economic Development in Mauritius.

**Vandna Jowaheer** is a Professor and Chair of Applied Statistics at the University of Mauritius, Mauritius. She is an elected Member of the International Statistical Institute and a Fellow of the African academic network. Dr. Jowaheer's research expertise lies in developing statistical models and methods which can be applied to analyze cross-sectional, multivariate, time-series and longitudinal data structures. She has published numerous papers in reputed international journals in the areas of theoretical and applied statistics.

# 1

# *Introduction*

## 1.1 Background

Bivariate time series are recorded in various studies, with an aim to understand the relationship between the co-movements of the two response variables over time. In case the joint distribution of the variables is normal, it is rather easy to estimate the correlation parameters and the regression effects of common explanatory variables on the responses which simplifies the prediction processes (Comte & Lieberman (2003), Ling & McAleer (2003), Lutkepohl (2005)).

However, in case of bivariate count responses, it is quite challenging to model the joint distribution of the response variables, to derive the estimates of the model parameters and to make predictions. Bivariate time series of counts are commonly observed, for example the number of stock transactions of two companies operating in the same line of business, the number of daytime and nighttime accidents in an area, the number of thefts and unemployed people in the city across years, the number of seizures along with the number of times drugs are taken per week and many others.

While the univariate time series of counts have been widely studied in literature (McKenzie (1985, 1986), Al-Osh & Alzaid (1987), McKenzie (1988), Joe (1996), Silva & Oliveira (2004), Kim & Park (2008), Zhang et al. (2009), Ristic et al. (2009), Borges et al. (2017)), the analysis of bivariate time series of counts has not been addressed adequately. The complexity in modeling and especially in estimating parameters of such bivariate series is affected not only by the interdependence of the two series but also by various time-independent or time-dependent explanatory variables that commonly influence both series, while inducing stationary or non-stationary correlations in the counting series. The next sections underpin the literature related to the univariate first-order integer-valued autoregressive (INAR(1)) and first-order integer-valued moving average (INMA(1)) models and their extension to the bivariate INAR(1) (BINAR(1)) and bivariate INMA(1) (BINMA(1)) models. The existing estimation methods are also discussed.

DOI: 10.1201/9781003600435-1

## 1.2 INAR(1) and BINAR(1) Models

### 1.2.1 INAR(1) Observation-Driven Models

Time series of counts are usually modeled using parameter-driven (PD) or observation-driven (OD) approaches as illustrated in Cox (1981). In the PD approaches, the serial-correlations are induced by some correlated randomly distributed effects (Zeger (1988), Shephard (1995), McDonald & Zucchini (1997), Jung & Tremayne (2003), FruHwirth-Schnatter & Wagner (2006), Ravishanker et al. (2014)) while in the OD approach, the time correlations are induced through the previous-lagged observations (Engle & Bollerslev (1986), McKenzie (1986, 1988), Zeger & Qaqish (1988), Alzaid & Al-Osh (1990), Jowaheer & Sutradhar (2002), Sutradhar (2003), Mamode Khan & Jowaheer (2013)). Under both approaches, time-varying covariates may be specified (Brannas (1995)), but the estimation of the regression parameters in the PD models is quite laborious since the specification of the likelihood function in such models is cumbersome while the construction of the likelihood function is more parsimonious in OD models (Chan & Ledolter (1995), Durbin & Koopman (2000)).

The simplest family of OD INAR(1) models developed initially by McKenzie (1985) and further extended by Al-Osh & Alzaid (1987) and McKenzie (1988) is specified as

$$Y_t = \rho \circ Y_{t-1} + R_t \quad \text{for} \quad t = 1, \dots, T, \quad Y_t \geq 0, \tag{1.1}$$

where '$\circ$' indicates the binomial thinning operator such that $\rho \circ Y_{t-1} = \sum_{j=1}^{Y_{t-1}} b_j(\rho) = z_t$ where $b_j(\rho)$ is the Bernoulli random variable with probability $\rho$, where $\rho \in [0,1]$ or simply $\{\rho \circ Y_{t-1} | Y_{t-1}\} \sim \text{Binomial}(Y_{t-1}, \rho)$ (Steutel & Van Harn (1979)) while $\{R_t\}$ is the corresponding innovation term at the $t^{th}$ time point independent of $\{Y_{t-1}\}$ such that $\text{Corr}(Y_{t-1}, R_t) = 0$ for any $t = 1, 2, \dots, T$. Alternatively, $\{\rho \circ Y_{t-1}\}$ may follow some other discrete distributions other than binomial distribution such as in the class of OD INAR(1) models introduced by Ristic et al. (2009), Nastic et al. (2016a) and Borges et al. (2017) where the conditional random variable $\{\rho \circ Y_{t-1} | Y_{t-1}\}$ was Negative Binomial (NB) or some generalized NB under the assumption that the counting series were Geometric and the innovations were mixed Geometric.

On the other hand, in the modeling of non-stationary INAR(1) series, several strategies have been used by researchers to consider non-stationarity: Some authors such as Brannas (1995), Jowaheer & Sutradhar (2005), Brijs et al. (2008), Sutradhar (2008) and Enciso-Mora et al. (2009) have reformulated the classical OD process of McKenzie (1985) by considering the non-stationarity through time-dependent covariates in the link predictor specification of the model. Nastic et al. (2016a) proposed modeling the non-stationary INAR(1) by assuming that the marginal means of the counting series vary with changes in the environmental conditions wherein these count responses are exposed to. These authors considered the counting series to follow the Geometric marginals based on the NB thinning process. However, in such INAR(1) process, some efforts are needed to determine the sequences of random environmental variates. On the other hand, an alternative strategy adopted by Kim & Park (2008) to treat non-stationarity was to simply remove the non-stationarity by using differencing together with the signed binomial thinning operator in order to cater for negative count observations, but such a method might influence the interpretation and significance of parameter estimates.

INAR(1) models have also been constructed under different innovation distributions notably Zero-Inflated Poisson (Jazi et al. (2012)), Compound Poisson (Schweer & Wei$\beta$ (2014)), Katz family (Kim & Lee (2017)) and Poisson-Lindley (Guikema & Coffelt (2009)), but all these models, so far, have been developed only under stationary moment conditions.

### 1.2.2 Extension to BINAR(1) Models

In analyzing BINAR(1) series, Pedeli & Karlis (2009, 2011) originally developed a constrained BINAR(1) model by extending the classical INAR(1) model (1.1) given by

$$Y_t^{[1]} = \rho_{11} \circ Y_{t-1}^{[1]} + R_t^{[1]} \tag{1.2}$$

$$Y_t^{[2]} = \rho_{22} \circ Y_{t-1}^{[2]} + R_t^{[2]} \tag{1.3}$$

or in matrix representation as

$$\begin{pmatrix} Y_t^{[1]} \\ Y_t^{[2]} \end{pmatrix} = \begin{pmatrix} \rho_{11} & 0 \\ 0 & \rho_{22} \end{pmatrix} \circ \begin{pmatrix} Y_{t-1}^{[1]} \\ Y_{t-1}^{[2]} \end{pmatrix} + \begin{pmatrix} R_t^{[1]} \\ R_t^{[2]} \end{pmatrix}. \tag{1.4}$$

In the above model, in addition to the serial-correlation structure in Equation (1.4), the two series $\{Y_t^{[1]}, Y_t^{[2]}\}$ are cross-correlated wherein this cross-correlation is induced by the correlated innovations $\{R_t^{[1]}, R_t^{[2]}\}$ such that

$$\text{Corr}(R_t^{[1]}, R_{t'}^{[2]}) = \begin{cases} \kappa_{12,t} & t = t', \\ 0 & t \neq t'. \end{cases} \tag{1.5}$$

In this model, the pair of innovations $\{R_t^{[1]}, R_t^{[2]}\}$ is first assumed to follow the bivariate Poisson distribution (Kocherlakota & Kocherlakota (2001)) with a time-independent cross-correlation $\kappa_{12,1}$ and in another instance, the bivariate NB (BVNB) distribution under Marshall & Olkin (1990) representation but under same level of over-dispersion.

However, the above model presents some limitations. This BINAR(1) model is based on the assumption that the matrix of the thinning coefficients is diagonal, which implies that the correlation between $\{R_t^{[1]}\}$ and $\{R_t^{[2]}\}$ is the only source of dependence between the bivariate series $\{Y_t^{[1]}, Y_t^{[2]}\}$. However, this assumption of diagonality of correlation may not be valid in some cases. For example, consider the joint modeling of the number of intra-day transactions of two correlated stocks. The current number of intra-day transactions of a specific stock is serially dependent on the number of intra-day transactions of this stock at the previous time point and also with the number of intra-day transactions of the other correlated stock at the previous time point, which leads to a second form of dependence. In this context, Pedeli & Karlis (2013a) proposed an extension of the BINAR(1) model (1.4) to a full or unconstrained BINAR(1) model:

$$Y_t^{[1]} = \rho_{11} \circ Y_{t-1}^{[1]} + \rho_{12} \circ Y_{t-1}^{[2]} + R_t^{[1]} \tag{1.6}$$

$$Y_t^{[2]} = \rho_{21} \circ Y_{t-1}^{[1]} + \rho_{22} \circ Y_{t-1}^{[2]} + R_t^{[2]} \tag{1.7}$$

or alternatively these two equations may be represented in matrix form as

$$\begin{pmatrix} Y_t^{[1]} \\ Y_t^{[2]} \end{pmatrix} = \begin{pmatrix} \rho_{11} & \rho_{12} \\ \rho_{21} & \rho_{22} \end{pmatrix} \circ \begin{pmatrix} Y_{t-1}^{[1]} \\ Y_{t-1}^{[2]} \end{pmatrix} + \begin{pmatrix} R_t^{[1]} \\ R_t^{[2]} \end{pmatrix}, \tag{1.8}$$

where $\rho_{ij}\circ$ are mutually independent and $\{\rho_{ij} \circ Y_{t-1}^{[j]} | Y_{t-1}^{[j]}\} \sim$ Binomial $(Y_{t-1}^{[j]}, \rho_{ij})$ (Steutel & Van Harn (1979)) for $i, j \in \{1, 2\}$, with similar properties of the binomial thinning operation as discussed in Subsection 1.2.1. This full BINAR(1) model fulfills the characteristic that each counting series is related with its previous counter-series lagged observation and also induces an extra cross-correlation between the series in addition to the correlated innovation terms. This two-way cross-correlation structure becomes complex to some extent in modeling the full BINAR(1) process and requires some multivariate thinning properties developed by Franke & Rao (1995) and Latour (1997) to derive its marginal and joint moments. Likewise, Pedeli & Karlis (2013a) assumed that the pair $\{R_t^{[1]}, R_t^{[2]}\}$ follows the bivariate Poisson and showed that the marginal counting series $\{Y_t^{[1]}\}$ and $\{Y_t^{[2]}\}$ follows the 4-parameter Hermite distribution. An important merit of the full BINAR(1) model is that it produces better Akaike Information Criterion (AIC) than the constrained BINAR(1) model (Pedeli & Karlis (2013a)). However, these authors drew attention that numerical maximization was used to optimize the conditional joint likelihood function $\{Y_t^{[1]}, Y_t^{[2]} | y_{t-1}^{[1]}, y_{t-1}^{[2]}\}$ since the expression was rather cumbersome. Moreover, under $\{R_t^{[1]}, R_t^{[2]}\}$ following the BVNB distribution, these authors pointed out that the marginal properties of $\{Y_t^{[1]}\}$ and $\{Y_t^{[2]}\}$ are hard to derive, thus making the construction of the conditional joint likelihood more difficult. Besides, under the bivariate Poisson innovation series and the Hermite counting series, only low or moderate over-dispersion is captured, thus making the model restricted for high over-dispersed series (Pedeli & Karlis (2013a)).

Another simple version of the full BINAR(1) model was proposed by Ristic et al. (2012) and Nastic et al. (2016b) where these authors assumed that the cross-correlation may be due to the relationship between the current series observation with the previous-lagged observation of the other series with the pair of innovations held independent. Ristic et al. (2012) and Nastic et al. (2016b) developed this new version of the full BINAR(1) model with Geometric marginal based on the NB thinning mechanism which indicate that this model was only suited for modeling series of restricted level of over-dispersion. Interestingly, Nastic et al. (2016b) showed that this novel full BINAR(1) model provides better AICs and Root Mean Square Error (RMSE) than the constrained BINAR(1) with NB innovations (Pedeli & Karlis (2009, 2011)) and full BINAR(1) with Poisson innovations by Pedeli & Karlis (2013a). Popovic et al. (2016) re-considered the model introduced by Ristic et al. (2012) with Geometric counting series using the binomial thinning operator. Under these assumptions, Popovic et al. (2016) re-derived new distributional parameters of the innovations and showed that the proposed model yields better AIC and RMSE than the constrained models of Pedeli & Karlis (2009, 2011). In another paper, Popovic (2016) proposed a novel constrained BINAR(1) model with the thinning coefficient randomly distributed as in Zheng et al. (2007). Similarly, this new BINAR(1) model as well has shown to provide better AIC and RMSE than the constrained and full BINAR(1) models by Pedeli & Karlis (2009, 2011, 2013a). However, up to this extent, these models were developed only under stationary moment conditions. Overall, such model developments contribute mostly to the modeling of over-dispersed real-life series but can only cater for situations of same level of over-dispersion of relatively low margin.

## 1.3 INMA(1) and BINMA(1) Models

### 1.3.1 INMA(1) Observation-Driven Models

Al-Osh & Alzaid (1988a) and McKenzie (1988) proposed the first INMA(1) model where the time series observation $Y_t$ is related with unrelated innovation terms $R_{t-1}$ and $R_t$ such that

$$Y_t = \rho \circ R_{t-1} + R_t, \tag{1.9}$$

where '$\circ$' is the binomial thinning operator with similar properties as in the INAR(1) model, that is, $\{\rho \circ R_{t-1} | R_{t-1}\} \sim \text{Binomial}(R_{t-1}, \rho)$. In addition, $\text{Corr}(Y_t, R_{t-1}) = 0$ since there is no dependence between $\{Y_t\}$ and $\{R_{t'}\}$ for $t \neq t', t = 1, \ldots, T$. In the initial phase, these authors developed the INMA(1) model and then extended to INMA model of order q (INMA(q)) under stationary moment conditions with Poisson innovation. Brannas & Hall (2001) extended the classical INMA models under different covariance structure representations. In order to estimate the effects of explanatory variables in INMA models, Brannas & Lonnbark (2006) and Brannas & Quoreshi (2010) introduced time-dependent covariates specification in the link predictor of the innovation terms while Jowaheer & Sutradhar (2005) considered the influence of the time-dependent covariates through the link function of the counting series. Under both cases, the resulting INMA process was non-stationary by considering time-variant covariates.

### 1.3.2 Extension to BINMA Models

The extension to BINMA model was undertaken by Quoreshi (2006a) and later the same author considered the multivariate INMA model (Quoreshi (2008)). Under both model developments, the innovation series were allowed marginally to follow some generalized Poisson moment structures without clearly specifying the joint distribution of the innovations under the dispersion cases. So far, Quoreshi (2006a) worked on a BINMA model of order $q_1$ and $q_2$ (BINMA($q_1$,$q_2$)) where

$$Y_t^{[1]} = R_t^{[1]} + \rho_{11} \circ R_{t-1}^{[1]} + \ldots + \rho_{1q_1} \circ R_{t-q_1}^{[1]} \tag{1.10}$$

$$Y_t^{[2]} = R_t^{[2]} + \rho_{21} \circ R_{t-1}^{[2]} + \ldots + \rho_{2q_2} \circ R_{t-q_2}^{[2]}. \tag{1.11}$$

From the above model specification, the cross-correlation between $Y_t^{[1]}$ and $Y_t^{[2]}$ is caused by the pairwise related innovations $\{R_t^{[1]}, R_t^{[2]}\}$ at the same time point such that $\text{Cov}(R_t^{[1]}, R_{t'}^{[2]}) = 0$ for $t \neq t'$ and Quoreshi (2006a) additionally assumed that the $\text{Cov}(R_t^{[1]}, R_t^{[2]})$ are constant over time. To consider the effects of the explanatory variables, Quoreshi (2006a, 2008) introduced the time-dependent covariates through the link predictor of the innovation series as in the INMA(1) model. The above BINMA model has been applied under different orders to analyze the real-life stock series of two companies quoted on the Stockholm Stock Exchange, Ericsson B and AstraZeneca.

## 1.4 Inference Procedures and Prediction in Time Series Models

### 1.4.1 Conditional Maximum Likelihood Estimation (CMLE)

CMLE is one of the most popular estimation approach used in INAR(1) models (Zheng et al. (2007), Zheng & Basawa (2008), Borges et al. (2017)). Based on the thinning and convolution properties, the conditional likelihood is rather easily constructed under the stationary conditions. In the BINAR(1) models (Pedeli & Karlis (2009, 2011, 2013a)), the CMLE equation was constructed by extending the conditional likelihood formulation to accommodate the bivariate Poisson innovation likelihood function. In Pedeli & Karlis (2009), the details pertaining to solving the CMLE including the Hessian entries are provided. However, we could remark that these entries are hugely cumbersome. In fact, these authors also reported same in their paper (Pedeli & Karlis (2013c)) and suggested an alternative composite likelihood approach but this latter approach still involves some intense operations. Similar complications were noted in Scotto et al. (2014) and Boudreault & Charpentier (2011). In addition, it is noteworthy that these complications occurred under stationary moment conditions and therefore, further computational difficulties are expected under the non-stationary bivariate processes. In addition, under non-stationary conditions, the derivation of marginal and joint probability generating function (pgf) is not always feasible, as illustrated by Nastic et al. (2016a).

### 1.4.2 Generalized Method of Moments (GMM)

This is a common alternative estimation technique popularized by Hansen (1982) to the likelihood-based approach especially when the formulation of the likelihood function is cumbersome. GMM is constructed from combining the various moment conditions to yield an objective function which are thereon optimized with respect to the parameters of interest. Some extensions of the GMM approach were made in longitudinal count data analysis by Qu et al. (2000) and Qu & Lindsay (2003), where these authors proposed extended GMM approach: Quadratic Inference Function and Adaptive Estimating Equations, respectively. Under these versions, the auto-covariance computation is carried out using the empirical covariance estimates. The GMM approach was also used by Pedeli & Karlis (2009, 2013b) in the BINAR(1) model where it was noted through the simulation experiment that the bias in the estimation of the mean and cross-covariance parameters was significantly higher compared to the CMLE approach.

### 1.4.3 Generalized Estimating Equations (GEE)

GEE is a simpler approach introduced by Zeger (1988) to analyze univariate time series of counts. This approach does not require the specification of the joint distribution or full likelihood but only the moments of the distribution and a working correlation structure are required. The GEE is specifically used to estimate the regression parameters while the autocorrelation parameters in the working structure are estimated by the Method of Moments (MoM). However, this approach suffers from major pitfalls as discussed by Crowder (1995) and Sutradhar & Das (1999). Crowder (1995) showed that in some simple cases of misspecified correlation structures, estimates of the "working" correlation parameters might not exist. By the same token, Sutradhar & Das (1999) demonstrated that in some simple cases, independence working structure-based GEE approach provides more efficient estimates than working correlation structures based-GEE approach.

### 1.4.4 Generalized Quasi-Likelihood (GQL)

To overcome the above limitations in the GEE approach, Sutradhar (2003) developed a GQL approach based on a robust autocorrelation structure which uniformly provides more efficient regression estimates as compared to GEE (Sutradhar et al. (2014)). The GQL method is based on an extension of the classical Wedderburn (1974) Quasi-Likelihood (QL) equation where only the correct specification of the mean and score function is required. The GQL approach has been developed to estimate parameters in univariate or longitudinal settings (Jowaheer & Sutradhar (2002), Sutradhar (2003), Sutradhar et al. (2014)), where it has shown to yield asymptotically equally efficient estimates as the likelihood techniques and in particular, at lesser computational costs (Sutradhar et al. (2014)). The GQL approach has also been developed to analyze non-stationary time series of counts in the univariate context and has proved to be an optimal and computationally feasible estimation method (Jowaheer & Sutradhar (2005), Mallick & Sutradhar (2008)). Note importantly that GQL has also shown to provide more efficient estimates than the empirical-based covariance entries GMM estimating equations (Jowaheer & Sutradhar (2009), Mamode Khan & Jowaheer (2013)).

### 1.4.5 Feasible Generalized Least Squares (FGLS)

The FGLS approach is constructed using empirical moment conditions that focuses on the unknown parameters of the conditional mean function. Since FGLS estimators have weighting based on individual observations, we normally expect the FGLS approach to perform better than GMM (Brannas (1995)). The first step of the FGLS approach is to create moment conditions based on predictor error, while in the second step, normal equations based on the conditional variance prediction error are used. Following these two steps, we obtain the criterion function of the FGLS and thereon it is minimized with respect to the parameters of interest. Quoreshi (2006a, 2008) applied the FGLS approach where the parameters of interest were jointly estimated using moment estimating equations based on the residual specification. It should be noted that in Quoreshi's work, the FGLS provided better estimates than GMM. However, one limitation of the FGLS approach is that it requires that the standardized residuals and the squared standardized residuals to be serially uncorrelated, which may not be the case in some real-life situations (Quoreshi (2006a, 2008), Brannas & Quoreshi (2010)).

## 1.5 Motivation

In the models described in Section 1.2, it is clear that so far, the emphasis has been laid on developing BINAR(1) models under stationary moment conditions only. Furthermore, under the existing BINAR(1) and BINMA(1) models, the distribution of the counting or innovation series has not been properly specified under the specific cases of under-dispersion or mixed-level of dispersion. In reality, there is a strong need to develop bivariate time series models that account for non-stationary serial- and cross-correlation influenced by time-dependent covariates, with the flexibility of modeling count series with different levels of dispersion.

Another pertinent point is that the estimation of parameters cannot be solely handled by the CMLE approach because this method is computationally intensive under the stationarity context and under non-stationarity, as argued by Nastic et al. (2016a), it is very tedious to formulate the joint pgf in spite of using the self-decomposability properties by Steutel & Van Harn (1986). An alternative approach for estimation of model parameters can be the FGLS, but it requires the elimination of serial correlation in the standardized residuals, which may not be feasible (Quoreshi (2006a, 2008), Brannas & Quoreshi (2010)). Although FGLS method have been able to provide efficient estimates in certain cases (Quoreshi (2006a, 2008), Brannas & Quoreshi (2010)), yet it will be of interest to formulate the GQL approach for bivariate time series models and compare it with existing estimation approaches in order to develop an estimation method that can yield efficient parameters as compared to the other methods of estimation and at relatively lower computational cost.

## 1.6 Aim and Objectives of the Book

Keeping in view the shortcomings of the existing models and estimation methods, the current book aims at developing models for bivariate time series of counts which can cater for non-stationarity, as well as different levels of dispersion. The major objectives are as follows:

1. To develop constrained and unconstrained BINAR(1) and BINMA(1) models with Poisson, NB and Conway-Maxwell-Poisson (COM-Poisson) innovations wherein the non-stationarity is considered through time-variant covariates which is introduced via the link predictor function.

2. To extend the univariate GQL developed by Sutradhar (2003) to a bivariate GQL to estimate the parameters of bivariate time series of counts and to compare these estimates with those obtained using CMLE and FGLS under simple stationary moment conditions in order to select a good estimation method for non-stationary case.

3. To assess the reliability of the performance of the proposed BINAR(1) and BINMA(1) models for stationary and non-stationary simulated

data and to design Monte-Carlo simulation algorithms to study the performance of proposed models under stationary and non-stationary data structures.

## 1.7 Organization of the Book

The subsequent chapters are organized as follows:

In Chapter 2, a constrained BINAR(1) model is introduced where the innovation series follow a bivariate Poisson distribution under some non-stationary distributional assumptions. The GQL estimation method is used to estimate the regression effects while the serial- and cross-correlation are estimated using a robust moment estimation technique. The application of the model and estimation method is made on simulated data. The GQL method is also compared with CMLE and FGLS approaches under the stationary moment conditions and their performance is compared through a simulation study. Under the non-stationary set-up, we propose the most feasible and optimal estimation method and its performance is assessed through a simulation study.

In Chapter 3, a constrained BINMA(1) model is developed where the innovation series follow the bivariate Poisson under time-varying moment assumptions. The GQL approach is used to estimate the regression effects and serial-correlation, while the cross-correlations are estimated by a robust MoM. The FGLS approach is formulated and compared through a simulation study in the stationary set-up. Under the non-stationary set-up, we use the more practical and optimal estimation method, which is assessed through a simulation study.

In Chapter 4, the constrained BINAR(1) model with Poisson innovations introduced in Chapter 2 is extended to an unconstrained structure with Poisson innovations under time-dependent moment conditions. The marginal and joint moments and covariances of the proposed model are derived under the distributional assumptions using the matrix multivariate thinning properties. Detail of the GQL estimating approach is provided.

In Chapter 5, the constrained BINMA(1) model with Poisson innovations introduced in Chapter 3 is extended to an unconstrained BINMA(1) model with Poisson innovations, by considering time-varying covariates. The marginal and joint moments are derived using the matrix multivariate binomial thinning mechanism. The GQL estimating equation is developed to estimate the regression effects and dependence parameters. A simulation study is conducted for different combinations of the model parameters to assess the performance of the GQL approach.

In Chapter 6, a constrained BINAR(1) model with the pair of innovations following a joint NB distribution that accommodates different over-dispersion indices is developed. The estimation of parameters is conducted using a GQL approach that operate in two phases. Monte-Carlo simulations are implemented to assess the performance of the proposed GQL under the wide range combinations of the model parameters.

In Chapter 7, a constrained BINMA(1) model with joint NB innovations is formulated that can cater for different levels of over-dispersion. The GQL approach is used to estimate the regression, over-dispersion and serial-correlation parameters. This is followed by a simulation study where the performance of the GQL approach is assessed.

In Chapter 8, the constrained BINAR(1) introduced in Chapter 6 is extended to an unconstrained BINAR(1) model with NB innovations under non-stationary moments. The estimation of the regression, over-dispersion and dependence parameters is conducted using a GQL approach and some lemmas are introduced to facilitate the computation of some higher-order moments.

In Chapter 9, the constrained BINMA(1) introduced in Chapter 7 is extended to an unconstrained BINMA(1) model with NB innovations under non-stationary moment conditions. The GQL approach is developed to estimate the regression, dependence and over-dispersion parameters, where some lemmas are used to compute some higher-order moments exactly. Monte-Carlo simulations are conducted to assess the feasibility of the GQL approach.

In Chapter 10, a constrained BINAR(1) model with COM-Poisson innovations is introduced. The regression and dependence effects are estimated using the GQL approach, while the serial- and cross-correlation parameters are estimated using some robust moment estimating equations. Simulation experiments are conducted to assess the performance of the estimation algorithms.

In Chapter 11, a constrained BINMA(1) model with COM-Poisson innovations under non-stationary moments is introduced. A trial experiment is conducted under the stationary set-up to investigate the validity of the proposed model to analyze equi-, under- or over-dispersed or mixed-dispersed data. The GQL approach is used to estimate the regression, dispersion and serial-correlation parameters, while the cross-correlation parameters are estimated using a robust moment estimating equation. A simulation study is carried out where equi-, over- and under-dispersed count time series are generated to assess the performance of the GQL approach.

In Chapter 12, the constrained BINAR(1) model introduced in Chapter 10 is extended to an unconstrained BINAR(1) model with COM-Poisson innovations under time-dependent moments that can accommodate different levels of over- and under-dispersion. The estimation of the model parameters is conducted via a GQL approach. We present some numerical experiments to illustrate the different scenarios of over-, equi- or under-dispersion or mixed-dispersion of the model.

In Chapter 13, the constrained BINMA(1) model introduced in Chapter 11 is extended to an unconstrained BINMA(1) model where the innovations are marginally COM-Poisson. A trial experiment is also presented to investigate the different scenarios under which the proposed model exhibit over-, equi- or under-dispersion or mixed-dispersion. A GQL approach is used to estimate the regression, dispersion and dependence parameters. Monte-Carlo experiments are implemented to assess the performance of the GQL approach.

Overall conclusion and some future directions are presented in Chapter 14.

# 2

# Constrained BINAR(1) Model with Correlated Poisson Innovations

## 2.1 Introduction

This chapter involves the modeling of the constrained non-stationary BINAR(1) process with inter-related Poisson innovations wherein the counting series are marginally Poisson with time-dependent moments induced by time-variant covariates. Constrained BINAR(1) model under bivariate Poisson related innovations has only been developed under stationary moment conditions (Pedeli & Karlis (2009, 2011), Karlis & Pedeli (2013)), but as discussed in Chapter 1, such bivariate time series model is not suitable to model real-life series with non-stationary conditions.

Secondly, this chapter considers the development of a novel estimation approach based on the QL procedures that do not require the knowledge of the conditional likelihood function. So far, CMLE has been used intensively to estimate the model parameters and has proved to provide more efficient estimates compared to other estimation methods (Pedeli & Karlis (2009, 2011)). However, this method involves heavy computations due to the complicated likelihood equations (Pedeli & Karlis (2013c)), and as illustrated by Nastic et al. (2016a), the derivation of the marginal and joint pgf is very difficult under the non-stationary set-up. Thus, we propose the GQL method of estimation in the bivariate setting to estimate the regression effects followed by a robust moment estimating equation to compute the serial-correlations of the series. GQL is based on a robust autocorrelation structure to represent the serial-correlations (Sutradhar & Das (1999)) and yields equally efficient estimates as CMLE (Sutradhar et al. (2014)). The bivariate dependence between the innovations is estimated using a robust moment estimating equation, similar to Sutradhar & Das (1999) and Pedeli & Karlis (2009, 2011). Hence, it is of substantive interest to develop GQL equations for BINAR(1) model. GQL equations are first developed for stationary bivariate time series with Poisson innovations and their performance is compared with CMLE and FGLS. The GQL approach is then extended to non-stationary set-up.

The plan of this chapter is as follows: In Section 2.2, a constrained BINAR(1) model under bivariate Poisson innovations is introduced and the marginal and joint moments and covariances are derived under the above

DOI: 10.1201/9781003600435-2

distributional assumptions. In Section 2.3, the GQL method of estimation is derived in the bivariate setting for stationary series and compared with CMLE and FGLS under the stationary moment conditions through a simulation study. In the same section, GQL is derived under the non-stationary moment conditions and a simulation study is performed where bivariate non-stationary Poisson data are generated under different serial-correlation and bivariate dependence values. The one-step ahead forecasting equations for the constrained BINAR(1) model with Poisson innovations are also developed. This chapter ends with the conclusion.

## 2.2 The Constrained BINAR(1) Model

The constrained BINAR(1) model is specified as

$$\begin{pmatrix} Y_t^{[1]} \\ Y_t^{[2]} \end{pmatrix} = \begin{pmatrix} \rho_{11} & 0 \\ 0 & \rho_{22} \end{pmatrix} \circ \begin{pmatrix} Y_{t-1}^{[1]} \\ Y_{t-1}^{[2]} \end{pmatrix} + \begin{pmatrix} R_t^{[1]} \\ R_t^{[2]} \end{pmatrix}, \tag{2.1}$$

where $\rho_{kk} \in [0,1]$ indicates the time correlation for the $k^{th}$ series such that $\{\rho_{kk} \circ Y_{t-1}^{[k]}\}$ indicates the binomial thinning operation (Steutel & Van Harn (1979), Aly & Bouzar (2005), Wei$\beta$ (2008b), Bourguignon & Vasconcellos (2015)), where

$$\{\rho_{kk} \circ Y_{t-1}^{[k]} | Y_{t-1}^{[k]}\} \sim \text{Binomial}(Y_{t-1}^{[k]}, \rho_{kk}), \quad \text{for} \quad k \in \{1,2\} \tag{2.2}$$

with the following properties:

1. $E(\rho_{kk} \circ Y_{t-1}^{[k]}) = \rho_{kk} E(Y_{t-1}^{[k]})$.
2. $\text{Var}(\rho_{kk} \circ Y_{t-1}^{[k]}) = \rho_{kk}(1-\rho_{kk})E(Y_{t-1}^{[k]}) + \rho_{kk}^2 \text{Var}(Y_{t-1}^{[k]})$.
3. $\text{Cov}(\rho_{11} \circ Y_{t-1}^{[1]}, \rho_{22} \circ Y_{t-1}^{[2]}) = \rho_{11}\rho_{22}\text{Cov}(Y_{t-1}^{[1]}, Y_{t-1}^{[2]})$.

In the above BINAR(1) model (2.1), $Y_{t-1}^{[k]} \sim \text{Poisson}(\mu_{t-1}^{[k]})$, where $\mu_t^{[k]} = \exp(\boldsymbol{x}_t' \boldsymbol{\beta}^{[k]})$. $\boldsymbol{x}_t = [x_{t1}, x_{t2}, \ldots, x_{tj}, \ldots, x_{tp}]'$ is a $(p \times 1)$ vector of covariates influencing both $Y_t^{[1]}$ and $Y_t^{[2]}$ with corresponding regression coefficients $\boldsymbol{\beta}^{[k]} = [\beta_1^{[k]}, \beta_2^{[k]}, \ldots, \beta_j^{[k]}, \ldots, \beta_p^{[k]}]'$ for $t = 1, 2, \ldots, T$ and $k \in \{1,2\}$.

The pair $\{R_t^{[1]}, R_t^{[2]}\}$ follows the bivariate Poisson distribution (Kocherlakota & Kocherlakota (2001)) such that

$$P_{(R_t^{[1]}, R_t^{[2]})}(u,v) = e^{-(\lambda_t^{[1]} + \lambda_t^{[2]} - \phi_t)} \frac{(\lambda_t^{[1]} - \phi_t)^u}{u!} \frac{(\lambda_t^{[2]} - \phi_t)^v}{v!}$$
$$\sum_{i=0}^{s} \left[ \binom{u}{i} \binom{v}{i} i! \left( \frac{\phi_t}{(\lambda_t^{[1]} - \phi_t)(\lambda_t^{[2]} - \phi_t)} \right)^i \right], \tag{2.3}$$

where $s = \min(u, v)$, $\lambda_t^{[1]}, \lambda_t^{[2]} > 0$ and $\phi_t \in [0, \min(\lambda_t^{[1]}, \lambda_t^{[2]})]$. $\phi_t$ denotes the covariance between the two innovation terms, such that

$$\text{Cov}(R_t^{[1]}, R_{t'}^{[2]}) = \begin{cases} \phi_t = \kappa_{12,t'}\sqrt{\lambda_t^{[1]}}\sqrt{\lambda_{t'}^{[2]}} & t = t', \\ 0 & t \neq t'. \end{cases} \tag{2.4}$$

where $\kappa_{12,t}$ denotes the cross-correlation coefficient between the innovation terms with $\kappa_{12,t} \in [0, 1]$. From the above joint distribution, $R_t^{[k]} \sim \text{Poisson}(\lambda_t^{[k]})$, where $\lambda_t^{[k]} = (\mu_t^{[k]} - \rho_{kk}\mu_{t-1}^{[k]})$ for $k \in \{1, 2\}$.

The random variables $Y_t^{[k]}$ and $R_u^{[l]}$ are independent for all $k, l \in \{1, 2\}$ and $t, u \in \mathbb{Z}$ for $t \neq u$. The relation between the counting series $Y_t^{[k]}$ and the innovation terms $R_t^{[k]}$ is thus summarized through the following lemma:

**Lemma 2.1.** *Suppose that the properties (1), (2) and (3) hold. Then,*

(a) $\text{Cov}(Y_t^{[k]}, R_t^{[k]}) = \text{Var}(R_t^{[k]})$ for $k \in \{1, 2\}$.

(b)
$$\text{Cov}(Y_t^{[k]}, R_{t+h}^{[j]}) = \begin{cases} \text{Cov}(R_t^{[k]}, R_t^{[j]}) & h = 0 \\ 0 & h \neq 0, \quad \text{for any} \quad k, j \in \{1, 2\}, k \neq j. \end{cases}$$

*Proof:*

(a)
$$\begin{aligned} \text{Cov}(Y_t^{[k]}, R_t^{[k]}) &= \text{Cov}(\rho_{kk} \circ Y_{t-1}^{[k]} + R_t^{[k]}, R_t^{[k]}) \\ &= \text{Cov}(\rho_{kk} \circ Y_{t-1}^{[k]}, R_t^{[k]}) + \text{Cov}(R_t^{[k]}, R_t^{[k]}) \\ &= \text{Var}(R_t^{[k]}). \end{aligned}$$

(b)
$$\begin{aligned} \text{Cov}(Y_t^{[k]}, R_{t+h}^{[j]}) &= \text{Cov}(\rho_{kk} \circ Y_{t-1}^{[k]} + R_t^{[k]}, R_{t+h}^{[j]}) \\ &= \text{Cov}(\rho_{kk} \circ Y_{t-1}^{[k]}, R_{t+h}^{[j]}) + \text{Cov}(R_t^{[k]}, R_{t+h}^{[j]}) \\ &= \text{Cov}(R_t^{[k]}, R_t^{[j]}) \quad \text{for} \quad h = 0. \end{aligned}$$

Based on the above conditions, it is proved that

$$\begin{aligned} E(Y_t^{[k]}) &= E_{Y_{t-1}^{[k]}} E[\rho_{kk} \circ Y_{t-1}^{[k]} + R_t^{[k]} | Y_{t-1}^{[k]}] \\ &= E_{Y_{t-1}^{[k]}} [\rho_{kk} Y_{t-1}^{[k]} + \mu_t^{[k]} - \rho_{kk}\mu_{t-1}^{[k]}] \\ &= \rho_{kk}\mu_{t-1}^{[k]} + \mu_t^{[k]} - \rho_{kk}\mu_{t-1}^{[k]} \\ &= \mu_t^{[k]} \end{aligned} \tag{2.5}$$

and

$$\begin{aligned}
\mathrm{Var}(Y_t^{[k]}) &= \mathrm{Var}[\rho_{kk} \circ Y_{t-1}^{[k]} + R_t^{[k]}] \\
&= E_{Y_{t-1}^{[k]}}[\mathrm{Var}(\rho_{kk} \circ Y_{t-1}^{[k]} | Y_{t-1}^{[k]})] + \mathrm{Var}_{Y_{t-1}^{[k]}}[E(\rho_{kk} \circ Y_{t-1}^{[k]} | Y_{t-1}^{[k]})] \\
&\quad + \mathrm{Var}[R_t^{[k]}] \\
&= E_{Y_{t-1}^{[k]}}[\rho_{kk}(1-\rho_{kk})Y_{t-1}^{[k]}] + \mathrm{Var}_{Y_{t-1}^{[k]}}[\rho_{kk} Y_{t-1}^{[k]}] + \mathrm{Var}[R_t^{[k]}] \\
&= \rho_{kk}(1-\rho_{kk})\mu_{t-1}^{[k]} + \rho_{kk}^2 \mu_{t-1}^{[k]} + \mu_t^{[k]} - \rho_{kk}\mu_{t-1}^{[k]} \\
&= \mu_t^{[k]}. \qquad (2.6)
\end{aligned}$$

Hence, $Y_t^{[k]} \sim \mathrm{Poisson}(\mu_t^{[k]})$.

The serial lag-$h$ correlation for the $k^{th}$ series is

$$\begin{aligned}
\mathrm{Cov}(Y_t^{[k]}, Y_{t+h}^{[k]}) &= \mathrm{Cov}[Y_t^{[k]}, (\rho_{kk} \circ Y_{t+h-1}^{[k]} + R_{t+h}^{[k]})] \\
&= \rho_{kk}\mathrm{Cov}[Y_t^{[k]}, Y_{t+h-1}^{[k]}] \\
&= \rho_{kk}\mathrm{Cov}[Y_t^{[k]}, (\rho_{kk} \circ Y_{t+h-2}^{[k]} + R_{t+h-1}^{[k]})] \\
&= \rho_{kk}^2\mathrm{Cov}[Y_t^{[k]}, Y_{t+h-2}^{[k]}] \\
&\vdots \\
&= \rho_{kk}^h\mathrm{Cov}[Y_t^{[k]}, Y_t^{[k]}] \\
&= \rho_{kk}^h \mu_t^{[k]} \qquad (2.7)
\end{aligned}$$

and hence

$$\mathrm{Corr}(Y_t^{[k]}, Y_{t+h}^{[k]}) = \rho_{kk}^h \sqrt{\frac{\mu_t^{[k]}}{\mu_{t+h}^{[k]}}}. \qquad (2.8)$$

As for the joint covariance $\mathrm{Cov}(Y_t^{[1]}, Y_t^{[2]})$, from Equation (2.1)

$$\begin{aligned}
\mathrm{Cov}(Y_t^{[1]}, Y_t^{[2]}) &= \mathrm{Cov}[(\rho_{11} \circ Y_{t-1}^{[1]} + R_t^{[1]}), (\rho_{22} \circ Y_{t-1}^{[2]} + R_t^{[2]})] \\
&= \rho_{11}\rho_{22}\mathrm{Cov}(Y_{t-1}^{[1]}, Y_{t-1}^{[2]}) + \mathrm{Cov}(R_t^{[1]}, R_t^{[2]}) \\
&= \rho_{11}\rho_{22}\mathrm{Cov}(Y_{t-1}^{[1]}, Y_{t-1}^{[2]}) \\
&\quad + \left[\kappa_{12,t}\sqrt{(\mu_t^{[1]} - \rho_{11}\mu_{t-1}^{[1]})}\sqrt{(\mu_t^{[2]} - \rho_{22}\mu_{t-1}^{[2]})}\right] \qquad (2.9)
\end{aligned}$$

and hence

$$\begin{aligned}
&\text{Corr}(Y_t^{[1]}, Y_t^{[2]}) \\
&= \frac{\rho_{11}\rho_{22}\sqrt{\mu_{t-1}^{[1]}}\sqrt{\mu_{t-1}^{[2]}}\text{Corr}(Y_{t-1}^{[1]}Y_{t-1}^{[2]}) + \kappa_{12,t}\sqrt{(\mu_t^{[1]} - \rho_{11}\mu_{t-1}^{[1]})}\sqrt{(\mu_t^{[2]} - \rho_{22}\mu_{t-1}^{[2]})}}{\sqrt{\mu_t^{[1]}}\sqrt{\mu_t^{[2]}}}.
\end{aligned} \tag{2.10}$$

As for the other off-diagonal cross-covariances,

$$\begin{aligned}
&\text{Cov}(Y_t^{[1]}, Y_{t+h}^{[2]}) \\
&= \text{Cov}[Y_t^{[1]}, (\rho_{22} \circ Y_{t+h-1}^{[2]} + R_{t+h}^{[2]})] \\
&= \rho_{22}\text{Cov}[Y_t^{[1]}, Y_{t+h-1}^{[2]}] \\
&= \rho_{22}\text{Cov}[Y_t^{[1]}, (\rho_{22} \circ Y_{t+h-2}^{[2]} + R_{t+h-1}^{[2]})] \\
&= \rho_{22}^2\text{Cov}[Y_t^{[1]}, Y_{t+h-2}^{[2]}] \\
&\vdots \\
&= \rho_{22}^h\text{Cov}[Y_t^{[1]}, Y_t^{[2]}] \\
&= \rho_{22}^h\left[\rho_{11}\rho_{22}\text{Cov}(Y_{t-1}^{[1]}, Y_{t-1}^{[2]}) + \kappa_{12,t}\sqrt{(\mu_t^{[1]} - \rho_{11}\mu_{t-1}^{[1]})}\sqrt{(\mu_t^{[2]} - \rho_{22}\mu_{t-1}^{[2]})}\right]
\end{aligned} \tag{2.11}$$

and

$$\begin{aligned}
&\text{Cov}(Y_{t+h}^{[1]}, Y_t^{[2]}) \\
&= \text{Cov}[Y_t^{[2]}, (\rho_{11} \circ Y_{t+h-1}^{[1]} + R_{t+h}^{[1]})] \\
&= \rho_{11}\text{Cov}[Y_t^{[2]}, Y_{t+h-1}^{[1]}] \\
&= \rho_{11}\text{Cov}[Y_t^{[2]}, (\rho_{11} \circ Y_{t+h-2}^{[1]} + R_{t+h-1}^{[1]})] \\
&= \rho_{11}^2\text{Cov}[Y_t^{[2]}, Y_{t+h-2}^{[1]}] \\
&\vdots \\
&= \rho_{11}^h\text{Cov}[Y_t^{[1]}, Y_t^{[2]}] \\
&= \rho_{11}^h\left[\rho_{11}\rho_{22}\text{Cov}(Y_{t-1}^{[1]}, Y_{t-1}^{[2]}) + \kappa_{12,t}\sqrt{(\mu_t^{[1]} - \rho_{11}\mu_{t-1}^{[1]})}\sqrt{(\mu_t^{[2]} - \rho_{22}\mu_{t-1}^{[2]})}\right].
\end{aligned} \tag{2.12}$$

Under the stationary moment assumption, where $\mu_t^{[k]} = \mu_1^{[k]}$,

$$\text{Corr}(Y_t^{[k]}, Y_{t+h}^{[k]}) = \rho_{kk}^h \tag{2.13}$$

for all same lag-$h$, whereas the cross-covariances at lag-$h$ reduce to

$$\text{Cov}(Y_{t+h}^{[1]}, Y_t^{[2]}) = \rho_{11}^h \text{Cov}(Y_t^{[1]}, Y_t^{[2]}) \text{ and } \text{Cov}(Y_t^{[1]}, Y_{t+h}^{[2]}) = \rho_{22}^h \text{Cov}(Y_t^{[1]}, Y_t^{[2]}), \tag{2.14}$$

where $\text{Cov}(Y_t^{[1]}, Y_t^{[2]})$ is given by

$$\text{Cov}(Y_t^{[1]}, Y_t^{[2]}) = \frac{\kappa_{12,1}\sqrt{(1-\rho_{11})\mu_1^{[1]}}\sqrt{(1-\rho_{22})\mu_1^{[2]}}}{(1-\rho_{11}\rho_{22})}. \tag{2.15}$$

## 2.3 Estimation of Parameters

In this section, the estimation methods for estimating regression, serial- and cross-dependence effects are first developed under stationary moments and then the most feasible method is developed under non-stationary moment conditions.

### 2.3.1 Stationary Conditions

In this subsection, the likelihood-based CMLE approach is developed. In addition, a more general non-likelihood-based FGLS approach is also developed. A GQL estimation approach is then formulated, which requires only the correct specification of the mean score and auto-covariance or a "robust" working auto-covariance structure.

#### 2.3.1.1 Likelihood-Based Approach

**Conditional Maximum Likelihood Estimation (CMLE)**

In order to construct the conditional likelihood, the marginal conditional probability for counting series $Y_t^{[1]}$ and $Y_t^{[2]}$ are expressed as:

$$f_1(y_t^{[1]} | y_{t-1}^{[1]}) = \begin{pmatrix} y_{t-1}^{[1]} \\ y_t^{[1]} \end{pmatrix} \rho_{11}^{y_t^{[1]}} (1-\rho_{11})^{y_{t-1}^{[1]} - y_t^{[1]}} \tag{2.16}$$

$$f_2(y_t^{[2]} | y_{t-1}^{[2]}) = \begin{pmatrix} y_{t-1}^{[2]} \\ y_t^{[2]} \end{pmatrix} \rho_{22}^{y_t^{[2]}} (1-\rho_{22})^{y_{t-1}^{[2]} - y_t^{[2]}}, \tag{2.17}$$

where the bivariate distribution of the innovation terms is of the form $f_3(u,v) = P_{(R_t^{[1]}, R_t^{[2]})}(u,v)$. Then, the conditional density is written as

$$
\begin{aligned}
&f((y_t^{[1]}, y_t^{[2]})|(y_{t-1}^{[1]}, y_{t-1}^{[2]}), \boldsymbol{\theta}) \\
&= \sum_u \sum_v f_1(y_t^{[1]} - u) f_2(y_t^{[2]} - v) f_3(u, v) \\
&= \sum_{u=0}^{g_1} \sum_{v=0}^{g_2} \begin{pmatrix} y_{t-1}^{[1]} \\ y_t^{[1]} - u \end{pmatrix} \rho_{11}^{y_t^{[1]} - u} (1 - \rho_{11})^{y_{t-1}^{[1]} - y_t^{[1]} + u} \\
&\times \begin{pmatrix} y_{t-1}^{[2]} \\ y_t^{[2]} - v \end{pmatrix} \rho_{22}^{y_t^{[2]} - v} (1 - \rho_{22})^{y_{t-1}^{[2]} - y_t^{[2]} + v} \\
&\times e^{-(\mu_1^{[1]} - \rho_{11}\mu_1^{[1]} + \mu_1^{[2]} - \rho_{22}\mu_1^{[2]} - \kappa_{12,1}\sqrt{\mu_1^{[1]} - \rho_{11}\mu_1^{[1]}}\sqrt{\mu_1^{[2]} - \rho_{22}\mu_1^{[2]}})} \\
&\sum_{m=0}^{\min(u,v)} \{[\mu_1^{[1]} - \rho_{11}\mu_1^{[1]} - \kappa_{12,1}\sqrt{\mu_1^{[1]} - \rho_{11}\mu_1^{[1]}}\sqrt{\mu_1^{[2]} - \rho_{22}\mu_1^{[2]}}]^{u-m} \\
&\times [\mu_1^{[2]} - \rho_{22}\mu_1^{[2]} - \kappa_{12,1}\sqrt{\mu_1^{[1]} - \rho_{11}\mu_1^{[1]}}\sqrt{\mu_1^{[2]} - \rho_{22}\mu_1^{[2]}}]^{v-m} \\
&\times [\kappa_{12,1}\sqrt{\mu_1^{[1]} - \rho_{11}\mu_1^{[1]}}\sqrt{\mu_1^{[2]} - \rho_{22}\mu_1^{[2]}}]^m\}/[(u-m)!(v-m)!m!],
\end{aligned} \tag{2.18}
$$

where $\boldsymbol{\theta} = [\rho_{11}, \rho_{22}, \beta_1^{[1]}, \beta_2^{[1]}, \beta_1^{[2]}, \beta_2^{[2]}, \kappa_{12,1}]$ is the vector of unknown parameters, $g_1 = \min(y_t^{[1]}, y_{t-1}^{[1]})$ and $g_2 = \min(y_t^{[2]}, y_{t-1}^{[2]})$.
The conditional likelihood function is thus given by

$$
L(\boldsymbol{\theta}|\boldsymbol{y}) = \prod_{t=1}^{T} f((y_t^{[1]}, y_t^{[2]})|(y_{t-1}^{[1]}, y_{t-1}^{[2]}), \boldsymbol{\theta}) \tag{2.19}
$$

for some initial value of $\boldsymbol{\theta}$ and by maximizing the conditional log-likelihood function

$$
\log[L(\boldsymbol{\theta}|\boldsymbol{y})] = \log\left[\sum_{t=1}^{T} f((y_t^{[1]}, y_t^{[2]})|(y_{t-1}^{[1]}, y_{t-1}^{[2]}), \boldsymbol{\theta})\right] \tag{2.20}
$$

with respect to $\boldsymbol{\theta}$, we obtain the following first-order derivatives:
Differentiating with respect to $\rho_{11}$, we have

$$
\begin{aligned}
&\sum_{t=1}^{T} \frac{\partial[\log L(\boldsymbol{\theta}|\boldsymbol{y})]}{\partial\rho_{11}} \\
&= \sum_{t=1}^{T} \frac{\sum_{u=0}^{g_1}\sum_{v=0}^{g_2} f_2(y_t^{[2]} - v) f_3(u,v) \frac{\partial}{\partial\rho_{11}} f_1(y_t^{[1]} - u) + f_1(y_t^{[1]} - u) f_2(y_t^{[2]} - v) \frac{\partial}{\partial\rho_{11}} f_3(u,v)}{\sum_{u=0}^{g_1}\sum_{v=0}^{g_2} f((y_t^{[1]}, y_t^{[2]})|(y_{t-1}^{[1]}, y_{t-1}^{[2]}), \boldsymbol{\theta})},
\end{aligned} \tag{2.21}
$$

where

$$
\begin{aligned}
&\frac{\partial}{\partial \rho_{11}} f_3(u,v) \\
&= \sum_{m=0}^{min(u,v)} \frac{e^{-(\mu_1^{[1]}-\rho_{11}\mu_1^{[1]}+\mu_1^{[2]}-\rho_{22}\mu_1^{[2]}-\kappa_{12,1}\sqrt{\mu_1^{[1]}-\rho_{11}\mu_1^{[1]}}\sqrt{\mu_1^{[2]}-\rho_{22}\mu_1^{[2]}})}}{(u-m)!(v-m)!m!} \\
&\quad \times \{[\mu_1^{[1]} - \frac{\kappa_{12,1}\mu_1^{[1]}\sqrt{\mu_1^{[2]}-\rho_{22}\mu_1^{[2]}}}{2\sqrt{\mu_1^{[1]}-\rho_{11}\mu_1^{[1]}}}] \\
&\quad \times [\mu_1^{[1]}-\rho_{11}\mu_1^{[1]}-\kappa_{12,1}\sqrt{\mu_1^{[1]}-\rho_{11}\mu_1^{[1]}}\sqrt{\mu_1^{[2]}-\rho_{22}\mu_1^{[2]}}]^{u-m} \\
&\quad \times [\mu_1^{[2]}-\rho_{22}\mu_1^{[2]}-\kappa_{12,1}\sqrt{\mu_1^{[1]}-\rho_{11}\mu_1^{[1]}}\sqrt{\mu_1^{[2]}-\rho_{22}\mu_1^{[2]}}]^{v-m} \\
&\quad \times [\kappa_{12,1}\sqrt{\mu_1^{[1]}-\rho_{11}\mu_1^{[1]}}\sqrt{\mu_1^{[2]}-\rho_{22}\mu_1^{[2]}}]^{m} \\
&\quad + \frac{1}{\mu_1^{[1]}-\rho_{11}\mu_1^{[1]}-\kappa_{12,1}\sqrt{\mu_1^{[1]}-\rho_{11}\mu_1^{[1]}}\sqrt{\mu_1^{[2]}-\rho_{22}\mu_1^{[2]}}} \\
&\quad \times [\mu_1^{[1]}-\rho_{11}\mu_1^{[1]}-\kappa_{12,1}\sqrt{\mu_1^{[1]}-\rho_{11}\mu_1^{[1]}}\sqrt{\mu_1^{[2]}-\rho_{22}\mu_1^{[2]}}]^{u-m} \\
&\quad \times [u-m][-\mu_1^{[1]} + \frac{\kappa_{12,1}\mu_1^{[1]}\sqrt{\mu_1^{[2]}-\rho_{22}\mu_1^{[2]}}}{2\sqrt{\mu_1^{[1]}-\rho_{11}\mu_1^{[1]}}}] \\
&\quad \times [\mu_1^{[2]}-\rho_{22}\mu_1^{[2]}-\kappa_{12,1}\sqrt{\mu_1^{[1]}-\rho_{11}\mu_1^{[1]}}\sqrt{\mu_1^{[2]}-\rho_{22}\mu_1^{[2]}}]^{v-m} \\
&\quad \times [\kappa_{12,1}\sqrt{\mu_1^{[1]}-\rho_{11}\mu_1^{[1]}}\sqrt{\mu_1^{[2]}-\rho_{22}\mu_1^{[2]}}]^{m} \\
&\quad + \frac{1}{2\sqrt{\mu_1^{[1]}-\rho_{11}\mu_1^{[1]}}[\mu_1^{[2]}-\rho_{22}\mu_1^{[2]}-\kappa_{12,1}\sqrt{\mu_1^{[1]}-\rho_{11}\mu_1^{[1]}}\sqrt{\mu_1^{[2]}-\rho_{22}\mu_1^{[2]}}]} \\
&\quad \times [\mu_1^{[1]}-\rho_{11}\mu_1^{[1]}-\kappa_{12,1}\sqrt{\mu_1^{[1]}-\rho_{11}\mu_1^{[1]}}\sqrt{\mu_1^{[2]}-\rho_{22}\mu_1^{[2]}}]^{u-m} \\
&\quad \times [\mu_1^{[2]}-\rho_{22}\mu_1^{[2]}-\kappa_{12,1}\sqrt{\mu_1^{[1]}-\rho_{11}\mu_1^{[1]}} \\
&\quad \times \sqrt{\mu_1^{[2]}-\rho_{22}\mu_1^{[2]}}]^{v-m}[v-m][\kappa_{12,1}\mu_1^{[1]}\sqrt{\mu_1^{[2]}-\rho_{22}\mu_1^{[2]}}] \\
&\quad \times [\kappa_{12,1}\sqrt{\mu_1^{[1]}-\rho_{11}\mu_1^{[1]}}\sqrt{\mu_1^{[2]}-\rho_{22}\mu_1^{[2]}}]^{m} - \frac{1}{2[\mu_1^{[1]}-\rho_{11}\mu_1^{[1]}]} \\
&\quad \times [\mu_1^{[1]}-\rho_{11}\mu_1^{[1]}-\kappa_{12,1}\sqrt{\mu_1^{[1]}-\rho_{11}\mu_1^{[1]}}\sqrt{\mu_1^{[2]}-\rho_{22}\mu_1^{[2]}}]^{u-m} \\
&\quad \times [\mu_1^{[2]}-\rho_{22}\mu_1^{[2]}-\kappa_{12,1}\sqrt{\mu_1^{[1]}-\rho_{11}\mu_1^{[1]}}\sqrt{\mu_1^{[2]}-\rho_{22}\mu_1^{[2]}}]^{v-m} \\
&\quad \times [\kappa_{12,1}\sqrt{\mu_1^{[1]}-\rho_{11}\mu_1^{[1]}}\sqrt{\mu_1^{[2]}-\rho_{22}\mu_1^{[2]}}]^{m}[m\mu_1^{[1]}]\} \qquad (2.22)
\end{aligned}
$$

and

$$\frac{\partial}{\partial \rho_{11}} f_1(y_t^{[1]} - u) = \left( \begin{array}{c} y_{t-1}^{[1]} \\ y_t^{[1]} - u \end{array} \right) [(y_t^{[1]} - u)\rho_{11}^{y_t^{[1]}-u-1}(1-\rho_{11})^{y_{t-1}^{[1]}-y_t^{[1]}+u} - (y_{t-1}^{[1]} - y_t^{[1]} + u)\rho_{11}^{y_t^{[1]}-u}(1-\rho_{11})^{y_{t-1}^{[1]}-y_t^{[1]}+u-1}]. \quad (2.23)$$

Differentiating with respect to $\rho_{22}$, we have

$$\sum_{t=1}^{T} \frac{\partial[\log \quad L(\boldsymbol{\theta}|\boldsymbol{y})]}{\partial \rho_{22}} = \sum_{t=1}^{T} \frac{\sum_{u=0}^{g_1}\sum_{v=0}^{g_2} f_1(y_t^{[1]}-u) f_3(u,v) \frac{\partial}{\partial \rho_{22}} f_2(y_t^{[2]}-v) + f_1(y_t^{[1]}-u) f_2(y_t^{[2]}-v) \frac{\partial}{\partial \rho_{22}} f_3(u,v)}{\sum_{u=0}^{g_1}\sum_{v=0}^{g_2} f((y_t^{[1]}, y_t^{[2]})|(y_{t-1}^{[1]}, y_{t-1}^{[2]}), \boldsymbol{\theta})}, \quad (2.24)$$

where

$$\frac{\partial}{\partial \rho_{22}} f_2(y_t^{[2]} - v) = \left( \begin{array}{c} y_{t-1}^{[2]} \\ y_t^{[2]} - v \end{array} \right) [(y_t^{[2]} - v)\rho_{22}^{y_t^{[2]}-v-1}(1-\rho_{22})^{y_{t-1}^{[2]}-y_t^{[2]}+v} - (y_{t-1}^{[2]} - y_t^{[2]} + v)\rho_{22}^{y_t^{[2]}-v}(1-\rho_{22})^{y_{t-1}^{[2]}-y_t^{[2]}+v-1}] \quad (2.25)$$

and

$$\begin{aligned}
&\frac{\partial}{\partial \rho_{22}} f_3(u,v) \\
&= \sum_{m=0}^{min(u,v)} \frac{e^{-(\mu_1^{[1]}-\rho_{11}\mu_1^{[1]}+\mu_1^{[2]}-\rho_{22}\mu_1^{[2]}-\kappa_{12,1}\sqrt{\mu_1^{[1]}-\rho_{11}\mu_1^{[1]}}\sqrt{\mu_1^{[2]}-\rho_{22}\mu_1^{[2]}})}}{(u-m)!(v-m)!m!} \\
&\times \{[\mu_1^{[2]} - \frac{\kappa_{12,1}\mu_1^{[2]}\sqrt{\mu_1^{[1]}-\rho_{11}\mu_1^{[1]}}}{2\sqrt{\mu_1^{[2]}-\rho_{22}\mu_1^{[2]}}}] \\
&\times [\mu_1^{[1]} - \rho_{11}\mu_1^{[1]} - \kappa_{12,1}\sqrt{\mu_1^{[1]}-\rho_{11}\mu_1^{[1]}}\sqrt{\mu_1^{[2]}-\rho_{22}\mu_1^{[2]}}]^{u-m} \\
&\times [\mu_1^{[2]} - \rho_{22}\mu_1^{[2]} - \kappa_{12,1}\sqrt{\mu_1^{[1]}-\rho_{11}\mu_1^{[1]}}\sqrt{\mu_1^{[2]}-\rho_{22}\mu_1^{[2]}}]^{v-m} \\
&\times [\kappa_{12,1}\sqrt{\mu_1^{[1]}-\rho_{11}\mu_1^{[1]}}\sqrt{\mu_1^{[2]}-\rho_{22}\mu_1^{[2]}}]^{m} \\
&+ \frac{1}{\mu_1^{[2]} - \rho_{22}\mu_1^{[2]} - \kappa_{12,1}\sqrt{\mu_1^{[1]}-\rho_{11}\mu_1^{[1]}}\sqrt{\mu_1^{[2]}-\rho_{22}\mu_1^{[2]}}} \\
&\times [\mu_1^{[1]} - \rho_{11}\mu_1^{[1]} - \kappa_{12,1}\sqrt{\mu_1^{[1]}-\rho_{11}\mu_1^{[1]}}\sqrt{\mu_1^{[2]}-\rho_{22}\mu_1^{[2]}}]^{u-m}[v-m] \\
&\times [-\mu_1^{[2]} + \frac{\kappa_{12,1}\mu_1^{[2]}\sqrt{\mu_1^{[1]}-\rho_{11}\mu_1^{[1]}}}{2\sqrt{\mu_1^{[2]}-\rho_{22}\mu_1^{[2]}}}]
\end{aligned}$$

$$\times [\mu_1^{[2]} - \rho_{22}\mu_1^{[2]} - \kappa_{12,1}\sqrt{\mu_1^{[1]} - \rho_{11}\mu_1^{[1]}}\sqrt{\mu_1^{[2]} - \rho_{22}\mu_1^{[2]}}]^{v-m}$$
$$\times [\kappa_{12,1}\sqrt{\mu_1^{[1]} - \rho_{11}\mu_1^{[1]}}\sqrt{\mu_1^{[2]} - \rho_{22}\mu_1^{[2]}}]^{m}$$
$$+ \frac{1}{2\sqrt{\mu_1^{[2]} - \rho_{22}\mu_1^{[2]}}[\mu_1^{[2]} - \rho_{22}\mu_1^{[2]} - \kappa_{12,1}\sqrt{\mu_1^{[1]} - \rho_{11}\mu_1^{[1]}}\sqrt{\mu_1^{[2]} - \rho_{22}\mu_1^{[2]}}]}$$
$$\times [\mu_1^{[1]} - \rho_{11}\mu_1^{[1]} - \kappa_{12,1}\sqrt{\mu_1^{[1]} - \rho_{11}\mu_1^{[1]}}\sqrt{\mu_1^{[2]} - \rho_{22}\mu_1^{[2]}}]^{u-m}$$
$$\times [\mu_1^{[2]} - \rho_{22}\mu_1^{[2]} - \kappa_{12,1}\sqrt{\mu_1^{[1]} - \rho_{11}\mu_1^{[1]}}$$
$$\times \sqrt{\mu_1^{[2]} - \rho_{22}\mu_1^{[2]}}]^{v-m}[u-m][\kappa_{12,1}\mu_1^{[2]}\sqrt{\mu_1^{[1]} - \rho_{11}\mu_1^{[1]}}]$$
$$\times [\kappa_{12,1}\sqrt{\mu_1^{[1]} - \rho_{11}\mu_1^{[1]}}\sqrt{\mu_1^{[2]} - \rho_{22}\mu_1^{[2]}}]^{m}$$
$$- \frac{1}{2[\mu_1^{[2]} - \rho_{22}\mu_1^{[2]}]}[\mu_1^{[1]} - \rho_{11}\mu_1^{[1]} - \kappa_{12,1}\sqrt{\mu_1^{[1]} - \rho_{11}\mu_1^{[1]}}\sqrt{\mu_1^{[2]} - \rho_{22}\mu_1^{[2]}}]^{u-m}$$
$$\times [\mu_1^{[2]} - \rho_{22}\mu_1^{[2]} - \kappa_{12,1}\sqrt{\mu_1^{[1]} - \rho_{11}\mu_1^{[1]}}\sqrt{\mu_1^{[2]} - \rho_{22}\mu_1^{[2]}}]^{v-m}$$
$$\times [\kappa_{12,1}\sqrt{\mu_1^{[1]} - \rho_{11}\mu_1^{[1]}}\sqrt{\mu_1^{[2]} - \rho_{22}\mu_1^{[2]}}]^{m}[m\mu_1^{[2]}]\}. \tag{2.26}$$

Differentiating with respect to $\kappa_{12,1}$, we have

$$\sum_{t=1}^{T}\frac{\partial[\log \quad L(\boldsymbol{\theta}|\boldsymbol{y})]}{\partial\kappa_{12,1}} = \sum_{t=1}^{T}\frac{\sum_{u=0}^{g_1}\sum_{v=0}^{g_2} f_1(y_t^{[1]}-u)f_2(y_t^{[2]}-v)\frac{\partial}{\partial\kappa_{12,1}}f_3(u,v)}{\sum_{u=0}^{g_1}\sum_{v=0}^{g_2} f((y_t^{[1]},y_t^{[2]})|(y_{t-1}^{[1]},y_{t-1}^{[2]}),\boldsymbol{\theta})}, \tag{2.27}$$

where

$$\frac{\partial}{\partial\kappa_{12,1}}f_3(u,v)$$
$$= \sum_{m=0}^{min(u,v)}\frac{e^{-(\mu_1^{[1]}-\rho_{11}\mu_1^{[1]}+\mu_1^{[2]}-\rho_{22}\mu_1^{[2]}-\kappa_{12,1}\sqrt{\mu_1^{[1]}-\rho_{11}\mu_1^{[1]}}\sqrt{\mu_1^{[2]}-\rho_{22}\mu_1^{[2]}})}}{(u-m)!(v-m)!m!}$$
$$\times \{[\sqrt{\mu_1^{[1]} - \rho_{11}\mu_1^{[1]}}\sqrt{\mu_1^{[2]} - \rho_{22}\mu_1^{[2]}}]$$
$$\times [\mu_1^{[1]} - \rho_{11}\mu_1^{[1]} - \kappa_{12,1}\sqrt{\mu_1^{[1]} - \rho_{11}\mu_1^{[1]}}\sqrt{\mu_1^{[2]} - \rho_{22}\mu_1^{[2]}}]^{u-m}$$
$$\times [\mu_1^{[2]} - \rho_{22}\mu_1^{[2]} - \kappa_{12,1}\sqrt{\mu_1^{[1]} - \rho_{11}\mu_1^{[1]}}\sqrt{\mu_1^{[2]} - \rho_{22}\mu_1^{[2]}}]^{v-m}$$
$$\times [\kappa_{12,1}\sqrt{\mu_1^{[1]} - \rho_{11}\mu_1^{[1]}}\sqrt{\mu_1^{[2]} - \rho_{22}\mu_1^{[2]}}]^{m}$$
$$- \frac{1}{\mu_1^{[1]} - \rho_{11}\mu_1^{[1]} - \kappa_{12,1}\sqrt{\mu_1^{[1]} - \rho_{11}\mu_1^{[1]}}\sqrt{\mu_1^{[2]} - \rho_{22}\mu_1^{[2]}}}$$
$$\times [\mu_1^{[1]} - \rho_{11}\mu_1^{[1]} - \kappa_{12,1}\sqrt{\mu_1^{[1]} - \rho_{11}\mu_1^{[1]}}\sqrt{\mu_1^{[2]} - \rho_{22}\mu_1^{[2]}}]^{u-m}[u-m]$$
$$\times [\sqrt{\mu_1^{[1]} - \rho_{11}\mu_1^{[1]}}\sqrt{\mu_1^{[2]} - \rho_{22}\mu_1^{[2]}}]$$
$$\times [\mu_1^{[2]} - \rho_{22}\mu_1^{[2]} - \kappa_{12,1}\sqrt{\mu_1^{[1]} - \rho_{11}\mu_1^{[1]}}\sqrt{\mu_1^{[2]} - \rho_{22}\mu_1^{[2]}}]^{v-m}$$

$$
\begin{aligned}
&\times [\kappa_{12,1}\sqrt{\mu_1^{[1]} - \rho_{11}\mu_1^{[1]}}\sqrt{\mu_1^{[2]} - \rho_{22}\mu_1^{[2]}}]^m \\
&- \frac{1}{\mu_1^{[2]} - \rho_{22}\mu_1^{[2]} - \kappa_{12,1}\sqrt{\mu_1^{[1]} - \rho_{11}\mu_1^{[1]}}\sqrt{\mu_1^{[2]} - \rho_{22}\mu_1^{[2]}}} \\
&\times [\mu_1^{[1]} - \rho_{11}\mu_1^{[1]} - \kappa_{12,1}\sqrt{\mu_1^{[1]} - \rho_{11}\mu_1^{[1]}}\sqrt{\mu_1^{[2]} - \rho_{22}\mu_1^{[2]}}]^{u-m}[v-m] \\
&\times [\sqrt{\mu_1^{[1]} - \rho_{11}\mu_1^{[1]}}\sqrt{\mu_1^{[2]} - \rho_{22}\mu_1^{[2]}}] \\
&\times [\mu_1^{[2]} - \rho_{22}\mu_1^{[2]} - \kappa_{12,1}\sqrt{\mu_1^{[1]} - \rho_{11}\mu_1^{[1]}}\sqrt{\mu_1^{[2]} - \rho_{22}\mu_1^{[2]}}]^{v-m} \\
&\times [\kappa_{12,1}\sqrt{\mu_1^{[1]} - \rho_{11}\mu_1^{[1]}}\sqrt{\mu_1^{[2]} - \rho_{22}\mu_1^{[2]}}]^m \\
&- \frac{1}{\kappa_{12,1}}[\mu_1^{[1]} - \rho_{11}\mu_1^{[1]} - \kappa_{12,1}\sqrt{\mu_1^{[1]} - \rho_{11}\mu_1^{[1]}}\sqrt{\mu_1^{[2]} - \rho_{22}\mu_1^{[2]}}]^{u-m}[m] \\
&\times [\mu_1^{[2]} - \rho_{22}\mu_1^{[2]} - \kappa_{12,1}\sqrt{\mu_1^{[1]} - \rho_{11}\mu_1^{[1]}}\sqrt{\mu_1^{[2]} - \rho_{22}\mu_1^{[2]}}]^{v-m} \\
&\times [\kappa_{12,1}\sqrt{\mu_1^{[1]} - \rho_{11}\mu_1^{[1]}}\sqrt{\mu_1^{[2]} - \rho_{22}\mu_1^{[2]}}]^m\}. \qquad (2.28)
\end{aligned}
$$

Differentiating with respect to $\beta_j^{[k]}$ for $k \in \{1, 2\}$, we have

$$
\sum_{t=1}^{T} \frac{\partial[\log \quad L(\boldsymbol{\theta}|\boldsymbol{y})]}{\partial \beta_j^{[k]}} = \sum_{t=1}^{T} \frac{\partial[\log \quad L(\boldsymbol{\theta}|\boldsymbol{y})]}{\partial \lambda_1^{[k]}} \times \frac{\partial \lambda_1^{[k]}}{\partial \beta_j^{[k]}}, \qquad (2.29)
$$

where $\lambda_1^{[k]} = \mu_1^{[k]} - \rho_{kk}\mu_1^{[k]}$, $\frac{\partial \lambda_1^{[k]}}{\partial \beta_j^{[k]}} = \mu_1^{[k]}x_{1j} - \rho_{kk}\mu_1^{[k]}x_{1j}$ and

$$
\sum_{t=1}^{T} \frac{\partial[\log \quad L(\boldsymbol{\theta}|\boldsymbol{y})]}{\partial \lambda_1^{[k]}} = \sum_{t=1}^{T} \frac{\sum_{u=0}^{g_1}\sum_{v=0}^{g_2} f_1(y_t^{[1]} - u) f_2(y_t^{[2]} - v) \frac{\partial}{\partial \lambda_1^{[k]}} f_3(u, v)}{\sum_{u=0}^{g_1}\sum_{v=0}^{g_2} f((y_t^{[1]}, y_t^{[2]})|(y_{t-1}^{[1]}, y_{t-1}^{[2]}), \boldsymbol{\theta})}, \qquad (2.30)
$$

where

$$
\begin{aligned}
&\frac{\partial}{\partial \lambda_1^{[1]}} f_3(u, v) \\
&= \sum_{m=0}^{min(u,v)} \frac{e^{-(\lambda_1^{[1]} + \lambda_1^{[2]} - \kappa_{12,1}\sqrt{\lambda_1^{[1]}}\sqrt{\lambda_1^{[2]}})}}{(u-m)!(v-m)!m!} \\
&\quad \times \{[-1 + \frac{\kappa_{12,1}\sqrt{\lambda_1^{[1]}}}{2\sqrt{\lambda_1^{[2]}}}][\lambda_1^{[1]} - \kappa_{12,1}\sqrt{\lambda_1^{[1]}}\sqrt{\lambda_1^{[2]}}]^{u-m}[\lambda_1^{[2]} - \kappa_{12,1}\sqrt{\lambda_1^{[1]}}\sqrt{\lambda_1^{[2]}}]^{v-m}[\kappa_{12,1}\sqrt{\lambda_1^{[1]}}\sqrt{\lambda_1^{[2]}}]^m \\
&\quad - \frac{[\lambda_1^{[1]} - \kappa_{12,1}\sqrt{\lambda_1^{[1]}}\sqrt{\lambda_1^{[2]}}]^{u-m}[u-m][\kappa_{12,1}\sqrt{\lambda_1^{[1]}}][\lambda_1^{[2]} - \kappa_{12,1}\sqrt{\lambda_1^{[1]}}\sqrt{\lambda_1^{[2]}}]^{v-m}[\kappa_{12,1}\sqrt{\lambda_1^{[1]}}\sqrt{\lambda_1^{[2]}}]^m}{2\sqrt{\lambda_1^{[2]}}[\lambda_1^{[1]} - \kappa_{12,1}\sqrt{\lambda_1^{[1]}}\sqrt{\lambda_1^{[2]}}]} \\
&\quad + [\frac{1}{\lambda_1^{[2]} - \kappa_{12,1}\sqrt{\lambda_1^{[1]}}\sqrt{\lambda_1^{[2]}}}][\lambda_1^{[1]} - \kappa_{12,1}\sqrt{\lambda_1^{[1]}}\sqrt{\lambda_1^{[2]}}]^{u-m}[\lambda_1^{[2]} - \kappa_{12,1}\sqrt{\lambda_1^{[1]}}\sqrt{\lambda_1^{[2]}}]^{v-m}[v-m]
\end{aligned}
$$

$$
\begin{aligned}
&\times [1 - \frac{\kappa_{12,1}\sqrt{\lambda_1^{[1]}}}{2\sqrt{\lambda_1^{[2]}}}][\kappa_{12,1}\sqrt{\lambda_1^{[1]}}\sqrt{\lambda_1^{[2]}}]^m \\
&+ \frac{[\lambda_1^{[1]} - \kappa_{12,1}\sqrt{\lambda_1^{[1]}}\sqrt{\lambda_1^{[2]}}]^{u-m}[\lambda_1^{[2]} - \kappa_{12,1}\sqrt{\lambda_1^{[1]}}\sqrt{\lambda_1^{[2]}}]^{v-m}[\kappa_{12,1}\sqrt{\lambda_1^{[1]}}\sqrt{\lambda_1^{[2]}}]^m[m]}{2\lambda_1^{[2]}}
\end{aligned} \tag{2.31}
$$

and

$$
\begin{aligned}
&\frac{\partial}{\partial\lambda_1^{[2]}} f_3(u, v) \\
&= \sum_{m=0}^{min(u,v)} \frac{e^{-(\lambda_1^{[1]}+\lambda_1^{[2]}-\kappa_{12,1}\sqrt{\lambda_1^{[1]}}\sqrt{\lambda_1^{[2]}})}}{(u-m)!(v-m)!m!} \\
&\times \{[-1+\frac{\kappa_{12,1}\sqrt{\lambda_1^{[2]}}}{2\sqrt{\lambda_1^{[1]}}}][\lambda_1^{[1]}-\kappa_{12,1}\sqrt{\lambda_1^{[1]}}\sqrt{\lambda_1^{[2]}}]^{u-m}[\lambda_1^{[2]}-\kappa_{12,1}\sqrt{\lambda_1^{[1]}}\sqrt{\lambda_1^{[2]}}]^{v-m}[\kappa_{12,1}\sqrt{\lambda_1^{[1]}}\sqrt{\lambda_1^{[2]}}]^m \\
&- \frac{[\lambda_1^{[1]}-\kappa_{12,1}\sqrt{\lambda_1^{[1]}}\sqrt{\lambda_1^{[2]}}]^{u-m}[v-m][\kappa_{12,1}\sqrt{\lambda_1^{[2]}}][\lambda_1^{[2]}-\kappa_{12,1}\sqrt{\lambda_1^{[1]}}\sqrt{\lambda_1^{[2]}}]^{v-m}[\kappa_{12,1}\sqrt{\lambda_1^{[1]}}\sqrt{\lambda_1^{[2]}}]^m}{2\sqrt{\lambda_1^{[1]}}[\lambda_1^{[2]}-\kappa_{12,1}\sqrt{\lambda_1^{[1]}}\sqrt{\lambda_1^{[2]}}]} \\
&+ [\frac{1}{\lambda_1^{[1]} - \kappa_{12,1}\sqrt{\lambda_1^{[1]}}\sqrt{\lambda_1^{[2]}}}][\lambda_1^{[1]} - \kappa_{12,1}\sqrt{\lambda_1^{[1]}}\sqrt{\lambda_1^{[2]}}]^{u-m}[u-m][\lambda_1^{[2]} - \kappa_{12,1}\sqrt{\lambda_1^{[1]}}\sqrt{\lambda_1^{[2]}}]^{v-m} \\
&\times [1 - \frac{\kappa_{12,1}\sqrt{\lambda_1^{[2]}}}{2\sqrt{\lambda_1^{[1]}}}][\kappa_{12,1}\sqrt{\lambda_1^{[1]}}\sqrt{\lambda_1^{[2]}}]^m \\
&+ \frac{[\lambda_1^{[1]} - \kappa_{12,1}\sqrt{\lambda_1^{[1]}}\sqrt{\lambda_1^{[2]}}]^{u-m}[\lambda_1^{[2]} - \kappa_{12,1}\sqrt{\lambda_1^{[1]}}\sqrt{\lambda_1^{[2]}}]^{v-m}[\kappa_{12,1}\sqrt{\lambda_1^{[1]}}\sqrt{\lambda_1^{[2]}}]^m[m]}{2\lambda_1^{[1]}}\}.
\end{aligned} \tag{2.32}
$$

Having obtained these scores, the unknown parameter $\boldsymbol{\theta}$ is estimated iteratively using

$$
\hat{\boldsymbol{\theta}}_{r+1} = \hat{\boldsymbol{\theta}}_r - [\log''[L(\hat{\boldsymbol{\theta}}_r)]]^{-1}[\log'[L(\hat{\boldsymbol{\theta}}_r)]], \tag{2.33}
$$

where $\log'[L(\hat{\boldsymbol{\theta}}_r)]$ is the vector of likelihood scores given by Equations (2.21), (2.24), (2.27) and (2.29), and $\log''[L(\hat{\boldsymbol{\theta}}_r)]$ is the vector of the double derivatives of the conditional log-likelihood function with respect to $\boldsymbol{\theta}$. To circumvent the task of solving this complicated system of non-linear equations, we use the numerical maximization which is found in most statistical and data analysis packages. Hence, the command *optim* from the R software system is used to solve the conditional log-likelihood function (2.20) by providing an initial value of the unknown parameter $\boldsymbol{\theta}$ and the time series data as follows:

$$
optim(\text{initial value of } \ \boldsymbol{\theta}, \quad \log[L(\boldsymbol{\theta}|\boldsymbol{y})], \quad \text{data}, \quad \text{hessian=true}) \tag{2.34}
$$

The *optim* command also provides the standard errors based on the observed Fisher information matrix by including Hessian in the command.

#### 2.3.1.2 Non–Likelihood-Based Approaches

**Feasible Generalized Least Squares (FGLS)**

The FGLS estimating equations to obtain the model parameters under the BINAR(1) are developed in this subsection. We require information on the specification of the conditional moments as follows:

$$e_{k1t} = y_t^{[k]} - E(Y_t^{[k]}|y_{t-1}^{[k]}), \quad \text{for} \quad k \in \{1, 2\}, \tag{2.35}$$

where $\mathrm{E}(Y_t^{[k]}|y_{t-1}^{[k]}) = \rho_{kk} y_{t-1}^{[k]} + (1 - \rho_{kk})\mu_1^{[k]}$.

To compute the unknown parameter $\boldsymbol{\theta} = [\rho_{11}, \rho_{22}, \beta_1^{[1]}, \beta_2^{[1]}, \beta_1^{[2]}, \beta_2^{[2]}, \kappa_{12,1}]$, the objective criterion function

$$\begin{aligned}
S_{FGLS} = &\sum_{t=1}^{T} \Big[\mathrm{Var}(Y_t^{[2]}|y_{t-1}^{[2]})(e_{11t}^2) + \mathrm{Var}(Y_t^{[1]}|y_{t-1}^{[1]})(e_{21t}^2) \\
&-2\mathrm{Cov}(Y_t^{[1]}, Y_t^{[2]}|y_{t-1}^{[1]}, y_{t-1}^{[2]})(e_{11t})(e_{21t})\Big] \\
&/ \Big[\mathrm{Var}(Y_t^{[1]}|y_{t-1}^{[1]})\mathrm{Var}(Y_t^{[2]}|y_{t-1}^{[2]}) - \mathrm{Cov}(Y_t^{[1]}, Y_t^{[2]}|y_{t-1}^{[1]}, y_{t-1}^{[2]})^2\Big] \\
= &\sum_{t=1}^{T} \{[\rho_{22}(1-\rho_{22})y_{t-1}^{[2]} + \mu_1^{[2]} - \rho_{22}\mu_1^{[2]}][y_t^{[1]} - (\rho_{11}y_{t-1}^{[1]} + \mu_1^{[1]} - \rho_{11}\mu_1^{[1]})]^2 \\
&\times [\rho_{11}(1 - \rho_{11})y_{t-1}^{[1]} + \mu_1^{[1]} - \rho_{11}\mu_1^{[1]}][y_t^{[2]} - (\rho_{22}y_{t-1}^{[2]} + \mu_1^{[2]} - \rho_{22}\mu_1^{[2]})]^2 \\
&- 2[\kappa_{12,1}\sqrt{\mu_1^{[1]} - \rho_{11}\mu_1^{[1]}}\sqrt{\mu_1^{[2]} - \rho_{22}\mu_1^{[2]}}][\rho_{11}y_{t-1}^{[1]} + \mu_1^{[1]} - \rho_{11}\mu_1^{[1]}] \\
&\times [\rho_{22}y_{t-1}^{[2]} + \mu_1^{[2]} - \rho_{22}\mu_1^{[2]}]\} \\
&/\{[\rho_{22}(1 - \rho_{22})y_{t-1}^{[2]} + \mu_1^{[2]} - \rho_{22}\mu_1^{[2]}][\rho_{11}(1 - \rho_{11})y_{t-1}^{[1]} + \mu_1^{[1]} - \rho_{11}\mu_1^{[1]}] \\
&- [\kappa_{12,1}\sqrt{\mu_1^{[1]} - \rho_{11}\mu_1^{[1]}}\sqrt{\mu_1^{[2]} - \rho_{22}\mu_1^{[2]}}]^2\}
\end{aligned} \tag{2.36}$$

is minimized with respect to the unknown parameter $\boldsymbol{\theta}$, where

1. $\mathrm{Var}(Y_t^{[k]}|y_{t-1}^{[k]}) = \rho_{kk}(1 - \rho_{kk})y_{t-1}^{[k]} + (1 - \rho_{kk})\mu_1^{[k]}$
2. $\mathrm{Cov}(Y_t^{[1]}, Y_t^{[2]}|y_{t-1}^{[1]}, y_{t-1}^{[2]}) = \mathrm{Cov}(R_t^{[1]}, R_t^{[2]}) = \kappa_{12,1}\sqrt{(1 - \rho_{11})\mu_1^{[1]}}$ $\sqrt{(1 - \rho_{22})\mu_1^{[2]}}$.

Hence, $\boldsymbol{\theta}$ is estimated iteratively using

$$\hat{\boldsymbol{\theta}}_{r+1} = \hat{\boldsymbol{\theta}}_r - [S''_{FGLS}(\hat{\boldsymbol{\theta}}_r)]^{-1}[S'_{FGLS}(\hat{\boldsymbol{\theta}}_r)]. \tag{2.37}$$

To facilitate the estimation, the command *optim*

$$\textit{optim}(\text{initial value of} \quad \boldsymbol{\theta}, \quad S_{FGLS}, \quad \text{data}, \quad \text{hessian=true}) \tag{2.38}$$

is used to solve the criterion $S_{FGLS}$ by providing the initial value of $\boldsymbol{\theta}$ and the time series data.

#### 2.3.1.3 Generalized Quasi-Likelihood (GQL)

In this subsection, we formulate the GQL method based on an extension of the classical Wedderburn (1974) QL equation where only the correct specification of the mean and score function is required (See also Sutradhar & Das (1999) and Sutradhar (2003)). In a recent paper, Sutradhar et al. (2014) showed that this approach yields asymptotically equally efficient estimates as likelihood-based methods. Hence, in this subsection, we develop a similar GQL approach for estimating the unknown parameters of the constrained BINAR(1) model. In fact, to facilitate the computation of the regression parameters, GQL equation is based on a simple mean score function. Note that the serial- and cross-correlation parameters are estimated using moment estimation approaches proposed by Sutradhar & Das (1999) and Jowaheer & Sutradhar (2005).

The GQL equation is expressed as

$$\boldsymbol{D}^{'}_{\boldsymbol{\beta}}\boldsymbol{\Sigma}^{-1}_{\boldsymbol{\beta}}(\boldsymbol{y}-\boldsymbol{\mu})=0, \tag{2.39}$$

where

$$\boldsymbol{y}=\begin{pmatrix}\boldsymbol{y}^{[1]}\\ \boldsymbol{y}^{[2]}\end{pmatrix}_{2T\times 1}, \boldsymbol{y}^{[1]}=[y_1^{[1]},y_2^{[1]},\dots,y_T^{[1]}]', \boldsymbol{y}^{[2]}=[y_1^{[2]},y_2^{[2]},\dots,y_T^{[2]}]' \tag{2.40}$$

and

$$\boldsymbol{\mu}=\begin{pmatrix}\boldsymbol{\mu}^{[1]}\\ \boldsymbol{\mu}^{[2]}\end{pmatrix}_{2T\times 1}, \boldsymbol{\mu}^{[1]}=[\mu_1^{[1]},\mu_1^{[1]},\dots,\mu_1^{[1]}]', \boldsymbol{\mu}^{[2]}=[\mu_1^{[2]},\mu_1^{[2]},\dots,\mu_1^{[2]}]', \tag{2.41}$$

where $[.]^{'}$ indicates the transpose of $[.]$.

The derivative matrix is a block diagonal $(2T\times 2p)$ matrix with $[\boldsymbol{D}^{[1]},\boldsymbol{D}^{[2]}]$ on the diagonals where

$$\boldsymbol{D}^{[k]}=\begin{pmatrix}\frac{\partial\mu_1^{[k]}}{\partial\beta_1^{[k]}} & \frac{\partial\mu_1^{[k]}}{\partial\beta_2^{[k]}} & \cdots & \frac{\partial\mu_1^{[k]}}{\partial\beta_p^{[k]}}\\ \frac{\partial\mu_1^{[k]}}{\partial\beta_1^{[k]}} & \frac{\partial\mu_1^{[k]}}{\partial\beta_2^{[k]}} & \cdots & \frac{\partial\mu_1^{[k]}}{\partial\beta_p^{[k]}}\\ \vdots & \vdots & \ddots & \vdots\\ \frac{\partial\mu_1^{[k]}}{\partial\beta_1^{[k]}} & \frac{\partial\mu_1^{[k]}}{\partial\beta_2^{[k]}} & \cdots & \frac{\partial\mu_1^{[k]}}{\partial\beta_p^{[k]}}\end{pmatrix}_{(T\times p)} \tag{2.42}$$

and $\frac{\partial\mu_1^{[k]}}{\partial\beta_j^{[k]}}=\mu_1^{[k]}x_{1j}$.

The covariance structure $\boldsymbol{\Sigma}_{\boldsymbol{\beta}}$ is

$$
\begin{array}{c} \\ Y_1^{[1]} \\ Y_2^{[1]} \\ \vdots \\ Y_T^{[1]} \\ \cdots \\ Y_1^{[2]} \\ Y_2^{[2]} \\ \vdots \\ Y_T^{[2]} \end{array}
\begin{array}{c}
\begin{array}{cccccccc} Y_1^{[1]} & Y_2^{[1]} & \cdots & Y_T^{[1]} & Y_1^{[2]} & Y_2^{[2]} & \cdots & Y_T^{[2]} \end{array} \\
\left(\begin{array}{ccccccccc}
V(Y_1^{[1]}) & Cov(Y_1^{[1]}, Y_2^{[1]}) & \cdots & Cov(Y_1^{[1]}, Y_T^{[1]}) & \vdots & Cov(Y_1^{[1]}, Y_1^{[2]}) & Cov(Y_1^{[1]}, Y_2^{[2]}) & \cdots & Cov(Y_1^{[1]}, Y_T^{[2]}) \\
Cov(Y_2^{[1]}, Y_1^{[1]}) & V(Y_2^{[1]}) & \cdots & Cov(Y_2^{[1]}, Y_T^{[1]}) & \vdots & Cov(Y_2^{[1]}, Y_1^{[2]}) & Cov(Y_2^{[1]}, Y_2^{[2]}) & \cdots & Cov(Y_2^{[1]}, Y_T^{[2]}) \\
\vdots & \vdots & \ddots & \vdots & \vdots & \vdots & \vdots & \ddots & \vdots \\
Cov(Y_T^{[1]}, Y_1^{[1]}) & Cov(Y_T^{[1]}, Y_2^{[1]}) & \cdots & V(Y_T^{[1]}) & \vdots & Cov(Y_T^{[1]}, Y_1^{[2]}) & Cov(Y_T^{[1]}, Y_2^{[2]}) & \cdots & Cov(Y_T^{[1]}, Y_T^{[2]}) \\
\cdots & \cdots & \cdots & \cdots & \vdots & \cdots & \cdots & \cdots & \cdots \\
Cov(Y_1^{[2]}, Y_1^{[1]}) & Cov(Y_1^{[2]}, Y_2^{[1]}) & \cdots & Cov(Y_1^{[2]}, Y_T^{[1]}) & \vdots & V(Y_1^{[2]}) & Cov(Y_1^{[2]}, Y_2^{[2]}) & \cdots & Cov(Y_1^{[2]}, Y_T^{[2]}) \\
Cov(Y_2^{[2]}, Y_1^{[1]}) & Cov(Y_2^{[2]}, Y_2^{[1]}) & \cdots & Cov(Y_2^{[2]}, Y_T^{[1]}) & \vdots & Cov(Y_2^{[2]}, Y_1^{[2]}) & V(Y_2^{[2]}) & \cdots & Cov(Y_2^{[2]}, Y_T^{[2]}) \\
\vdots & \vdots & \ddots & \vdots & \vdots & \vdots & \vdots & \ddots & \vdots \\
Cov(Y_T^{[2]}, Y_1^{[1]}) & Cov(Y_T^{[2]}, Y_2^{[1]}) & \cdots & Cov(Y_T^{[2]}, Y_T^{[1]}) & \vdots & Cov(Y_T^{[2]}, Y_1^{[2]}) & Cov(Y_T^{[2]}, Y_2^{[2]}) & \cdots & V(Y_T^{[2]})
\end{array}\right)
\end{array}
\tag{2.43}
$$

where the entries are computed using Equations (2.13)–(2.15).

Under the GQL approach, the stationary serial-correlation in Equation (2.13) is calculated using Sutradhar & Das (1999) and Jowaheer & Sutradhar (2005) moment estimation approach

$$\hat{\rho}_{kk,h} = \frac{\sum_{t=1}^{T-h} \tilde{y_t}^{[k]} \tilde{y}^{[k]}_{(t+h)}/(T-h)}{\sum_{t=1}^{T} (\tilde{y}_t^{[k]})^2/T}, \tag{2.44}$$

where $\hat{\rho}_{kk,h}$ is the estimated serial-correlation for the $k^{th}$ series at lag-$h$ and $\tilde{y_t}^{[k]} = \frac{y_t^{[k]} - \hat{\mu}_1^{[k]}}{\sqrt{\hat{\mu}_1^{[k]}}}$. From here, the cross-correlation $\kappa_{12,1}$ is estimated using Equation (2.15) as follows:

$$\hat{\kappa}_{12,1} = \frac{(1-\hat{\rho}_{11}\hat{\rho}_{22})\tilde{\text{Cov}}(Y_t^{[1]}, Y_t^{[2]})}{\sqrt{(\hat{\mu}_1^{[1]} - \hat{\rho}_{11}\hat{\mu}_1^{[1]})}\sqrt{(\hat{\mu}_1^{[2]} - \hat{\rho}_{22}\hat{\mu}_1^{[2]})}}, \tag{2.45}$$

where $\tilde{\text{Cov}}(Y_t^{[1]}, Y_t^{[2]}) = \frac{1}{T} \sum_{t=1}^{T} (y_t^{[1]} - \hat{\mu}_1^{[1]})(y_t^{[2]} - \hat{\mu}_1^{[2]})$.

The Newton-Raphson iterative technique is used to solve the GQL in Equation (2.39) and this yields

$$\begin{pmatrix} \hat{\boldsymbol{\beta}}^{[1]}_{r+1} \\ \hat{\boldsymbol{\beta}}^{[2]}_{r+1} \end{pmatrix} = \begin{pmatrix} \hat{\boldsymbol{\beta}}^{[1]}_{r} \\ \hat{\boldsymbol{\beta}}^{[2]}_{r} \end{pmatrix} + [\boldsymbol{D}'_{\boldsymbol{\beta}} \boldsymbol{\Sigma}^{-1}_{\boldsymbol{\beta}} \boldsymbol{D}_{\boldsymbol{\beta}}]^{-1}_r [\boldsymbol{D}'_{\boldsymbol{\beta}} \boldsymbol{\Sigma}^{-1}_{\boldsymbol{\beta}} (\boldsymbol{y} - \boldsymbol{\mu})]_r, \tag{2.46}$$

where $\hat{\boldsymbol{\beta}}^{[k]}_r$ are the estimates at the $r^{th}$ iteration and $[.]_r$ are the values of the expression at the $r^{th}$ iteration.

The algorithm works as follows: For an initial value of $\hat{\boldsymbol{\beta}}^{[k]}$, $[\beta_1^{[1]} = \beta_2^{[1]} = 0.01, \beta_1^{[2]} = \beta_2^{[2]} = 0.02]$, the serial-correlation (2.44) and cross-correlation (2.45) are calculated. These estimates are then substituted in Equations (2.13)–(2.15) to estimate the covariance structure (2.43). Then, using Equation (2.46), an updated estimate of $\hat{\boldsymbol{\beta}}^{[k]}$ is obtained. The new set of regression parameters is then used to re-calculate components (2.44) and (2.45). This iterative process continues until the absolute difference between the updated and previous regression estimates is less than a tolerance $10^{-5}$. Under mild regulatory conditions, the asymptotic normality and consistency of $\hat{\boldsymbol{\beta}}$ is derived as follows:
From Equation (2.39), for the true $\boldsymbol{\beta}$, let us define

$$g(\boldsymbol{\beta}) = \boldsymbol{D}' [\boldsymbol{\Sigma}(\boldsymbol{\beta})]^{-1} (\boldsymbol{y} - \boldsymbol{\mu}(\boldsymbol{\beta})), \tag{2.47}$$

where $\text{E}(\boldsymbol{Y}) = \boldsymbol{\mu}(\boldsymbol{\beta})$ and the exact auto-covariance structure is $\boldsymbol{\Sigma}(\boldsymbol{\beta})$.
By the Central Limit Theorem (CLT), $\boldsymbol{Y} \sim \text{Normal}(\boldsymbol{\mu}(\boldsymbol{\beta}), \boldsymbol{\Sigma}(\boldsymbol{\beta}))$ and it follows that $\text{E}[g(\boldsymbol{\beta})]=0$ and

$$\text{Cov}[g(\boldsymbol{\beta})] = \boldsymbol{D}' [\boldsymbol{\Sigma}(\boldsymbol{\beta})]^{-1} \boldsymbol{D} = \text{Var}^*(\boldsymbol{\beta}). \tag{2.48}$$

Further to Lindeberg-Feller CLT (Amemiya (1985), McDonald (2005)),

$$H = [\text{Cov}(g(\boldsymbol{\beta}))]^{-1/2} g(\boldsymbol{\beta}) \to \text{Normal}(0, I_p), \tag{2.49}$$

where $(p \times 1)$ is the dimension of $g(\boldsymbol{\beta})$.
By first-order Taylor series expansion, $g(\boldsymbol{\beta}) + (\hat{\boldsymbol{\beta}} - \boldsymbol{\beta})g^{'}(\boldsymbol{\beta}) = 0$. Then,

$$\begin{aligned}(\hat{\boldsymbol{\beta}} - \boldsymbol{\beta}) &= -[g^{'}(\boldsymbol{\beta})]^{-1} g(\boldsymbol{\beta}) \\ &= [\text{Cov}(g(\boldsymbol{\beta}))]^{-1} g(\boldsymbol{\beta}) \\ &= [\text{Cov}(g(\boldsymbol{\beta}))]^{-1/2} [\text{Cov}(g(\boldsymbol{\beta}))]^{-1/2} g(\boldsymbol{\beta}) \\ &= [\text{Cov}(g(\boldsymbol{\beta}))]^{-1/2} H \to \text{Normal}(\mathbf{0}, \text{Cov}^{-1}(g(\boldsymbol{\beta})))\end{aligned} \tag{2.50}$$

and

$$\hat{\boldsymbol{\beta}} \to \text{Normal}\left(\boldsymbol{\beta}, \text{Var}^{*-1}(\boldsymbol{\beta})\right). \tag{2.51}$$

Hence, the standard errors of the regression estimates are calculated using the hessian part $[\boldsymbol{D}^{'}\boldsymbol{\Sigma}^{-1}\boldsymbol{D}]^{-1}$.

#### 2.3.1.4 Simulation

This section presents some simulation experiments to generate time series of stationary BINAR(1) data. The CMLE, FGLS and GQL estimation methods explained in the Subsections 2.3.1.1 and 2.3.1.2 are applied to estimate the regression and dependence parameters. For this simulation study, the responses are generated using the OD model (2.1). In this experiment, the bivariate innovation series are generated using the multivariate Poisson R code developed by Yahav & Shmueli (2011) where for each time point, a pair of simulated innovation term based on the stationary means described in Equation (2.3) is generated for $\kappa_{12,1} = 0.1, 0.3, 0.5, 0.9$, $\beta_1^{[k]} = \beta_2^{[k]} = 1$ for $k \in \{1, 2\}$, within different combinations of the serial-correlations $[\rho_{11}, \rho_{22}] = [0.9, 0.9], [0.9, 0.3]$ and [0.3,0.3]. The simulation of the innovation terms is implemented in the software R via the library MASS as follows:

- *Generate* $\{R_t^{[1]}, R_t^{[1]}\} \leftarrow function(2, T, R, lambda)$
- $\{$ $normal_{mu} = rep(0, 2)$
  $normal = mvrnorm(T, normal_{mu}, R)$
  $pois = normal$
  $p1 = pnorm(normal)$
  $for(s \quad in 1:2) pois[, s] = qpois(p1[, s], lambda[s])$
  $return(pois)$
- $\}$

where $R = matrix(c(1, \kappa_{12,1}, \kappa_{12,1}, 1), 2, 2), lambda = matrix((\mu_1^{[1]} - \rho_{11}\mu_1^{[1]}), (\mu_1^{[2]} - \rho_{22}\mu_1^{[2]}), 2, 1)$.

Hence, using the simulated pair of values $(r_t^{[1]}, r_t^{[2]})$ and $Y_{t-1}^{[k]} \sim$ Poisson$(\mu_1^{[k]})$, we generate observations $y_t^{[1]}$ and $y_t^{[2]}$ using Equation (2.1).

Five thousand samples of data for different values of $\kappa_{12,1}$ under $T = 60, 100$ and $600$ are generated based on a two-dimensional covariate vector where the two elements are

$$x_{t1} = \begin{cases} -1 & (t = 1, \ldots, T/4), \\ 0 & (t = (T/4) + 1, \ldots, 3T/4), \\ 1 & (t = (3T/4) + 1, \ldots, T), \end{cases}$$

$$x_{t2} = \begin{cases} -2 & (t = 1, \ldots, T/4), \\ 1 & (t = (T/4) + 1, \ldots, 3T/4), \\ -2 & (t = (3T/4) + 1, \ldots, T). \end{cases}$$

The results are shown in Tables 2.1–2.6:

**TABLE 2.1**

Estimates of the Parameters and Standard Errors under Stationary BINAR(1) Process for $\rho_{kk} = [\rho_{11} = 0.9, \rho_{22} = 0.9]$ and $\kappa_{12,1} = [0.1, 0.3]$

| $\kappa_{12,1}$ | $T$ | Method | $\hat\beta_1^{[1]}$ | $\hat\beta_2^{[1]}$ | $\hat\beta_1^{[2]}$ | $\hat\beta_2^{[2]}$ | $\hat\rho_{11}$ | $\hat\rho_{22}$ | $\hat\kappa_{12,1}$ |
|---|---|---|---|---|---|---|---|---|---|
| 0.1 | 60 | CMLE | 0.9845 | 1.0614 | 0.9811 | 1.0112 | 0.8843 | 0.9178 | 0.1134 |
| | | | (0.1131) | (0.1111) | (0.1114) | (0.1123) | (0.0885) | (0.0838) | (0.0511) |
| | | FGLS | 0.9820 | 1.0155 | 0.9809 | 1.0172 | 0.8815 | 0.9159 | 0.1279 |
| | | | (0.1294) | (0.1290) | (0.1297) | (0.1286) | (0.0992) | (0.0998) | (0.0610) |
| | | GQL | 0.9832 | 1.0148 | 0.9812 | 1.0152 | 0.8822 | 0.9170 | 0.1168 |
| | | | (0.1161) | (0.1147) | (0.1145) | (0.1152) | | | |
| | 100 | CMLE | 0.9874 | 1.0298 | 0.9918 | 0.9881 | 0.8855 | 0.8941 | 0.1019 |
| | | | (0.0911) | (0.0920) | (0.0925) | (0.0940) | (0.0642) | (0.0616) | (0.0429) |
| | | FGLS | 0.9878 | 0.9964 | 0.9947 | 0.9901 | 0.8901 | 0.8912 | 0.1249 |
| | | | (0.1055) | (0.1066) | (0.1072) | (0.1089) | (0.0779) | (0.0755) | (0.0577) |
| | | GQL | 0.9888 | 1.0092 | 0.9911 | 0.9854 | 0.8848 | 0.8908 | 0.1048 |
| | | | (0.0930) | (0.0943) | (0.0940) | (0.0966) | | | |
| | 600 | CMLE | 0.9922 | 0.9930 | 1.0011 | 1.0042 | 0.8915 | 0.9052 | 0.1017 |
| | | | (0.0317) | (0.0325) | (0.0315) | (0.0313) | (0.0296) | (0.0244) | (0.0182) |
| | | FGLS | 0.9969 | 0.9961 | 1.0087 | 1.0058 | 0.8916 | 0.9094 | 0.1138 |
| | | | (0.0439) | (0.0449) | (0.0442) | (0.0438) | (0.0355) | (0.0287) | (0.0245) |
| | | GQL | 0.9918 | 0.9921 | 1.0002 | 1.0052 | 0.8990 | 0.9048 | 0.1034 |
| | | | (0.0318) | (0.0325) | (0.0315) | (0.0314) | | | |
| 0.3 | 60 | CMLE | 1.0181 | 0.9802 | 1.0189 | 0.9895 | 0.9181 | 0.8810 | 0.3173 |
| | | | (0.1165) | (0.1140) | (0.1145) | (0.1173) | (0.0812) | (0.0825) | (0.1210) |
| | | FGLS | 1.0315 | 0.9849 | 1.0114 | 0.9855 | 0.9118 | 0.8822 | 0.3294 |
| | | | (0.1220) | (0.1291) | (0.1295) | (0.1222) | (0.0948) | (0.0993) | (0.1389) |
| | | GQL | 1.0363 | 0.9827 | 1.0170 | 0.9840 | 0.9138 | 0.8885 | 0.3192 |
| | | | (0.1195) | (0.1174) | (0.1171) | (0.1203) | | | |
| | 100 | CMLE | 1.0079 | 0.9950 | 1.0069 | 0.9927 | 0.9095 | 0.9077 | 0.3053 |
| | | | (0.0953) | (0.0925) | (0.0936) | (0.0951) | (0.0632) | (0.0621) | (0.1170) |
| | | FGLS | 1.0289 | 0.9916 | 1.0099 | 0.9915 | 0.9053 | 0.9038 | 0.3183 |
| | | | (0.1098) | (0.1070) | (0.1088) | (0.1093) | (0.0755) | (0.0721) | (0.1284) |
| | | GQL | 1.0589 | 0.9939 | 1.0071 | 0.9901 | 0.9085 | 0.9058 | 0.3082 |
| | | | (0.0975) | (0.0947) | (0.0976) | (0.0970) | | | |

*(Continued)*

**TABLE 2.1**
(*Continued*)

| $\kappa_{12,1}$ | $T$ | Method | $\hat{\beta}_1^{[1]}$ | $\hat{\beta}_2^{[1]}$ | $\hat{\beta}_1^{[2]}$ | $\hat{\beta}_2^{[2]}$ | $\hat{\rho}_{11}$ | $\hat{\rho}_{22}$ | $\hat{\kappa}_{12,1}$ |
|---|---|---|---|---|---|---|---|---|---|
| | 600 | CMLE | 0.9978 | 1.0078 | 0.9922 | 1.0071 | 0.8945 | 0.8981 | 0.3011 |
| | | | (0.0350) | (0.0371) | (0.0366) | (0.0326) | (0.0203) | (0.0294) | (0.0520) |
| | | FGLS | 0.9973 | 1.0098 | 0.9969 | 1.0096 | 0.8961 | 0.8909 | 0.3131 |
| | | | (0.0478) | (0.0492) | (0.0490) | (0.0451) | (0.0373) | (0.0333) | (0.0679) |
| | | GQL | 0.9918 | 1.0090 | 0.9979 | 1.0012 | 0.8957 | 0.8946 | 0.3024 |
| | | | (0.0350) | (0.0372) | (0.0367) | (0.0326) | | | |

**TABLE 2.2**
Estimates of the Parameters and Standard Errors under Stationary BINAR(1) Process for $\rho_{kk} = [\rho_{11} = 0.9, \rho_{22} = 0.9]$ and $\kappa_{12,1} = [0.5, 0.9]$

| $\kappa_{12,1}$ | $T$ | Method | $\hat{\beta}_1^{[1]}$ | $\hat{\beta}_2^{[1]}$ | $\hat{\beta}_1^{[2]}$ | $\hat{\beta}_2^{[2]}$ | $\hat{\rho}_{11}$ | $\hat{\rho}_{22}$ | $\hat{\kappa}_{12,1}$ |
|---|---|---|---|---|---|---|---|---|---|
| 0.5 | 60 | CMLE | 1.0181 | 0.9835 | 1.0172 | 0.9818 | 0.8866 | 0.8851 | 0.5152 |
| | | | (0.1152) | (0.1180) | (0.1144) | (0.1120) | (0.0876) | (0.0898) | (0.1204) |
| | | FGLS | 1.0113 | 0.9846 | 1.0191 | 0.9881 | 0.8884 | 0.8867 | 0.5171 |
| | | | (0.1318) | (0.1340) | (0.1290) | (0.1280) | (0.1030) | (0.1014) | (0.1468) |
| | | GQL | 1.0111 | 0.9809 | 1.0164 | 0.9816 | 0.8858 | 0.8861 | 0.5194 |
| | | | (0.1182) | (0.1208) | (0.1177) | (0.1148) | | | |
| | 100 | CMLE | 0.9911 | 1.0098 | 0.9916 | 1.0085 | 0.8929 | 0.8912 | 0.4949 |
| | | | (0.0919) | (0.0965) | (0.0924) | (0.0985) | (0.0627) | (0.0663) | (0.1120) |
| | | FGLS | 0.9915 | 1.0076 | 0.9901 | 1.0099 | 0.8926 | 0.8919 | 0.4925 |
| | | | (0.1071) | (0.1095) | (0.1069) | (0.1128) | (0.0786) | (0.0722) | (0.1278) |
| | | GQL | 0.9917 | 1.0086 | 0.9921 | 1.0096 | 0.8920 | 0.8934 | 0.4929 |
| | | | (0.0940) | (0.0983) | (0.0946) | (0.1005) | | | |
| | 600 | CMLE | 1.0059 | 1.0018 | 0.9993 | 0.9996 | 0.9032 | 0.8944 | 0.5015 |
| | | | (0.0320) | (0.0330) | (0.0351) | (0.0321) | (0.0293) | (0.0262) | (0.0546) |
| | | FGLS | 1.0024 | 1.0015 | 0.9967 | 0.9949 | 0.9014 | 0.8991 | 0.5039 |
| | | | (0.0452) | (0.0462) | (0.0479) | (0.0454) | (0.0392) | (0.0381) | (0.0674) |
| | | GQL | 1.0012 | 1.0011 | 0.9997 | 0.9990 | 0.9010 | 0.8947 | 0.5031 |
| | | | (0.0322) | (0.0331) | (0.0353) | (0.0321) | | | |
| 0.9 | 60 | CMLE | 0.9855 | 0.9854 | 0.9811 | 1.0448 | 0.8871 | 0.8814 | 0.9131 |
| | | | (0.1133) | (0.1111) | (0.1141) | (0.1161) | (0.0869) | (0.0834) | (0.1292) |
| | | FGLS | 0.9859 | 0.9869 | 0.9841 | 1.0198 | 0.8860 | 0.8802 | 0.9197 |
| | | | (0.1295) | (0.1292) | (0.1298) | (0.1320) | (0.0932) | (0.0912) | (0.1491) |
| | | GQL | 0.9807 | 0.9803 | 0.9833 | 1.0156 | 0.8825 | 0.8865 | 0.9151 |
| | | | (0.1162) | (0.1141) | (0.1177) | (0.1195) | | | |
| | 100 | CMLE | 1.0073 | 0.9912 | 1.0085 | 0.9921 | 0.8912 | 0.8922 | 0.9067 |
| | | | (0.0936) | (0.0911) | (0.0942) | (0.0912) | (0.0677) | (0.0660) | (0.1174) |
| | | FGLS | 1.0086 | 0.9911 | 1.0072 | 0.9922 | 0.8911 | 0.8928 | 0.9091 |
| | | | (0.1080) | (0.1060) | (0.1096) | (0.1059) | (0.0755) | (0.0716) | (0.1290) |
| | | GQL | 1.0081 | 0.9935 | 1.0077 | 0.9927 | 0.8905 | 0.8939 | 0.9090 |
| | | | (0.0957) | (0.0936) | (0.0960) | (0.0935) | | | |
| | 600 | CMLE | 0.9985 | 1.0019 | 1.0011 | 0.9957 | 0.9021 | 0.8981 | 0.9018 |
| | | | (0.0310) | (0.0321) | (0.0311) | (0.0333) | (0.0250) | (0.0228) | (0.0595) |
| | | FGLS | 0.9954 | 1.0037 | 1.0053 | 0.9943 | 0.9011 | 0.8963 | 0.9035 |
| | | | (0.0442) | (0.0453) | (0.0448) | (0.0467) | (0.0315) | (0.0386) | (0.0660) |
| | | GQL | 0.9971 | 1.0016 | 1.0013 | 0.9981 | 0.9026 | 0.8989 | 0.9034 |
| | | | (0.0315) | (0.0323) | (0.0316) | (0.0335) | | | |

**TABLE 2.3**
Estimates of the Parameters and Standard Errors under Stationary BINAR(1) Process for $\rho_{kk} = [\rho_{11} = 0.9, \rho_{22} = 0.3]$ and $\kappa_{12,1} = [0.1, 0.3]$

| $\kappa_{12,1}$ | $T$ | Method | $\hat{\beta}_1^{[1]}$ | $\hat{\beta}_2^{[1]}$ | $\hat{\beta}_1^{[2]}$ | $\hat{\beta}_2^{[2]}$ | $\hat{\rho}_{11}$ | $\hat{\rho}_{22}$ | $\hat{\kappa}_{12,1}$ |
|---|---|---|---|---|---|---|---|---|---|
| 0.1 | 60 | CMLE | 0.9858 | 0.9827 | 0.9852 | 0.9859 | 0.9111 | 0.3152 | 0.1123 |
| | | | (0.1142) | (0.1112) | (0.1121) | (0.1120) | (0.0811) | (0.0855) | (0.0526) |
| | | FGLS | 0.9810 | 0.9819 | 0.9855 | 0.9885 | 0.9133 | 0.3132 | 0.1291 |
| | | | (0.1293) | (0.1278) | (0.1282) | (0.1286) | (0.0938) | (0.0943) | (0.0638) |
| | | GQL | 0.9851 | 0.9871 | 0.9881 | 0.9847 | 0.9136 | 0.3126 | 0.1187 |
| | | | (0.1170) | (0.1145) | (0.1150) | (0.1153) | | | |
| | 100 | CMLE | 0.9914 | 1.0133 | 1.0131 | 0.9936 | 0.8911 | 0.2978 | 0.1105 |
| | | | (0.0916) | (0.0911) | (0.0903) | (0.0905) | (0.0624) | (0.0641) | (0.0410) |
| | | FGLS | 0.9911 | 1.0113 | 1.0134 | 0.9915 | 0.8988 | 0.2913 | 0.1252 |
| | | | (0.1057) | (0.1063) | (0.1053) | (0.1055) | (0.0710) | (0.0756) | (0.0564) |
| | | GQL | 0.9943 | 1.0131 | 1.0122 | 0.9941 | 0.8909 | 0.2990 | 0.1129 |
| | | | (0.0934) | (0.0930) | (0.0921) | (0.0922) | | | |
| | 600 | CMLE | 0.9975 | 0.9969 | 0.9959 | 1.0028 | 0.8949 | 0.2983 | 0.1019 |
| | | | (0.0325) | (0.0345) | (0.0316) | (0.0317) | (0.0237) | (0.0223) | (0.0130) |
| | | FGLS | 0.9978 | 0.9959 | 0.9965 | 1.0011 | 0.8990 | 0.2988 | 0.1145 |
| | | | (0.0456) | (0.0479) | (0.0459) | (0.0468) | (0.0351) | (0.0366) | (0.0224) |
| | | GQL | 0.9978 | 0.9976 | 0.9980 | 1.0012 | 0.8957 | 0.2994 | 0.1040 |
| | | | (0.0331) | (0.0355) | (0.0318) | (0.0319) | | | |
| 0.3 | 60 | CMLE | 0.9878 | 0.9885 | 1.0210 | 1.0272 | 0.8887 | 0.2869 | 0.3189 |
| | | | (0.1163) | (0.1140) | (0.1150) | (0.1138) | (0.0837) | (0.0817) | (0.1240) |
| | | FGLS | 0.9825 | 0.9861 | 1.0131 | 1.0145 | 0.8856 | 0.2868 | 0.3281 |
| | | | (0.1329) | (0.1295) | (0.1314) | (0.1299) | (0.0942) | (0.0972) | (0.1420) |
| | | GQL | 0.9831 | 0.9815 | 1.0121 | 1.0189 | 0.8827 | 0.2871 | 0.3141 |
| | | | (0.1194) | (0.1170) | (0.1180) | (0.1165) | | | |
| | 100 | CMLE | 0.9913 | 1.0108 | 0.9987 | 1.0107 | 0.9101 | 0.2913 | 0.3014 |
| | | | (0.0924) | (0.0906) | (0.0928) | (0.0921) | (0.0675) | (0.0617) | (0.1186) |
| | | FGLS | 0.9914 | 1.0112 | 0.9919 | 1.0109 | 0.9111 | 0.2949 | 0.2895 |
| | | | (0.1077) | (0.1055) | (0.1073) | (0.1075) | (0.0782) | (0.0798) | (0.1274) |
| | | GQL | 0.9919 | 1.0111 | 0.9987 | 1.0131 | 0.9126 | 0.2955 | 0.2971 |
| | | | (0.0942) | (0.0920) | (0.0940) | (0.0941) | | | |
| | 600 | CMLE | 1.0045 | 0.9951 | 1.0071 | 0.9979 | 0.8950 | 0.2966 | 0.2993 |
| | | | (0.0315) | (0.0349) | (0.0345) | (0.0317) | (0.0218) | (0.0250) | (0.0528) |
| | | FGLS | 1.0012 | 0.9982 | 1.0066 | 0.9962 | 0.8959 | 0.2972 | 0.2898 |
| | | | (0.0445) | (0.0480) | (0.0475) | (0.0448) | (0.0331) | (0.0344) | (0.0671) |
| | | GQL | 1.0002 | 0.9981 | 1.0015 | 0.9974 | 0.8984 | 0.2956 | 0.2997 |
| | | | (0.0320) | (0.0358) | (0.0346) | (0.0317) | | | |

It can be seen from Tables 2.1–2.6 that both GQL and CMLE yield consistent estimates of the regression and dependence effects irrespective of the combination of low and high serial- or cross-correlation coefficients. Notably, there are no huge discrepancies between the standard errors of the estimates under both approaches with CMLE yielding estimates with slightly lower standard errors. As the time point increases, both methods yield estimates of same level of asymptotic efficiency, which is a similar conclusion as in Sutradhar et al. (2014). It is also noted that GQL gives more efficient estimates than FGLS. The next table provides the number of non-convergent simulations

**TABLE 2.4**
Estimates of the Parameters and Standard Errors under Stationary BINAR(1) Process for $\rho_{kk} = [\rho_{11} = 0.9, \rho_{22} = 0.3]$ and $\kappa_{12,1} = [0.5, 0.9]$

| $\kappa_{12,1}$ | $T$ | Method | $\hat{\beta}_1^{[1]}$ | $\hat{\beta}_2^{[1]}$ | $\hat{\beta}_1^{[2]}$ | $\hat{\beta}_2^{[2]}$ | $\hat{\rho}_{11}$ | $\hat{\rho}_{22}$ | $\hat{\kappa}_{12,1}$ |
|---|---|---|---|---|---|---|---|---|---|
| 0.5 | 60 | CMLE | 0.9836 | 0.9813 | 1.0135 | 1.0153 | 0.8833 | 0.3187 | 0.4880 |
| | | | (0.1127) | (0.1128) | (0.1116) | (0.1144) | (0.0802) | (0.0886) | (0.1249) |
| | | FGLS | 0.9889 | 0.9829 | 1.0112 | 1.0134 | 0.8874 | 0.3179 | 0.4866 |
| | | | (0.1289) | (0.1290) | (0.1286) | (0.1298) | (0.0998) | (0.0942) | (0.1345) |
| | | GQL | 0.9831 | 0.9869 | 1.0187 | 1.0168 | 0.8854 | 0.3133 | 0.4846 |
| | | | (0.1157) | (0.1158) | (0.1145) | (0.1144) | | | |
| | 100 | CMLE | 1.0111 | 1.0112 | 0.9923 | 0.9929 | 0.8918 | 0.3048 | 0.5065 |
| | | | (0.0912) | (0.0913) | (0.0923) | (0.0951) | (0.0640) | (0.0630) | (0.1146) |
| | | FGLS | 1.0116 | 1.0112 | 0.9969 | 0.9931 | 0.8955 | 0.3051 | 0.5092 |
| | | | (0.1063) | (0.1064) | (0.1074) | (0.1092) | (0.0751) | (0.0741) | (0.1266) |
| | | GQL | 1.0119 | 1.0111 | 0.9966 | 0.9938 | 0.8971 | 0.3041 | 0.5082 |
| | | | (0.0930) | (0.0931) | (0.0950) | (0.0951) | | | |
| | 600 | CMLE | 1.0021 | 1.0092 | 0.9939 | 0.9942 | 0.9022 | 0.3018 | 0.4985 |
| | | | (0.0312) | (0.0322) | (0.0313) | (0.0345) | (0.0235) | (0.0222) | (0.0519) |
| | | FGLS | 1.0082 | 1.0025 | 0.9973 | 0.9979 | 0.9029 | 0.3027 | 0.4976 |
| | | | (0.0435) | (0.0459) | (0.0444) | (0.0479) | (0.0381) | (0.0317) | (0.0648) |
| | | GQL | 1.0065 | 1.0021 | 0.9989 | 0.9971 | 0.9013 | 0.3026 | 0.4962 |
| | | | (0.0313) | (0.0325) | (0.0324) | (0.0347) | | | |
| 0.9 | 60 | CMLE | 0.9848 | 0.9854 | 1.0145 | 1.0187 | 0.8876 | 0.3124 | 0.9152 |
| | | | (0.1133) | (0.1140) | (0.1121) | (0.1128) | (0.0835) | (0.0828) | (0.1247) |
| | | FGLS | 0.9816 | 0.9890 | 1.0116 | 1.0116 | 0.8817 | 0.3174 | 0.9117 |
| | | | (0.1295) | (0.1294) | (0.1289) | (0.1285) | (0.0977) | (0.0979) | (0.1331) |
| | | GQL | 0.9876 | 0.9831 | 1.0104 | 1.0170 | 0.8801 | 0.3121 | 0.9116 |
| | | | (0.1162) | (0.1170) | (0.1150) | (0.1158) | | | |
| | 100 | CMLE | 1.0122 | 0.9925 | 1.0125 | 0.9931 | 0.9015 | 0.3013 | 0.9044 |
| | | | (0.0915) | (0.0971) | (0.0914) | (0.0999) | (0.0649) | (0.0613) | (0.1162) |
| | | FGLS | 1.0115 | 0.9942 | 1.0111 | 0.9922 | 0.9085 | 0.3019 | 0.9061 |
| | | | (0.1058) | (0.1095) | (0.1065) | (0.1150) | (0.0725) | (0.0771) | (0.1286) |
| | | GQL | 1.0101 | 0.9954 | 1.0118 | 0.9901 | 0.9058 | 0.3094 | 0.9054 |
| | | | (0.0935) | (0.0990) | (0.0931) | (0.1015) | | | |
| | 600 | CMLE | 0.9981 | 1.0098 | 1.0080 | 0.9973 | 0.9035 | 0.3019 | 0.9016 |
| | | | (0.0311) | (0.0368) | (0.0333) | (0.0317) | (0.0296) | (0.0279) | (0.0570) |
| | | FGLS | 0.9962 | 1.0098 | 1.0022 | 0.9975 | 0.9029 | 0.3017 | 0.9045 |
| | | | (0.0471) | (0.0499) | (0.0478) | (0.0473) | (0.0372) | (0.0383) | (0.0660) |
| | | GQL | 0.9975 | 1.0049 | 1.0031 | 0.9998 | 0.9025 | 0.3018 | 0.9030 |
| | | | (0.0320) | (0.0369) | (0.0333) | (0.0320) | | | |

under GQL, CMLE and FGLS approaches for the different combinations for the different time size.

From Table 2.7, we conclude that GQL is computationally more appealing since it yields fewer non-convergent simulations as compared to CMLE and FGLS. The problem with CMLE is mainly its double derivative log-likelihood matrix is rather ill-conditioned in the intermediate computations, which leads to convergence failures whereas the Hessian component in the GQL is rather less intensive though some failures were noted. In addition, the discrepancies in standard errors between GQL and CMLE decrease as sample size increases,

**TABLE 2.5**
Estimates of the Parameters and Standard Errors under Stationary BINAR(1) Process for $\rho_{kk} = [\rho_{11} = 0.3, \rho_{22} = 0.3]$ and $\kappa_{12,1} = [0.1, 0.3]$

| $\kappa_{12,1}$ | $T$ | Method | $\hat{\beta}_1^{[1]}$ | $\hat{\beta}_2^{[1]}$ | $\hat{\beta}_1^{[2]}$ | $\hat{\beta}_2^{[2]}$ | $\hat{\rho}_{11}$ | $\hat{\rho}_{22}$ | $\hat{\kappa}_{12,1}$ |
|---|---|---|---|---|---|---|---|---|---|
| 0.1 | 60 | CMLE | 1.0129 | 0.9842 | 0.9879 | 1.0115 | 0.3121 | 0.3101 | 0.1114 |
| | | | (0.1115) | (0.1134) | (0.1136) | (0.1127) | (0.0822) | (0.0894) | (0.0408) |
| | | FGLS | 1.0151 | 0.9869 | 0.9815 | 1.0149 | 0.3162 | 0.3139 | 0.1259 |
| | | | (0.1275) | (0.1288) | (0.1290) | (0.1289) | (0.0915) | (0.0996) | (0.0613) |
| | | GQL | 1.0102 | 0.9809 | 0.9883 | 1.0171 | 0.3109 | 0.3178 | 0.1148 |
| | | | (0.1140) | (0.1160) | (0.1162) | (0.1155) | | | |
| | 100 | CMLE | 1.0124 | 0.9919 | 1.0111 | 0.9923 | 0.2914 | 0.3059 | 0.1110 |
| | | | (0.0932) | (0.0935) | (0.0944) | (0.0924) | (0.0611) | (0.0647) | (0.0331) |
| | | FGLS | 1.0117 | 0.9912 | 1.0111 | 0.9914 | 0.2982 | 0.3010 | 0.1239 |
| | | | (0.1083) | (0.1095) | (0.1090) | (0.1075) | (0.0773) | (0.0798) | (0.0590) |
| | | GQL | 1.1121 | 0.9912 | 1.0111 | 0.9912 | 0.2933 | 0.3047 | 0.1131 |
| | | | (0.0950) | (0.0955) | (0.0963) | (0.0941) | | | |
| | 600 | CMLE | 0.9989 | 1.0015 | 1.0049 | 0.9966 | 0.2959 | 0.3019 | 0.1021 |
| | | | (0.0302) | (0.0310) | (0.0343) | (0.0322) | (0.0243) | (0.0208) | (0.0040) |
| | | FGLS | 0.9969 | 1.0029 | 1.0011 | 0.9944 | 0.2991 | 0.3003 | 0.1141 |
| | | | (0.0435) | (0.0445) | (0.0478) | (0.0456) | (0.0364) | (0.0374) | (0.0120) |
| | | GQL | 0.9962 | 1.0008 | 1.0070 | 0.9971 | 0.2973 | 0.3011 | 0.1047 |
| | | | (0.0310) | (0.0311) | (0.0352) | (0.0325) | | | |
| 0.3 | 60 | CMLE | 0.9818 | 0.9869 | 1.0179 | 1.0115 | 0.2849 | 0.3169 | 0.3119 |
| | | | (0.1112) | (0.1123) | (0.1147) | (0.1105) | (0.0889) | (0.0876) | (0.1272) |
| | | FGLS | 0.9872 | 0.9810 | 1.0151 | 1.0101 | 0.2816 | 0.3189 | 0.3271 |
| | | | (0.1274) | (0.1277) | (0.1299) | (0.1268) | (0.0998) | (0.0986) | (0.1388) |
| | | GQL | 0.9857 | 0.9813 | 1.0117 | 1.0157 | 0.2838 | 0.3166 | 0.3159 |
| | | | (0.1141) | (0.1152) | (0.1176) | (0.1135) | | | |
| | 100 | CMLE | 0.9914 | 1.0112 | 1.0117 | 0.9905 | 0.3011 | 0.3019 | 0.3010 |
| | | | (0.0910) | (0.0925) | (0.0914) | (0.0926) | (0.0660) | (0.0649) | (0.1189) |
| | | FGLS | 0.9914 | 1.0119 | 1.0115 | 0.9969 | 0.3079 | 0.3025 | 0.3252 |
| | | | (0.1066) | (0.1079) | (0.1056) | (0.1078) | (0.0721) | (0.0790) | (0.1222) |
| | | GQL | 0.9901 | 1.0117 | 1.0101 | 0.9917 | 0.3065 | 0.3061 | 0.3042 |
| | | | (0.0930) | (0.0954) | (0.0942) | (0.0955) | | | |
| | 600 | CMLE | 1.0010 | 0.9959 | 0.9956 | 1.0013 | 0.3006 | 0.3003 | 0.3012 |
| | | | (0.0349) | (0.0331) | (0.0311) | (0.0312) | (0.0260) | (0.0244) | (0.0507) |
| | | FGLS | 1.0083 | 0.9970 | 0.9913 | 1.0092 | 0.3045 | 0.3016 | 0.3132 |
| | | | (0.0479) | (0.0463) | (0.0444) | (0.0446) | (0.0355) | (0.0311) | (0.0609) |
| | | GQL | 1.0008 | 0.9921 | 0.9952 | 1.0086 | 0.3039 | 0.3025 | 0.3022 |
| | | | (0.0350) | (0.0340) | (0.0314) | (0.0315) | | | |

such that for $T = 600$, the GQL estimates are asymptotically equally efficient as CMLE. Hence, based on these findings, we recommend only the GQL approach under the non-stationary set-up.

### 2.3.2 Non-Stationary Conditions

Based on the findings of Subsection 2.3.1.3, it is concluded that GQL provides more efficient estimates than FGLS while asymptotically equally efficient estimates than CMLE under the stationary moment condition. Also, the number

**TABLE 2.6**
Estimates of the Parameters and Standard Errors under Stationary BINAR(1) Process for $\rho_{kk} = [\rho_{11} = 0.3, \rho_{22} = 0.3]$ and $\kappa_{12,1} = [0.5, 0.9]$

| $\kappa_{12,1}$ | $T$ | Method | $\hat{\beta}_1^{[1]}$ | $\hat{\beta}_2^{[1]}$ | $\hat{\beta}_1^{[2]}$ | $\hat{\beta}_2^{[2]}$ | $\hat{\rho}_{11}$ | $\hat{\rho}_{22}$ | $\hat{\kappa}_{12,1}$ |
|---|---|---|---|---|---|---|---|---|---|
| 0.5 | 60 | CMLE | 0.9870 | 0.9810 | 1.0149 | 1.0199 | 0.2891 | 0.3102 | 0.4898 |
| | | | (0.1110) | (0.1120) | (0.1147) | (0.1118) | (0.0889) | (0.0828) | (0.1210) |
| | | FGLS | 0.9813 | 0.9817 | 1.0110 | 1.0186 | 0.2801 | 0.3148 | 0.4855 |
| | | | (0.1275) | (0.1284) | (0.1299) | (0.1279) | (0.0913) | (0.0964) | (0.1386) |
| | | GQL | 0.9806 | 0.9828 | 1.0112 | 1.0121 | 0.2806 | 0.3135 | 0.4874 |
| | | | (0.1140) | (0.1150) | (0.1176) | (0.1148) | | | |
| | 100 | CMLE | 1.0125 | 1.0134 | 0.9936 | 0.9921 | 0.3012 | 0.2922 | 0.5078 |
| | | | (0.0948) | (0.0930) | (0.0920) | (0.0910) | (0.0672) | (0.0645) | (0.1107) |
| | | FGLS | 1.0115 | 1.0155 | 0.9952 | 0.9916 | 0.3069 | 0.2918 | 0.5084 |
| | | | (0.1089) | (0.1080) | (0.1070) | (0.1063) | (0.0720) | (0.0751) | (0.1255) |
| | | GQL | 1.0141 | 1.0174 | 0.9901 | 0.9978 | 0.3063 | 0.2926 | 0.5096 |
| | | | (0.0966) | (0.0947) | (0.0936) | (0.0930) | | | |
| | 600 | CMLE | 1.0049 | 1.0098 | 0.9994 | 0.9959 | 0.3011 | 0.2992 | 0.4985 |
| | | | (0.0328) | (0.0317) | (0.0349) | (0.0330) | (0.0282) | (0.0258) | (0.0536) |
| | | FGLS | 1.0033 | 1.0042 | 0.9981 | 0.9941 | 0.3015 | 0.2952 | 0.4968 |
| | | | (0.0459) | (0.0448) | (0.0479) | (0.0463) | (0.0397) | (0.0390) | (0.0620) |
| | | GQL | 1.0054 | 1.0058 | 0.9984 | 0.9988 | 0.3025 | 0.2977 | 0.4974 |
| | | | (0.0330) | (0.0321) | (0.0357) | (0.0333) | | | |
| 0.9 | 60 | CMLE | 0.9812 | 1.0115 | 1.0125 | 0.9885 | 0.3113 | 0.3104 | 0.9108 |
| | | | (0.1120) | (0.1111) | (0.1115) | (0.1127) | (0.0823) | (0.0886) | (0.1229) |
| | | FGLS | 0.9891 | 1.0115 | 1.0192 | 0.9855 | 0.3119 | 0.3109 | 0.9141 |
| | | | (0.1285) | (0.1276) | (0.1274) | (0.1288) | (0.0958) | (0.0930) | (0.1371) |
| | | GQL | 0.9848 | 1.0125 | 1.0113 | 0.9821 | 0.3135 | 0.3101 | 0.9133 |
| | | | (0.1150) | (0.1140) | (0.1140) | (0.1155) | | | |
| | 100 | CMLE | 0.9946 | 0.9978 | 1.0127 | 1.0117 | 0.3071 | 0.3011 | 0.9031 |
| | | | (0.0927) | (0.0935) | (0.0946) | (0.0954) | (0.0635) | (0.0614) | (0.1118) |
| | | FGLS | 0.9918 | 0.9982 | 1.0111 | 1.0119 | 0.3058 | 0.3099 | 0.9055 |
| | | | (0.1078) | (0.1088) | (0.1097) | (0.1096) | (0.0717) | (0.0714) | (0.1242) |
| | | GQL | 0.9957 | 0.9917 | 1.0108 | 1.0107 | 0.3058 | 0.3053 | 0.9048 |
| | | | (0.0946) | (0.0955) | (0.0964) | (0.0972) | | | |
| | 600 | CMLE | 0.9964 | 0.9989 | 1.0012 | 1.0077 | 0.3049 | 0.3002 | 0.9002 |
| | | | (0.0311) | (0.0342) | (0.0336) | (0.0353) | (0.0211) | (0.0297) | (0.0572) |
| | | FGLS | 0.9953 | 0.9989 | 1.0015 | 1.0097 | 0.3042 | 0.3018 | 0.9021 |
| | | | (0.0442) | (0.0475) | (0.0469) | (0.0484) | (0.0345) | (0.0350) | (0.0677) |
| | | GQL | 0.9963 | 0.9977 | 1.0014 | 1.0045 | 0.3038 | 0.3013 | 0.9011 |
| | | | (0.0314) | (0.0345) | (0.0340) | (0.0359) | | | |

of non-convergent simulations for CMLE is very high as compared to GQL and under the non-stationary moment assumption, we expect more computational problems in implementing the CMLE algorithm. Hence, only the GQL estimation method is derived under the non-stationary moment assumption.

#### 2.3.2.1 Generalized Quasi-Likelihood (GQL)

In this subsection, a GQL equation is developed to estimate the regression effects in the non-stationary BINAR(1) process. The GQL equation is expressed

**TABLE 2.7**
Number of Non-Convergent Simulations under GQL, CMLE and FGLS for the Different Combinations

| $\hat{\rho}_{11}$ | $\hat{\rho}_{22}$ | $\hat{\kappa}_{12,1}$ | T | GQL | CMLE | FGLS | $\hat{\rho}_{11}$ | $\hat{\rho}_{22}$ | $\hat{\kappa}_{12,1}$ | T | GQL | CMLE | FGLS |
|---|---|---|---|---|---|---|---|---|---|---|---|---|---|
| 0.9 | 0.9 | 0.1 | 60 | 175 | 600 | 320 | 0.9 | 0.3 | 0.5 | 60 | 190 | 625 | 325 |
| | | | 100 | 140 | 535 | 300 | | | | 100 | 160 | 540 | 280 |
| | | | 600 | 60 | 370 | 240 | | | | 600 | 80 | 390 | 215 |
| 0.9 | 0.9 | 0.3 | 60 | 185 | 615 | 340 | 0.9 | 0.3 | 0.9 | 60 | 200 | 630 | 335 |
| | | | 100 | 150 | 540 | 315 | | | | 100 | 160 | 570 | 300 |
| | | | 600 | 80 | 380 | 260 | | | | 600 | 90 | 400 | 230 |
| 0.9 | 0.9 | 0.5 | 60 | 195 | 630 | 355 | 0.3 | 0.3 | 0.1 | 60 | 160 | 575 | 290 |
| | | | 100 | 165 | 555 | 335 | | | | 100 | 120 | 520 | 270 |
| | | | 600 | 90 | 395 | 265 | | | | 600 | 50 | 350 | 200 |
| 0.9 | 0.9 | 0.9 | 60 | 205 | 640 | 360 | 0.3 | 0.3 | 0.3 | 60 | 170 | 600 | 300 |
| | | | 100 | 170 | 575 | 310 | | | | 100 | 130 | 525 | 275 |
| | | | 600 | 100 | 410 | 270 | | | | 600 | 70 | 360 | 220 |
| 0.9 | 0.3 | 0.1 | 60 | 165 | 585 | 310 | 0.3 | 0.3 | 0.5 | 60 | 180 | 620 | 315 |
| | | | 100 | 130 | 530 | 270 | | | | 100 | 150 | 530 | 285 |
| | | | 600 | 55 | 360 | 215 | | | | 600 | 75 | 380 | 230 |
| 0.9 | 0.3 | 0.3 | 60 | 180 | 605 | 320 | 0.3 | 0.3 | 0.9 | 60 | 190 | 625 | 320 |
| | | | 100 | 140 | 530 | 275 | | | | 100 | 150 | 560 | 280 |
| | | | 600 | 75 | 370 | 220 | | | | 600 | 80 | 390 | 235 |

as

$$\boldsymbol{D_\beta}'\boldsymbol{\Sigma_\beta}^{-1}(\boldsymbol{y}-\boldsymbol{\mu}) = 0, \tag{2.52}$$

where

$$\boldsymbol{y} = \begin{pmatrix} \boldsymbol{y}^{[1]} \\ \boldsymbol{y}^{[2]} \end{pmatrix}_{2T\times 1}, \boldsymbol{y}^{[1]} = [y_1^{[1]}, y_2^{[1]}, \dots, y_T^{[1]}]', \boldsymbol{y}^{[2]} = [y_1^{[2]}, y_2^{[2]}, \dots, y_T^{[2]}]' \tag{2.53}$$

and

$$\boldsymbol{\mu} = \begin{pmatrix} \boldsymbol{\mu}^{[1]} \\ \boldsymbol{\mu}^{[2]} \end{pmatrix}_{2T\times 1}, \boldsymbol{\mu}^{[1]} = [\mu_1^{[1]}, \mu_2^{[1]}, \dots, \mu_T^{[1]}]', \boldsymbol{\mu}^{[2]} = [\mu_1^{[2]}, \mu_2^{[2]}, \dots, \mu_T^{[2]}]'. \tag{2.54}$$

The derivative matrix is a block diagonal $(2T \times 2p)$ matrix with $[\boldsymbol{D_\beta}^{[1]}, \boldsymbol{D_\beta}^{[2]}]$ on the diagonals where

$$\boldsymbol{D_\beta}^{[k]} = \begin{pmatrix} \frac{\partial \mu_1^{[k]}}{\partial \beta_1^{[k]}} & \frac{\partial \mu_1^{[k]}}{\partial \beta_2^{[k]}} & \cdots & \frac{\partial \mu_1^{[k]}}{\partial \beta_p^{[k]}} \\ \frac{\partial \mu_2^{[k]}}{\partial \beta_1^{[k]}} & \frac{\partial \mu_2^{[k]}}{\partial \beta_2^{[k]}} & \cdots & \frac{\partial \mu_2^{[k]}}{\partial \beta_p^{[k]}} \\ \vdots & \vdots & \ddots & \vdots \\ \frac{\partial \mu_T^{[k]}}{\partial \beta_1^{[k]}} & \frac{\partial \mu_T^{[k]}}{\partial \beta_2^{[k]}} & \cdots & \frac{\partial \mu_T^{[k]}}{\partial \beta_p^{[k]}} \end{pmatrix}_{(T\times p)} \tag{2.55}$$

and $\frac{\partial \mu_t^{[k]}}{\partial \beta_j^{[k]}} = \mu_t^{[k]} x_{tj}$.

The covariance structure $\boldsymbol{\Sigma}_{\boldsymbol{\beta}}$ is similar to Equation (2.43) with the entries computed by Equations (2.6)–(2.12). However, note that for $t = 1$, $\text{Cov}(Y_t^{[1]}, Y_t^{[2]})$ is computed using Equation (2.15), while for $t = 2, \ldots, T$, $\text{Cov}(Y_t^{[1]}, Y_t^{[2]})$ is computed iteratively using Equation (2.9).

Under the GQL approach, the serial-correlation in Equation (2.8) is calculated using Sutradhar & Das (1999) and Jowaheer & Sutradhar (2005) moment estimation approach

$$\frac{\sum_{t=1}^{T-1} \tilde{y_t}^{[k]} \tilde{y}_{(t+1)}^{[k]}/(T-1)}{\sum_{t=1}^{T} (\tilde{y}_t^{[k]})^2/T} = \hat{\rho}_{kk} \frac{\sum_{t=1}^{T-1} \sqrt{\frac{\mu_t^{[k]}}{\mu_{t+1}^{[k]}}}}{(T-1)} \tag{2.56}$$

which yields

$$\hat{\rho}_{kk} = \frac{T \sum_{t=1}^{T-1} \tilde{y_t}^{[k]} \tilde{y}_{(t+1)}^{[k]}}{\sum_{t=1}^{T} (\tilde{y}_t^{[k]})^2 \sum_{t=1}^{T-1} \sqrt{\frac{\mu_t^{[k]}}{\mu_{t+1}^{[k]}}}} \tag{2.57}$$

where $\tilde{y_t}^{[k]} = \frac{y_t^{[k]} - \hat{\mu}_t^{[k]}}{\sqrt{\hat{\mu}_t^{[k]}}}$.

The cross-correlation $\kappa_{12,t}$ is estimated using Equation (2.9), that is,

$$\hat{\kappa}_{12,t} = \frac{\tilde{\text{Cov}}(Y_t^{[1]}, Y_t^{[2]}) - \hat{\rho}_{11}\hat{\rho}_{22}\tilde{\text{Cov}}(Y_{t-1}^{[1]}, Y_{t-1}^{[2]})}{\sqrt{(\hat{\mu}_t^{[1]} - \hat{\rho}_{11}\hat{\mu}_{t-1}^{[1]})}\sqrt{(\hat{\mu}_t^{[2]} - \hat{\rho}_{22}\hat{\mu}_{t-1}^{[2]})}}, \tag{2.58}$$

where $\tilde{\text{Cov}}(Y_t^{[1]}, Y_t^{[2]}) = \frac{1}{T} \sum_{t=1}^{T} (y_t^{[1]} - \hat{\mu}_t^{[1]})(y_t^{[2]} - \hat{\mu}_t^{[2]})$, $\tilde{\text{Cov}}(Y_{t-1}^{[1]}, Y_{t-1}^{[2]}) = \frac{1}{T-1} \sum_{t=2}^{T} (y_{t-1}^{[1]} - \hat{\mu}_{t-1}^{[1]})(y_{t-1}^{[2]} - \hat{\mu}_{t-1}^{[2]})$ and $\hat{\mu}_0^{[k]} = \hat{\mu}_1^{[k]}$.

The Newton-Raphson iterative technique is used to solve the GQL in Equation (2.52) and this yields

$$\begin{pmatrix} \hat{\boldsymbol{\beta}}_{r+1}^{[1]} \\ \hat{\boldsymbol{\beta}}_{r+1}^{[2]} \end{pmatrix} = \begin{pmatrix} \hat{\boldsymbol{\beta}}_{r}^{[1]} \\ \hat{\boldsymbol{\beta}}_{r}^{[2]} \end{pmatrix} + [\boldsymbol{D}_{\boldsymbol{\beta}}{}' \boldsymbol{\Sigma}_{\boldsymbol{\beta}}{}^{-1} \boldsymbol{D}_{\boldsymbol{\beta}}]_r^{-1} [\boldsymbol{D}_{\boldsymbol{\beta}}{}' \boldsymbol{\Sigma}_{\boldsymbol{\beta}}{}^{-1} (\boldsymbol{y} - \boldsymbol{\mu})]_r, \tag{2.59}$$

where $\hat{\boldsymbol{\beta}}_r^{[k]}$ are the estimates at the $r^{th}$ iteration and $[.]_r$ are the values of the expression at the $r^{th}$ iteration.

The algorithm works as follows: For an initial value of regression parameters, $\rho_{kk}$, $\kappa_{12,t}$ and the covariance matrix are computed. Using these estimates, the iterative Equation (2.59) is solved to obtain an updated regression coefficients. These updated regression estimates are used to re-compute $\rho_{kk}$, $\kappa_{12,t}$ and the entries of the covariance matrix. Then, using the iterative Equation (2.59), we obtain another updated regression coefficients, which are used to recalculate $\rho_{kk}$, $\kappa_{12,t}$ and the covariance matrix. This cycle of iteration continues until convergence. Under mild regulatory conditions, the regression estimators are consistent and $((\hat{\boldsymbol{\beta}}^{[1]}, \hat{\boldsymbol{\beta}}^{[2]}) - (\boldsymbol{\beta}^{[1]}, \boldsymbol{\beta}^{[2]}))'$ follows an asymptotic normal distribution with mean $\mathbf{0}$ and covariance matrix $[\boldsymbol{D_\beta}'\boldsymbol{\Sigma_\beta}^{-1}\boldsymbol{D_\beta}]^{-1}$ (Refer to Subsection 2.3.1.2).

Following Sutradhar (2008), the conditional expectation and variance of the one-step ahead forecast $Y_{t+1}^{[k]}$ given $y_t^{[k]}$ are as follows:

$$\begin{aligned} E(Y_{t+1}^{[k]}|y_t^{[k]}) &= E(\rho_{kk} \circ Y_t^{[k]}|y_t^{[k]}) + E(R_{t+1}^{[k]}) \\ &= \rho_{kk}y_t^{[k]} + \mu_{t+1}^{[k]} - \rho_{kk}\mu_t^{[k]} \\ &= \mu_{t+1}^{[k]} + \rho_{kk}(y_t^{[k]} - \mu_t^{[k]}) \end{aligned} \tag{2.60}$$

and

$$\begin{aligned} \text{Var}(Y_{t+1}^{[k]}|y_t^{[k]}) &= \text{Var}(\rho_{kk} \circ Y_t^{[k]}|y_t^{[k]}) + \text{Var}(R_{t+1}^{[k]}) \\ &= \rho_{kk}(1-\rho_{kk})y_t^{[k]} + \mu_{t+1}^{[k]} - \rho_{kk}\mu_t^{[k]} \\ &= \mu_{t+1}^{[k]} + \rho_{kk}(y_t^{[k]} - \mu_t^{[k]}) - \rho_{kk}^2 y_t^{[k]}. \end{aligned} \tag{2.61}$$

#### 2.3.2.2 Simulation

In this section, we generate time series of non-stationary BINAR(1) data using the OD model (2.1) and the GQL method under non-stationary moment assumption is applied to estimate the regression parameters. The bivariate innovation series are generated using the multivariate Poisson R code developed in Subsection 2.3.1.3, where $R = matrix(c(1, \kappa_{12,1}, \kappa_{12,1}, 1), 2, 2)$ and $lambda = matrix((\mu_t^{[1]} - \rho_{11}\mu_{t-1}^{[1]}), (\mu_t^{[2]} - \rho_{22}\mu_{t-1}^{[2]}), 2, 1)$.

Hence, for each time point, a pair of simulated innovation term based on the non-stationary means described in Equation (2.3) is generated, but using $\kappa_{12,t} = \kappa_{12,1}$ for facilitating the simulation process where $\kappa_{12,1} = 0.1, 0.3, 0.5, 0.9$, $\beta_1^{[k]} = \beta_2^{[k]} = 1$ for $k \in \{1, 2\}$ and $[\rho_{11}, \rho_{22}]$=[0.9,0.9], [0.9,0.3], [0.3,0.3]. Note that the non-stationarity will still be accounted in spite of a fixed $\kappa_{12,1}$ in the simulation process since the mean at each time point varies for both series, hence the covariance $\phi_t$ will still vary. Hence, the series of values $y_t^{[1]}$ and $y_t^{[2]}$ are generated using the simulated pair of innovations $(r_t^{[1]}, r_t^{[2]})$ and $Y_{t-1}^{[k]} \sim \text{Poisson}(\mu_{t-1}^{[k]})$.

Five thousand samples of data for different values of $\kappa_{12,1}$ under $T = 60, 100$ and $600$ are generated based on a two-dimensional covariate vector where the first element is

$$x_{t1} = \begin{cases} -1 + t & (t = 1, \ldots, T/4), \\ rnorm(1, 0, 1) & (t = (T/4) + 1, \ldots, 3T/4), \\ 1 + t & (t = (3T/4) + 1, \ldots, T), \end{cases}$$

where *rnorm* generates random numbers from normal distribution with mean 0 and variance 1 in R and the second element $x_{t2}$ is random Poisson with parameter 2. The results are shown in Tables 2.8–2.10.

**TABLE 2.8**
Estimates of the Parameters and Standard Errors under Non-Stationary BINAR(1) Process for $\rho_{kk} = [\rho_{11} = 0.9, \rho_{22} = 0.9]$

| $\kappa_{12,1}$ | $T$ | $\hat{\beta}_1^{[1]}$ | $\hat{\beta}_2^{[1]}$ | $\hat{\beta}_1^{[2]}$ | $\hat{\beta}_2^{[2]}$ | $\hat{\rho}_{11}$ | $\hat{\rho}_{22}$ | $\hat{\kappa}_{12,1}$ |
|---|---|---|---|---|---|---|---|---|
| 0.1 | 60 | 0.9818 | 0.9855 | 0.9866 | 0.9838 | 0.8847 | 0.8859 | 0.1180 |
| | | (0.0910) | (0.0920) | (0.0918) | (0.0990) | | | |
| | 100 | 1.0106 | 0.9909 | 0.9901 | 0.9913 | 0.9087 | 0.8921 | 0.1040 |
| | | (0.0700) | (0.0711) | (0.0712) | (0.0713) | | | |
| | 600 | 1.0050 | 0.9975 | 1.0010 | 1.0015 | 0.9036 | 0.9008 | 0.1011 |
| | | (0.0111) | (0.0110) | (0.0111) | (0.0102) | | | |
| 0.3 | 60 | 1.0179 | 1.0154 | 1.0109 | 1.0135 | 0.9149 | 0.9105 | 0.3154 |
| | | (0.1046) | (0.0940) | (0.1021) | (0.0950) | | | |
| | 100 | 1.0135 | 0.9916 | 0.9923 | 1.0182 | 0.8968 | 0.9055 | 0.3025 |
| | | (0.0730) | (0.0834) | (0.0815) | (0.0710) | | | |
| | 600 | 1.0018 | 0.9990 | 1.0047 | 1.0027 | 0.9014 | 0.9002 | 0.3015 |
| | | (0.0223) | (0.0160) | (0.0157) | (0.0250) | | | |
| 0.5 | 60 | 1.0176 | 1.0122 | 0.9826 | 0.9864 | 0.9155 | 0.8849 | 0.4872 |
| | | (0.1104) | (0.1014) | (0.1102) | (0.1112) | | | |
| | 100 | 0.9919 | 0.9902 | 1.0099 | 1.0086 | 0.9077 | 0.9066 | 0.4961 |
| | | (0.0812) | (0.0853) | (0.0711) | (0.0978) | | | |
| | 600 | 0.9986 | 0.9993 | 1.0025 | 1.0012 | 0.9012 | 0.9018 | 0.5009 |
| | | (0.0178) | (0.0390) | (0.0249) | (0.0190) | | | |
| 0.9 | 60 | 1.0132 | 0.9841 | 1.0181 | 0.9853 | 0.8831 | 0.8896 | 0.8839 |
| | | (0.1211) | (0.1250) | (0.1130) | (0.1326) | | | |
| | 100 | 0.9958 | 1.0099 | 1.0085 | 1.0088 | 0.8953 | 0.8989 | 0.8912 |
| | | (0.0905) | (0.0910) | (0.0720) | (0.0841) | | | |
| | 600 | 1.0009 | 1.0021 | 1.0003 | 1.0034 | 0.9006 | 0.9012 | 0.8989 |
| | | (0.0320) | (0.0370) | (0.0311) | (0.0428) | | | |

The results in Tables 2.8–2.10 represent the simulated mean regression estimates for the two series and their corresponding standard errors are displayed in (.). These estimates are obtained using the iterative process (2.59) with small initial values. From the simulation tables, it can be deduced that the GQL regression, serial and bivariate correlation coefficients are consistent and close to the true values. As the number of time points increases to $T = 100$ and $T = 600$, the standard errors decrease. However, there are some failures in a number of simulations for almost all cluster sizes irrespective of high or low serial-correlations in the non-stationary case. In fact, for $T = 60$ and $\kappa_{12,1} = 0.5$, the GQL algorithm fails in almost 215 simulations and as the number of time points increases to $T = 100$ for same $\kappa_{12,1}$, the number of non-convergent simulations decreases to 190 when $\rho_{11} = \rho_{22} = 0.9$. For same values of serial-correlations and T=100 but $\kappa_{12,1} = 0.9$, the number of non-convergent simulations increases to 210. This indicates that this GQL algorithm is working quite smoothly for extremely low and high bivariate dependence irrespective of high or low serial-correlations. The non-convergent simulations were due to either the ill-conditioned covariance matrix or the Hessian matrix (2.59). Note that, in these tables, the non-stationary serial-correlations

**TABLE 2.9**
Estimates of the Parameters and Standard Errors under Non-Stationary BINAR(1) Process for $\rho_{kk} = [\rho_{11} = 0.9, \rho_{22} = 0.3]$

| $\kappa_{12,1}$ | $T$ | $\hat{\beta}_1^{[1]}$ | $\hat{\beta}_2^{[1]}$ | $\hat{\beta}_1^{[2]}$ | $\hat{\beta}_2^{[2]}$ | $\hat{\rho}_{11}$ | $\hat{\rho}_{22}$ | $\hat{\kappa}_{12,1}$ |
|---|---|---|---|---|---|---|---|---|
| 0.1 | 60 | 1.0115 | 1.0126 | 1.0171 | 1.0117 | 0.8829 | 0.2809 | 0.1093 |
| | | (0.0925) | (0.0930) | (0.0910) | (0.0950) | | | |
| | 100 | 0.9923 | 0.9978 | 0.9916 | 1.0108 | 0.8994 | 0.2980 | 0.1052 |
| | | (0.0710) | (0.0750) | (0.0708) | (0.0711) | | | |
| | 600 | 1.0091 | 0.9984 | 0.9962 | 1.0086 | 0.9001 | 0.3013 | 0.1015 |
| | | (0.0111) | (0.0131) | (0.0113) | (0.0114) | | | |
| 0.3 | 60 | 0.9849 | 0.9879 | 1.0141 | 1.0114 | 0.8886 | 0.2831 | 0.3152 |
| | | (0.0945) | (0.0930) | (0.0911) | (0.0921) | | | |
| | 100 | 1.0111 | 1.0116 | 1.0115 | 0.9911 | 0.8994 | 0.3052 | 0.3058 |
| | | (0.0710) | (0.0750) | (0.0790) | (0.0763) | | | |
| | 600 | 0.9925 | 0.9961 | 1.0096 | 1.0092 | 0.8931 | 0.2969 | 0.3024 |
| | | (0.0151) | (0.0117) | (0.0175) | (0.0116) | | | |
| 0.5 | 60 | 0.9841 | 1.0182 | 1.0159 | 1.0198 | 0.8803 | 0.3128 | 0.5141 |
| | | (0.1020) | (0.1125) | (0.1130) | (0.0920) | | | |
| | 100 | 1.0195 | 1.0138 | 1.0190 | 0.9925 | 0.8992 | 0.2919 | 0.5031 |
| | | (0.0880) | (0.0710) | (0.0810) | (0.0740) | | | |
| | 600 | 0.9992 | 0.9955 | 1.0097 | 1.0036 | 0.8993 | 0.3029 | 0.5027 |
| | | (0.0245) | (0.0330) | (0.0116) | (0.0125) | | | |
| 0.9 | 60 | 1.0164 | 1.0148 | 1.0147 | 1.0115 | 0.9137 | 0.3103 | 0.8818 |
| | | (0.1225) | (0.1126) | (0.0911) | (0.1225) | | | |
| | 100 | 1.0115 | 1.0113 | 0.9991 | 0.9905 | 0.9073 | 0.3019 | 0.8978 |
| | | (0.0970) | (0.0915) | (0.0710) | (0.0813) | | | |
| | 600 | 1.0057 | 1.0093 | 0.9992 | 0.9975 | 0.8987 | 0.2987 | 0.8989 |
| | | (0.0364) | (0.0151) | (0.0322) | (0.0112) | | | |

**TABLE 2.10**
Estimates of the Parameters and Standard Errors under Non-Stationary BINAR(1) Process for $\rho_{kk} = [\rho_{11} = 0.3, \rho_{22} = 0.3]$

| $\kappa_{12,1}$ | $T$ | $\hat{\beta}_1^{[1]}$ | $\hat{\beta}_2^{[1]}$ | $\hat{\beta}_1^{[2]}$ | $\hat{\beta}_2^{[2]}$ | $\hat{\rho}_{11}$ | $\hat{\rho}_{22}$ | $\hat{\kappa}_{12,1}$ |
|---|---|---|---|---|---|---|---|---|
| 0.1 | 60 | 1.0162 | 1.0113 | 0.9803 | 1.0136 | 0.2855 | 0.3177 | 0.1124 |
| | | (0.0913) | (0.0946) | (0.0921) | (0.0916) | | | |
| | 100 | 1.0110 | 1.0104 | 1.0114 | 1.0125 | 0.3063 | 0.2960 | 0.1053 |
| | | (0.0720) | (0.0742) | (0.0711) | (0.0760) | | | |
| | 600 | 1.0025 | 1.0056 | 1.0095 | 0.9951 | 0.2997 | 0.3044 | 0.1036 |
| | | (0.0114) | (0.0103) | (0.0123) | (0.0138) | | | |
| 0.3 | 60 | 0.9893 | 0.9877 | 1.0138 | 1.0159 | 0.3164 | 0.2891 | 0.3118 |
| | | (0.0908) | (0.0918) | (0.1035) | (0.0906) | | | |
| | 100 | 1.0111 | 1.0102 | 1.0129 | 1.0133 | 0.2984 | 0.3049 | 0.3031 |
| | | (0.0750) | (0.0821) | (0.0711) | (0.0905) | | | |
| | 600 | 1.0021 | 0.9921 | 0.9905 | 1.0024 | 0.2965 | 0.3086 | 0.3012 |
| | | (0.0255) | (0.0122) | (0.0311) | (0.0112) | | | |

(*Continued*)

**TABLE 2.10**
*(Continued)*

| $\kappa_{12,1}$ | $T$ | $\hat{\beta}_1^{[1]}$ | $\hat{\beta}_2^{[1]}$ | $\hat{\beta}_1^{[2]}$ | $\hat{\beta}_2^{[2]}$ | $\hat{\rho}_{11}$ | $\hat{\rho}_{22}$ | $\hat{\kappa}_{12,1}$ |
|---|---|---|---|---|---|---|---|---|
| 0.5 | 60 | 1.0187 | 1.0145 | 1.0156 | 0.9857 | 0.2887 | 0.3116 | 0.4897 |
| | | (0.1015) | (0.1215) | (0.1231) | (0.1110) | | | |
| | 100 | 0.9911 | 1.0124 | 1.0146 | 1.0105 | 0.2956 | 0.2958 | 0.5076 |
| | | (0.0814) | (0.0849) | (0.0711) | (0.0751) | | | |
| | 600 | 1.0093 | 1.0089 | 1.0003 | 1.0021 | 0.3009 | 0.3033 | 0.5025 |
| | | (0.0216) | (0.0113) | (0.0235) | (0.0117) | | | |
| 0.9 | 60 | 1.0191 | 0.9887 | 0.9825 | 1.0145 | 0.2809 | 0.3194 | 0.8889 |
| | | (0.1254) | (0.1306) | (0.0912) | (0.1012) | | | |
| | 100 | 1.0118 | 0.9972 | 0.9973 | 1.0126 | 0.2958 | 0.3052 | 0.8978 |
| | | (0.0770) | (0.0942) | (0.0730) | (0.0825) | | | |
| | 600 | 1.0054 | 1.0021 | 1.0019 | 1.0024 | 0.2962 | 0.3015 | 0.8985 |
| | | (0.0311) | (0.0117) | (0.0214) | (0.0110) | | | |

for the different lags as computed in Equation (2.57) and the bivariate dependence $\kappa_{12,t}$ are not displayed due to their high dimensions but were reliable values in the interval (0,1).

## 2.4 Summary

This chapter has considered the modeling of the constrained BINAR(1) process under the innovation series following stationary and non-stationary moment conditions. In the literature, the BINAR(1) model was only implemented under stationary moment conditions of the Poisson innovations wherein the parameters were estimated mostly using the CMLE approach. In this book, we have re-analyzed the BINAR(1) process under a new parametrization of the innovations series that has been extended easily to the non-stationary case. In addition of the CMLE and FGLS, we have proposed a novel QL estimation method: the bivariate GQL to estimate the regression parameters while the serial- and cross-correlation parameters are estimated using a robust moment estimating equation. We have illustrated through Monte-Carlo simulation that GQL provides estimates with lower bias than FGLS while asymptotically equally efficient estimates as CMLE for the constrained stationary BINAR(1) series under Poisson innovations. Moreover, GQL has been shown to yield fewer non-convergent simulations than the computationally intensive CMLE which has motivated to explore this new approach to analyze non-stationary BINAR(1) process in Subsection 2.3.2. The simulation results under the non-stationary set-up indicate that GQL is a commendable approach as it has shown to provide consistent estimates with lesser computational failures. The next chapter aims at developing a constrained BINMA(1) model following the same approach.

# 3

# *Constrained BINMA(1) Model with Correlated Poisson Innovations*

## 3.1 Introduction

In this chapter, we develop a constrained BINMA(1) model based on a similar diagonal cross-correlation structure as in Chapter 2 where the cross-correlation between the series is induced by the dependence between the innovation series only. The innovation series here are defined under simpler distributional parametrization of the moments. However, in the BINMA case, covariates are introduced via the link predictor function of the innovation series similar to Quoreshi (2006a, 2008).

As for the inferential procedures, it is common in INMA and BINMA processes to apply the FGLS approach (Quoreshi (2006a, 2008)). This chapter as well explores this method under the stationary moment conditions for the newly proposed BINMA(1) model. A GQL estimation approach is also proposed to estimate the parameters of the model. It is noted that the GQL approach has proved to yield more efficient estimates than FGLS in the constrained stationary BINAR(1) series. Hence, in the proposed model, the GQL is used to estimate the regression and serial-correlation parameters while the cross-correlation between the innovations is estimated via a robust moment estimating equation approach developed by Sutradhar & Das (1999). The performance of these estimation methods is compared under stationary BINMA(1) series and a unified estimation approach is proposed for the analysis of non-stationary BINMA(1) series with Poisson innovations.

The plan of this chapter is therefore as follows: In Section 3.2, a constrained BINMA(1) model with Poisson innovations is introduced and the marginal and joint moments and covariances are derived under the given distributional assumptions. In Section 3.3, the GQL and FGLS methods of estimation are derived in the bivariate context and through a simulation study, their performance is assessed under stationary moment conditions. In the same section, GQL is derived under non-stationary moment conditions and its performance is assessed through a simulation study. The one-step ahead forecasting equations for the BINMA(1) model with Poisson innovations are also developed. This chapter ends with a conclusion.

 DOI: 10.1201/9781003600435-3

## 3.2 The Constrained BINMA(1) Model

The constrained BINMA(1) model is expressed as

$$\begin{pmatrix} Y_t^{[1]} \\ Y_t^{[2]} \end{pmatrix} = \begin{pmatrix} \rho_{11} & 0 \\ 0 & \rho_{22} \end{pmatrix} \circ \begin{pmatrix} R_{t-1}^{[1]} \\ R_{t-1}^{[2]} \end{pmatrix} + \begin{pmatrix} R_t^{[1]} \\ R_t^{[2]} \end{pmatrix}, \tag{3.1}$$

where $\rho_{kk} \in [0,1]$ for $k \in \{1,2\}$. $\{\rho_{kk} \circ R_{t-1}^{[k]}\}$ indicates the binomial thinning operation such that $\{\rho_{kk} \circ R_{t-1}^{[k]} | R_{t-1}^{[k]}\} \sim \text{Binomial}(R_{t-1}^{[k]}, \rho_{kk})$ (Steutel & Van Harn (1979)), with the following properties:

1. $E(\rho_{kk} \circ R_{t-1}^{[k]}) = \rho_{kk} E(R_{t-1}^{[k]})$
2. $\text{Var}(\rho_{kk} \circ R_{t-1}^{[k]}) = \rho_{kk}(1-\rho_{kk}) E(R_{t-1}^{[k]}) + \rho_{kk}^2 \text{Var}(R_{t-1}^{[k]})$
3. $\text{Cov}(\rho_{11} \circ R_{t-1}^{[1]}, \rho_{22} \circ R_{t-1}^{[2]}) = \rho_{11}\rho_{22} \text{Cov}(R_{t-1}^{[1]}, R_{t-1}^{[2]})$

for $k \in \{1,2\}$. $R_t^{[k]}$ and $R_{t-1}^{[k]}$ are independent but $R_t^{[1]}$ and $R_t^{[2]}$ follow the bivariate Poisson distribution as in Equation (2.3) with same assumptions, but $R_t^{[k]} \sim \text{Poisson}(\lambda_t^{[k]})$ where $\lambda_t^{[k]} = \exp(\boldsymbol{x}_t^{'} \boldsymbol{\beta}^{[k]})$, with $\boldsymbol{x}_t = [x_{t1}, x_{t2}, \ldots, x_{tj}, \ldots, x_{tp}]^{'}$ and $\boldsymbol{\beta}^{[k]} = [\beta_1^{[k]}, \beta_2^{[k]}, \ldots, \beta_j^{[k]}, \ldots, \beta_p^{[k]}]^{'}$ for $t = 1, 2, \ldots, T$ and $k \in \{1,2\}$.

Based on the above conditions and using Lemma 2.1, it can be shown that

$$\begin{aligned} E(Y_t^{[k]}) &= E_{R_{t-1}^{[k]}} E(\rho_{kk} \circ R_{t-1}^{[k]} | R_{t-1}^{[k]}) + E(R_t^{[k]}) \\ &= \rho_{kk} \lambda_{t-1}^{[k]} + \lambda_t^{[k]} \end{aligned} \tag{3.2}$$

and

$$\begin{aligned} \text{Var}(Y_t^{[k]}) &= \text{Var}[\rho_{kk} \circ R_{t-1}^{[k]} + R_t^{[k]}] \\ &= E_{R_{t-1}^{[k]}}[\text{Var}(\rho_{kk} \circ R_{t-1}^{[k]} | R_{t-1}^{[k]})] + \text{Var}_{R_{t-1}^{[k]}}[E(\rho_{kk} \circ R_{t-1}^{[k]} | R_{t-1}^{[k]})] \\ &\quad + \text{Var}[R_t^{[k]}] \\ &= \rho_{kk}(1-\rho_{kk}) \lambda_{t-1}^{[k]} + \rho_{kk}^2 \lambda_{t-1}^{[k]} + \lambda_t^{[k]} \\ &= \rho_{kk} \lambda_{t-1}^{[k]} + \lambda_t^{[k]}. \end{aligned} \tag{3.3}$$

Hence, $Y_t^{[k]} \sim \text{Poisson}(\lambda_t^{[k]} + \rho_{kk} \lambda_{t-1}^{[k]})$.

The serial lag-$h$ correlation for the $k^{th}$ series is

$$\begin{aligned}
\mathrm{Cov}(Y_t^{[k]}, Y_{t+h}^{[k]}) &= \mathrm{Cov}(Y_t^{[k]}, \rho_{kk} \circ R_{t+h-1}^{[k]} + R_{t+h}^{[k]}) \\
&= \rho_{kk}\mathrm{Cov}(Y_t^{[k]}, R_{t+h-1}^{[k]}) \\
&= \rho_{kk}\mathrm{Cov}(\rho_{kk} \circ R_{t-1}^{[k]} + R_t^{[k]}, R_{t+h-1}^{[k]}) \\
&= \rho_{kk}\mathrm{Cov}(R_t^{[k]}, R_{t+h-1}^{[k]}) \\
&= \begin{cases} \rho_{kk}\lambda_t^{[k]} & h = 1 \\ 0 & h > 1 \end{cases}
\end{aligned} \tag{3.4}$$

and hence

$$\mathrm{Corr}(Y_t^{[k]}, Y_{t+h}^{[k]}) = \begin{cases} \frac{\rho_{kk}\lambda_t^{[k]}}{\sqrt{\rho_{kk}\lambda_{t-1}^{[k]}+\lambda_t^{[k]}}\sqrt{\rho_{kk}\lambda_{t+h-1}^{[k]}+\lambda_{t+h}^{[k]}}} & h = 1, \\ 0 & h > 1. \end{cases} \tag{3.5}$$

As for the joint covariance $\mathrm{Cov}(Y_t^{[1]}, Y_t^{[2]})$, from Equation (3.1)

$$\begin{aligned}
\mathrm{Cov}(Y_t^{[1]}, Y_t^{[2]}) &= \mathrm{Cov}(\rho_{11} \circ R_{t-1}^{[1]} + R_t^{[1]}, \rho_{22} \circ R_{t-1}^{[2]} + R_t^{[2]}) \\
&= \rho_{11}\rho_{22}\mathrm{Cov}(R_{t-1}^{[1]}, R_{t-1}^{[2]}) + \mathrm{Cov}(R_t^{[1]}, R_t^{[2]}) \\
&= \rho_{11}\rho_{22}\kappa_{12,t-1}\sqrt{\lambda_{t-1}^{[1]}}\sqrt{\lambda_{t-1}^{[2]}} + \kappa_{12,t}\sqrt{\lambda_t^{[1]}}\sqrt{\lambda_t^{[2]}}
\end{aligned} \tag{3.6}$$

and hence

$$\mathrm{Corr}(Y_t^{[1]}, Y_t^{[2]}) = \frac{\rho_{11}\rho_{22}\kappa_{12,t-1}\sqrt{\lambda_{t-1}^{[1]}}\sqrt{\lambda_{t-1}^{[2]}} + \kappa_{12,t}\sqrt{\lambda_t^{[1]}}\sqrt{\lambda_t^{[2]}}}{\sqrt{\lambda_t^{[1]} + \rho_{11}\lambda_{t-1}^{[1]}}\sqrt{\lambda_t^{[2]} + \rho_{22}\lambda_{t-1}^{[2]}}}. \tag{3.7}$$

As for the other off-diagonal cross-covariances,

$$\begin{aligned}
\mathrm{Cov}(Y_t^{[1]}, Y_{t+h}^{[2]}) &= \mathrm{Cov}(Y_t^{[1]}, \rho_{22} \circ R_{t+h-1}^{[2]} + R_{t+h}^{[2]}) \\
&= \rho_{22}\mathrm{Cov}(Y_t^{[1]}, R_{t+h-1}^{[2]}) \\
&= \rho_{22}\mathrm{Cov}(\rho_{11} \circ R_{t-1}^{[1]} + R_t^{[1]}, R_{t+h-1}^{[2]}) \\
&= \rho_{22}\mathrm{Cov}(R_t^{[1]}, R_{t+h-1}^{[2]}) \\
&= \begin{cases} \rho_{22}\kappa_{12,t}\sqrt{\lambda_t^{[1]}}\sqrt{\lambda_t^{[2]}} & h = 1 \\ 0 & h > 1 \end{cases}
\end{aligned} \tag{3.8}$$

and similarly

$$\mathrm{Cov}(Y_{t+h}^{[1]}, Y_t^{[2]}) = \begin{cases} \rho_{11}\kappa_{12,t}\sqrt{\lambda_t^{[1]}}\sqrt{\lambda_t^{[2]}} & h = 1, \\ 0 & h > 1. \end{cases} \tag{3.9}$$

Under the stationary moment assumption, where $\lambda_t^{[k]} = \lambda_1^{[k]}$,

$$E(Y_t^{[k]}) = \text{Var}(Y_t^{[k]}) = (1+\rho_{kk})\lambda_1^{[k]}, \tag{3.10}$$

$$\text{Corr}(Y_t^{[k]}, Y_{t+1}^{[k]}) = \frac{\rho_{kk}}{1+\rho_{kk}} \tag{3.11}$$

and the cross-covariances are

$$\text{Cov}(Y_t^{[1]}, Y_t^{[2]}) = (1+\rho_{11}\rho_{22})\kappa_{12,1}\sqrt{\lambda_1^{[1]}}\sqrt{\lambda_1^{[2]}} \tag{3.12}$$

with

$$\begin{aligned}\text{Cov}(Y_{t+1}^{[1]}, Y_t^{[2]}) = \rho_{11}\kappa_{12,1}\sqrt{\lambda_1^{[1]}}\sqrt{\lambda_1^{[2]}} \quad \text{and} \quad \text{Cov}(Y_t^{[1]}, Y_{t+1}^{[2]})\\ = \rho_{22}\kappa_{12,1}\sqrt{\lambda_1^{[1]}}\sqrt{\lambda_1^{[2]}}.\end{aligned} \tag{3.13}$$

## 3.3 Estimation of Parameters

Similar to Section 2.3, the estimation part is divided into two subsections: Firstly estimation of model parameters under stationary moments and secondly estimation of model parameters under non-stationary moment conditions.

### 3.3.1 Stationary Conditions

#### 3.3.1.1 Feasible Generalized Least Squares (FGLS)

In the case of BINMA(1) process,

$$e_{k1t} = y_t^{[k]} - E(Y_t^{[k]}|r_{t-1}^{[k]}) \tag{3.14}$$

where $\text{E}(Y_t^{[k]}|r_{t-1}^{[k]}) = \rho_{kk}r_{t-1}^{[k]} + \lambda_1^{[k]}$. Note that $r_{t-1}^{[k]}$ are approximated by their respective means.

To compute the unknown parameter $\boldsymbol{\theta} = [\rho_{11}, \rho_{22}, \beta_1^{[1]}, \beta_2^{[1]}, \beta_1^{[2]}, \beta_2^{[2]}, \kappa_{12,1}]$, we formulate the objective criterion function

$$\begin{aligned}S_{FGLS} = \sum_{t=1}^{T}\Big[&\text{Var}(Y_t^{[2]}|r_{t-1}^{[2]})(e_{11t}^2) + \text{Var}(Y_t^{[1]}|r_{t-1}^{[1]})(e_{21t}^2)\\ &-2\text{Cov}(Y_t^{[1]}, Y_t^{[2]}|r_{t-1}^{[1]}, r_{t-1}^{[2]})(e_{11t})(e_{21t})\Big]\\ &/\Big[\text{Var}(Y_t^{[1]}|r_{t-1}^{[1]})\text{Var}(Y_t^{[2]}|r_{t-1}^{[2]}) - \text{Cov}(Y_t^{[1]}, Y_t^{[2]}|r_{t-1}^{[1]}, r_{t-1}^{[2]})^2\Big]\end{aligned}$$

$$
\begin{aligned}
&= \sum_{t=1}^{T}\{[\rho_{22}(1-\rho_{22})r_{t-1}^{[2]} + \lambda_1^{[2]}][y_t^{[1]} - (\rho_{11}r_{t-1}^{[1]} + \lambda_1^{[1]})]^2 \\
&\times [\rho_{11}(1-\rho_{11})r_{t-1}^{[1]} + \lambda_1^{[1]}][y_t^{[2]} - (\rho_{22}r_{t-1}^{[2]} + \lambda_1^{[2]})]^2 \\
&- 2[\kappa_{12,1}\sqrt{\lambda_1^{[1]}}\sqrt{\lambda_1^{[2]}}][\rho_{11}r_{t-1}^{[1]} + \lambda_1^{[1]}][\rho_{22}r_{t-1}^{[2]} + \lambda_1^{[2]}]\} \\
&/\{[\rho_{22}(1-\rho_{22})r_{t-1}^{[2]} + \lambda_1^{[2]}][\rho_{11}(1-\rho_{11})r_{t-1}^{[1]} + \lambda_1^{[1]}] \\
&- [\kappa_{12,1}\sqrt{\lambda_1^{[1]}}\sqrt{\lambda_1^{[2]}}]^2\},
\end{aligned} \tag{3.15}
$$

where

1. $\text{Var}(Y_t^{[k]}|r_{t-1}^{[k]}) = \rho_{kk}(1-\rho_{kk})r_{t-1}^{[k]} + \lambda_1^{[k]}$.
2. $\text{Cov}(Y_t^{[1]}, Y_t^{[2]}|r_{t-1}^{[1]}, r_{t-1}^{[2]}) = \text{Cov}(R_t^{[1]}, R_t^{[2]}) = \kappa_{12,1}\sqrt{\lambda_1^{[1]}}\sqrt{\lambda_1^{[2]}}$.

The objective criterion function is minimized with respect to the unknown parameter $\boldsymbol{\theta}$ using the *optim* function as follows:

$$
\textit{optim}(\text{initial value of } \quad \boldsymbol{\theta}, \quad S_{FGLS}, \quad \text{data}, \quad \text{hessian=true}) \tag{3.16}
$$

(Refer to Subsection 2.3.1.2 for more details).

#### 3.3.1.2 Generalized Quasi-Likelihood (GQL)

In this subsection, we develop the GQL equation to estimate the regression and serial-correlation parameters in the BINMA(1) process, where only the marginal moments and joint covariances of the two series are required. The GQL equation is expressed as

$$
\boldsymbol{D}_{\boldsymbol{\beta},\boldsymbol{\rho}}^{'}\boldsymbol{\Sigma}_{\boldsymbol{\beta},\boldsymbol{\rho}}^{-1}(\boldsymbol{y}-\boldsymbol{\mu}) = 0 \tag{3.17}
$$

where

$$
\boldsymbol{y} = \begin{pmatrix} \boldsymbol{y}^{[1]} \\ \boldsymbol{y}^{[2]} \end{pmatrix}_{2T\times 1}, \boldsymbol{y}^{[1]} = [y_1^{[1]}, y_2^{[1]}, \ldots, y_T^{[1]}]', \boldsymbol{y}^{[2]} = [y_1^{[2]}, y_2^{[2]}, \ldots, y_T^{[2]}]' \tag{3.18}
$$

and

$$
\boldsymbol{\mu} = \begin{pmatrix} \boldsymbol{\mu}^{[1]} \\ \boldsymbol{\mu}^{[2]} \end{pmatrix}_{2T\times 1}, \boldsymbol{\mu}^{[1]} = [\mu_1^{[1]}, \mu_1^{[1]}, \ldots, \mu_1^{[1]}]', \boldsymbol{\mu}^{[2]} = [\mu_1^{[2]}, \mu_1^{[2]}, \ldots, \mu_1^{[2]}]'. \tag{3.19}
$$

The tri-diagonal covariance matrix $\boldsymbol{\Sigma}_{\boldsymbol{\beta},\boldsymbol{\rho}}$ is given by

$$
\begin{array}{c}
 \\
Y_1^{[1]} \\
Y_2^{[1]} \\
\vdots \\
Y_T^{[1]} \\
\cdots \\
Y_1^{[2]} \\
Y_2^{[2]} \\
\vdots \\
Y_T^{[2]}
\end{array}
\begin{array}{c}
\begin{array}{cccccccc}
Y_1^{[1]} & Y_2^{[1]} & \cdots & Y_T^{[1]} & Y_1^{[2]} & Y_2^{[2]} & \cdots & Y_T^{[2]}
\end{array} \\
\left(\begin{array}{cccc:cccc}
V(Y_1^{[1]}) & C(Y_1^{[1]},Y_2^{[1]}) & & & C(Y_1^{[1]},Y_1^{[2]}) & C(Y_1^{[1]},Y_2^{[2]}) & & \\
C(Y_2^{[1]},Y_1^{[1]}) & V(Y_2^{[1]}) & C(Y_2^{[1]},Y_3^{[1]}) & & C(Y_2^{[1]},Y_1^{[2]}) & C(Y_2^{[1]},Y_2^{[2]}) & C(Y_2^{[1]},Y_3^{[2]}) & \\
 & & \ddots & & & & \ddots & \\
 & & C(Y_T^{[1]},Y_{T-1}^{[1]}) & V(Y_T^{[1]}) & & & C(Y_T^{[1]},Y_{T-1}^{[2]}) & C(Y_T^{[1]},Y_T^{[2]}) \\
\cdots & \cdots & \cdots & \cdots & \cdots & \cdots & \cdots & \cdots \\
C(Y_1^{[2]},Y_1^{[1]}) & C(Y_1^{[2]},Y_2^{[1]}) & & & V(Y_1^{[2]}) & C(Y_1^{[2]},Y_2^{[2]}) & & \\
C(Y_2^{[2]},Y_1^{[1]}) & C(Y_2^{[2]},Y_2^{[1]}) & C(Y_2^{[2]},Y_3^{[1]}) & & C(Y_2^{[2]},Y_1^{[2]}) & V(Y_2^{[2]}) & C(Y_2^{[2]},Y_3^{[2]}) & \\
 & & \ddots & & & & \ddots & \\
 & & C(Y_T^{[2]},Y_{T-1}^{[1]}) & C(Y_T^{[2]},Y_T^{[1]}) & & & C(Y_T^{[2]},Y_{T-1}^{[2]}) & V(Y_T^{[2]})
\end{array}\right)
\end{array}
\tag{3.20}
$$

where $V(Y_t^{[k]}) = \text{Var}(Y_t^{[k]})$ and $C(Y_t^{[k]}, Y_{t-1}^{[k']}) = \text{Cov}(Y_t^{[k]}, Y_{t-1}^{[k']})$, for $k, k' \in \{1, 2\}$. The entries of the above covariance matrix are computed using Equations (3.10)–(3.13).
As for the derivative matrix, $\boldsymbol{D}_{\beta,\rho} = [\boldsymbol{D}_1, \boldsymbol{D}_2, \ldots, \boldsymbol{D}_t, \ldots, \boldsymbol{D}_T]'$ with

$$\boldsymbol{D}_t = \begin{pmatrix} \frac{\partial \mu_1^{[1]}}{\partial \boldsymbol{\beta}^{[1]}} & 0 \\ 0 & \frac{\partial \mu_1^{[2]}}{\partial \boldsymbol{\beta}^{[2]}} \\ \frac{\partial \mu_1^{[1]}}{\partial \rho_{11}} & 0 \\ 0 & \frac{\partial \mu_1^{[2]}}{\partial \rho_{22}} \end{pmatrix}_{(2p+2)\times 2}, \tag{3.21}$$

where $\mu_1^{[k]} = E(Y_t^{[k]}) = (1 + \rho_{kk})\lambda_1^{[k]}$.
The entries of the derivative matrix are:

1. $\frac{\partial \mu_1^{[k]}}{\partial \beta_j^{[k]}} = \rho_{kk}\lambda_1^{[k]} x_{1j} + \lambda_1^{[k]} x_{1j}$.

2. $\frac{\partial \mu_1^{[k]}}{\partial \rho_{kk}} = \lambda_1^{[k]}$.

Following Equation (3.12), the dependence parameter $\kappa_{12,1}$ is estimated as follows:

$$\kappa_{12,1} = \frac{\tilde{\text{Cov}}(Y_t^{[1]}, Y_t^{[2]})}{(1 + \hat{\rho}_{11}\hat{\rho}_{22})\sqrt{\hat{\lambda}_1^{[1]}}\sqrt{\hat{\lambda}_1^{[2]}}} \tag{3.22}$$

where $\tilde{\text{Cov}}(Y_t^{[1]}, Y_t^{[2]}) = \frac{1}{T}\sum_{t=1}^{T}(y_t^{[1]} - \hat{\mu}_1^{[1]})(y_t^{[2]} - \hat{\mu}_1^{[2]})$.

The Newton-Raphson iterative scheme to estimate $\boldsymbol{\beta}^{[k]}$ and $\rho_{kk}$ is of the form

$$(\ \hat{\boldsymbol{\psi}}_{r+1}\ ) = (\ \hat{\boldsymbol{\psi}}_r\ ) + [\boldsymbol{D}_{\psi}{}'\boldsymbol{\Sigma}_{\psi}{}^{-1}\boldsymbol{D}_{\psi}]_r^{-1}[\boldsymbol{D}_{\psi}{}'\boldsymbol{\Sigma}_{\psi}{}^{-1}(\boldsymbol{y} - \boldsymbol{\mu})]_r, \tag{3.23}$$

where $\hat{\boldsymbol{\psi}} = [\hat{\beta}_1^{[1]}, \hat{\beta}_2^{[1]}, \hat{\beta}_1^{[2]}, \hat{\beta}_2^{[2]}, \hat{\rho}_{11}, \hat{\rho}_{22}]$ and $\hat{\boldsymbol{\psi}}_r$ are the estimates at the $r^{th}$ iteration and $[.]_r$ are the values of the expression at the $r^{th}$ iteration.

The estimation procedure works as follows: For an initial $\hat{\boldsymbol{\psi}}$, $[\beta_1^{[1]} = \beta_2^{[1]} = 0.01, \beta_1^{[2]} = \beta_2^{[2]} = 0.02, \rho_{11} = 0.1, \rho_{22} = 0.15]$, the dependence parameter $\kappa_{12,1}$, the score and Hessian components are calculated and replaced in Equation (3.23) to obtain an updated $\hat{\boldsymbol{\psi}}$. The new set of regressions and serial-correlation parameters are then used to re-calculate the dependence, score and Hessian components, which are again replaced in Equation (3.23) to obtain a new set of $\hat{\boldsymbol{\psi}}$. This cycle of iteration continues until convergence. Under mild regulatory conditions, the regression and dependence estimators are consistent and $(\hat{\boldsymbol{\psi}} - \boldsymbol{\psi})'$ follows an asymptotic normal distribution with mean $\boldsymbol{0}$ and covariance matrix $[\boldsymbol{D}_{\psi}{}'\boldsymbol{\Sigma}_{\psi}{}^{-1}\boldsymbol{D}_{\psi}]^{-1}$, as discussed in Subsection 2.3.1.2.

#### 3.3.1.3 Simulation

This section consists of a simulation study where stationary BINMA(1) Poisson time series are generated for different values of $T$, that is, $T = 60, 100$ and 600 and where the BINMA(1) observations are subject to a $(2 \times 1)$ vector of covariates specified as

$$x_{t1} = \begin{cases} -1 & (t = 1, \ldots, T/4), \\ 0 & (t = (T/4) + 1, \ldots, 3T/4), \\ 1 & (t = (3T/4) + 1, \ldots, T), \end{cases}$$

$$x_{t2} = \begin{cases} -2 & (t = 1, \ldots, T/4), \\ 1 & (t = (T/4) + 1, \ldots, 3T/4), \\ 2 & (t = (3T/4) + 1, \ldots, T). \end{cases}$$

For this simulation process, $\beta_1^{[k]} = \beta_2^{[k]} = 0.3$ and the pair of innovation terms is generated using the R codes developed in Subsection 2.3.1.3 for generating multivariate Poisson counts via the library MASS function, where $R = matrix(c(1, \kappa_{12,1}, \kappa_{12,1}, 1), 2, 2)$ and $lambda = matrix((\lambda_1^{[1]}, \lambda_1^{[2]}), 2, 1)$.

$\kappa_{12,1}$ takes values $0.1, 0.3, 0.5, 0.9$ along with different combinations of the serial-correlation $[\rho_{11}, \rho_{22}]$=[0.9,0.9],[0.9,0.3] and [0.3,0.3]. In the simulations, different combinations of these serial-correlation values are taken into account to assess the performance of FGLS and GQL under high and low cross-correlation. For each time point, 5000 simulations are run and the mean estimates of the regression and correlation parameters are tabled (Tables 3.1–3.3) under stationary set-up:

**TABLE 3.1**
Estimates of the Parameters and Standard Errors under Stationary BINMA(1) Process for $\rho_{kk} = [\rho_{11} = 0.9, \rho_{22} = 0.9]$

| $\kappa_{12,1}$ | $T$ | Method | $\hat{\beta}_1^{[1]}$ | $\hat{\beta}_2^{[1]}$ | $\hat{\beta}_1^{[2]}$ | $\hat{\beta}_2^{[2]}$ | $\hat{\rho}_{11}$ | $\hat{\rho}_{22}$ | $\hat{\kappa}_{12,1}$ |
|---|---|---|---|---|---|---|---|---|---|
| 0.1 | 60 | FGLS | 0.2811 | 0.3134 | 0.3169 | 0.2815 | 0.8805 | 0.9147 | 0.1276 |
| | | | (0.1091) | (0.1056) | (0.1074) | (0.1010) | (0.0945) | (0.0933) | (0.0626) |
| | | GQL | 0.2835 | 0.3103 | 0.3132 | 0.2868 | 0.8868 | 0.9133 | 0.1182 |
| | | | (0.0942) | (0.0941) | (0.0998) | (0.0974) | (0.0860) | (0.0868) | |
| | 100 | FGLS | 0.2872 | 0.3129 | 0.3171 | 0.3101 | 0.8884 | 0.8887 | 0.1233 |
| | | | (0.0991) | (0.0912) | (0.0913) | (0.0964) | (0.0803) | (0.0891) | (0.0545) |
| | | GQL | 0.2918 | 0.3081 | 0.3063 | 0.3042 | 0.8917 | 0.8914 | 0.1155 |
| | | | (0.0723) | (0.0730) | (0.0744) | (0.0740) | (0.0633) | (0.0620) | |
| | 600 | FGLS | 0.2911 | 0.3098 | 0.3052 | 0.2944 | 0.8820 | 0.9104 | 0.1066 |
| | | | (0.0387) | (0.0361) | (0.0338) | (0.0321) | (0.0242) | (0.0297) | (0.0251) |
| | | GQL | 0.2968 | 0.3068 | 0.3062 | 0.2958 | 0.8971 | 0.9040 | 0.1018 |
| | | | (0.0265) | (0.0236) | (0.0260) | (0.0285) | (0.0130) | (0.0137) | |

(*Continued*)

**TABLE 3.1**
*(Continued)*

| $\kappa_{12,1}$ | $T$ | Method | $\hat{\beta}_1^{[1]}$ | $\hat{\beta}_2^{[1]}$ | $\hat{\beta}_1^{[2]}$ | $\hat{\beta}_2^{[2]}$ | $\hat{\rho}_{11}$ | $\hat{\rho}_{22}$ | $\hat{\kappa}_{12,1}$ |
|---|---|---|---|---|---|---|---|---|---|
| 0.3 | 60 | FGLS | 0.2788 | 0.3234 | 0.2818 | 0.2813 | 0.8804 | 0.8798 | 0.3264 |
| | | | (0.1161) | (0.1162) | (0.1178) | (0.1176) | (0.1025) | (0.1055) | (0.1127) |
| | | GQL | 0.2812 | 0.3110 | 0.2831 | 0.2847 | 0.8804 | 0.8816 | 0.3183 |
| | | | (0.0935) | (0.0950) | (0.0935) | (0.0935) | (0.0845) | (0.0910) | |
| | 100 | FGLS | 0.3116 | 0.2903 | 0.3110 | 0.2910 | 0.9122 | 0.8933 | 0.3132 |
| | | | (0.0959) | (0.0951) | (0.0956) | (0.0953) | (0.0889) | (0.0872) | (0.1062) |
| | | GQL | 0.3030 | 0.2979 | 0.3024 | 0.2967 | 0.9099 | 0.8968 | 0.3094 |
| | | | (0.0720) | (0.0730) | (0.0722) | (0.0732) | (0.0630) | (0.0688) | |
| | 600 | FGLS | 0.2978 | 0.3107 | 0.2915 | 0.2934 | 0.9111 | 0.9068 | 0.2984 |
| | | | (0.0379) | (0.0327) | (0.0324) | (0.0397) | (0.0284) | (0.0206) | (0.0581) |
| | | GQL | 0.2993 | 0.3029 | 0.2958 | 0.2980 | 0.9086 | 0.9067 | 0.2943 |
| | | | (0.0207) | (0.0225) | (0.0208) | (0.0230) | (0.0114) | (0.0140) | |
| 0.5 | 60 | FGLS | 0.2765 | 0.3316 | 0.2748 | 0.3389 | 0.9344 | 0.9387 | 0.5371 |
| | | | (0.1166) | (0.1181) | (0.1165) | (0.1183) | (0.1095) | (0.1055) | (0.1274) |
| | | GQL | 0.2804 | 0.3165 | 0.2858 | 0.3173 | 0.9100 | 0.9134 | 0.5182 |
| | | | (0.0935) | (0.0950) | (0.0966) | (0.0951) | (0.0899) | (0.0831) | |
| | 100 | FGLS | 0.3161 | 0.3151 | 0.2896 | 0.2874 | 0.9140 | 0.8890 | 0.4870 |
| | | | (0.0954) | (0.0958) | (0.0961) | (0.0963) | (0.0851) | (0.0821) | (0.1032) |
| | | GQL | 0.3041 | 0.3058 | 0.2942 | 0.2959 | 0.9011 | 0.8968 | 0.4954 |
| | | | (0.0720) | (0.0730) | (0.0770) | (0.0721) | (0.0695) | (0.0629) | |
| | 600 | FGLS | 0.3105 | 0.3111 | 0.2932 | 0.3101 | 0.8888 | 0.9107 | 0.5112 |
| | | | (0.0391) | (0.0368) | (0.0335) | (0.0338) | (0.0286) | (0.0240) | (0.0530) |
| | | GQL | 0.3022 | 0.3011 | 0.2992 | 0.3009 | 0.8998 | 0.9016 | 0.5038 |
| | | | (0.0205) | (0.0225) | (0.0208) | (0.0230) | (0.0190) | (0.0170) | |
| 0.9 | 60 | FGLS | 0.2790 | 0.3234 | 0.2759 | 0.3270 | 0.8773 | 0.9234 | 0.9296 |
| | | | (0.1180) | (0.1184) | (0.1187) | (0.1179) | (0.1032) | (0.1027) | (0.1242) |
| | | GQL | 0.2845 | 0.3150 | 0.2881 | 0.3124 | 0.8869 | 0.9169 | 0.9162 |
| | | | (0.0950) | (0.0946) | (0.0965) | (0.0953) | (0.0898) | (0.0872) | |
| | 100 | FGLS | 0.3191 | 0.3185 | 0.3188 | 0.2905 | 0.9187 | 0.8906 | 0.9195 |
| | | | (0.0964) | (0.0966) | (0.0971) | (0.0968) | (0.0845) | (0.0893) | (0.1081) |
| | | GQL | 0.3023 | 0.3054 | 0.3039 | 0.2985 | 0.9028 | 0.8963 | 0.9039 |
| | | | (0.0799) | (0.0733) | (0.0810) | (0.0730) | (0.0685) | (0.0650) | |
| | 600 | FGLS | 0.2939 | 0.3138 | 0.2948 | 0.3050 | 0.8953 | 0.8908 | 0.9050 |
| | | | (0.0344) | (0.0349) | (0.0347) | (0.0350) | (0.0292) | (0.0272) | (0.0591) |
| | | GQL | 0.2976 | 0.3039 | 0.2976 | 0.3009 | 0.8991 | 0.8994 | 0.9022 |
| | | | (0.0208) | (0.0225) | (0.0220) | (0.0226) | (0.0193) | (0.0113) | |

Tables 3.1–3.3 display the mean estimates of the regression, serial- and cross-correlation parameters. The values in the brackets (.) indicate the standard errors for the regression and dependence estimates. As it can be seen, the GQL approach yields more efficient estimates of the regression and correlation parameters than FGLS. We also note that as the number of time points increases, the standard errors of the regression parameters decrease across the estimation methods which is expected. However, some non-convergent simulations were noted and are displayed in Table 3.4.

**TABLE 3.2**
Estimates of the Parameters and Standard Errors under Stationary BINMA(1) Process for $\rho_{kk} = [\rho_{11} = 0.9, \rho_{22} = 0.3]$

| $\kappa_{12,1}$ | $T$ | Method | $\hat{\beta}_1^{[1]}$ | $\hat{\beta}_2^{[1]}$ | $\hat{\beta}_1^{[2]}$ | $\hat{\beta}_2^{[2]}$ | $\hat{\rho}_{11}$ | $\hat{\rho}_{22}$ | $\hat{\kappa}_{12,1}$ |
|---|---|---|---|---|---|---|---|---|---|
| 0.1 | 60 | FGLS | 0.2782 | 0.3203 | 0.3223 | 0.3273 | 0.9224 | 0.3283 | 0.1231 |
| | | | (0.1110) | (0.1195) | (0.1147) | (0.1119) | (0.1010) | (0.1092) | (0.0539) |
| | | GQL | 0.2818 | 0.3132 | 0.3142 | 0.3103 | 0.9101 | 0.3172 | 0.1130 |
| | | | (0.0960) | (0.0982) | (0.0945) | (0.0951) | (0.0870) | (0.0862) | |
| | 100 | FGLS | 0.2847 | 0.3111 | 0.3148 | 0.2823 | 0.8826 | 0.2877 | 0.1212 |
| | | | (0.0992) | (0.0922) | (0.0927) | (0.0971) | (0.0804) | (0.0833) | (0.0428) |
| | | GQL | 0.2955 | 0.3012 | 0.3070 | 0.2987 | 0.8910 | 0.2971 | 0.1106 |
| | | | (0.0760) | (0.0731) | (0.0740) | (0.0747) | (0.0665) | (0.0689) | |
| | 600 | FGLS | 0.2929 | 0.3061 | 0.3076 | 0.2929 | 0.9081 | 0.3043 | 0.1143 |
| | | | (0.0362) | (0.0337) | (0.0314) | (0.0375) | (0.0265) | (0.0242) | (0.0102) |
| | | GQL | 0.2968 | 0.3003 | 0.3033 | 0.2990 | 0.9047 | 0.3019 | 0.1026 |
| | | | (0.0217) | (0.0227) | (0.0205) | (0.0210) | (0.0135) | (0.0124) | |
| 0.3 | 60 | FGLS | 0.3241 | 0.2756 | 0.2782 | 0.3220 | 0.8744 | 0.3261 | 0.3147 |
| | | | (0.1125) | (0.1193) | (0.1195) | (0.1110) | (0.1050) | (0.1064) | (0.1235) |
| | | GQL | 0.3170 | 0.2868 | 0.2844 | 0.3112 | 0.8839 | 0.3151 | 0.2993 |
| | | | (0.0997) | (0.0980) | (0.0999) | (0.0981) | (0.0897) | (0.0847) | |
| | 100 | FGLS | 0.3153 | 0.2893 | 0.3155 | 0.2825 | 0.9151 | 0.3182 | 0.3119 |
| | | | (0.0923) | (0.0932) | (0.0901) | (0.0959) | (0.0819) | (0.0871) | (0.1014) |
| | | GQL | 0.3045 | 0.2926 | 0.3088 | 0.2989 | 0.9018 | 0.3095 | 0.3016 |
| | | | (0.0760) | (0.0791) | (0.0732) | (0.0799) | (0.0660) | (0.0656) | |
| | 600 | FGLS | 0.3066 | 0.3064 | 0.2872 | 0.3075 | 0.8869 | 0.3094 | 0.3092 |
| | | | (0.0375) | (0.0313) | (0.0348) | (0.0356) | (0.0283) | (0.0288) | (0.0581) |
| | | GQL | 0.3009 | 0.3021 | 0.2945 | 0.3017 | 0.8982 | 0.3047 | 0.3005 |
| | | | (0.0218) | (0.0225) | (0.0220) | (0.0203) | (0.0190) | (0.0165) | |
| 0.5 | 60 | FGLS | 0.2814 | 0.2830 | 0.3208 | 0.3267 | 0.9228 | 0.3201 | 0.5251 |
| | | | (0.1185) | (0.1122) | (0.1184) | (0.1188) | (0.1068) | (0.1045) | (0.1291) |
| | | GQL | 0.2864 | 0.2892 | 0.3127 | 0.3157 | 0.9116 | 0.3108 | 0.5187 |
| | | | (0.0941) | (0.0948) | (0.0950) | (0.0966) | (0.0899) | (0.0896) | |
| | 100 | FGLS | 0.2950 | 0.3161 | 0.2902 | 0.3112 | 0.8881 | 0.3116 | 0.5166 |
| | | | (0.0904) | (0.0993) | (0.0912) | (0.0977) | (0.0881) | (0.0820) | (0.1040) |
| | | GQL | 0.2985 | 0.3111 | 0.2991 | 0.3081 | 0.8909 | 0.3098 | 0.5092 |
| | | | (0.0745) | (0.0788) | (0.0795) | (0.0736) | (0.0699) | (0.0698) | |
| | 600 | FGLS | 0.2885 | 0.3160 | 0.2849 | 0.3130 | 0.8894 | 0.3139 | 0.5113 |
| | | | (0.0351) | (0.0319) | (0.0319) | (0.0373) | (0.0263) | (0.0213) | (0.0515) |
| | | GQL | 0.2991 | 0.3013 | 0.2993 | 0.3065 | 0.8931 | 0.3007 | 0.5012 |
| | | | (0.0215) | (0.0224) | (0.0203) | (0.0213) | (0.0133) | (0.0120) | |
| 0.9 | 60 | FGLS | 0.2725 | 0.2754 | 0.3274 | 0.2744 | 0.9274 | 0.2788 | 0.9273 |
| | | | (0.1191) | (0.1152) | (0.1120) | (0.1172) | (0.1016) | (0.1010) | (0.1267) |
| | | GQL | 0.2811 | 0.2893 | 0.3101 | 0.2801 | 0.9147 | 0.2810 | 0.9123 |
| | | | (0.0975) | (0.0947) | (0.0955) | (0.0934) | (0.0857) | (0.0865) | |
| | 100 | FGLS | 0.3115 | 0.3132 | 0.2892 | 0.2862 | 0.8893 | 0.3137 | 0.9176 |
| | | | (0.0980) | (0.0904) | (0.0948) | (0.0973) | (0.0867) | (0.0812) | (0.1045) |
| | | GQL | 0.3093 | 0.3067 | 0.2934 | 0.2998 | 0.8971 | 0.3010 | 0.9076 |
| | | | (0.0760) | (0.0741) | (0.0735) | (0.0738) | (0.0667) | (0.0699) | |
| | 600 | FGLS | 0.2925 | 0.2895 | 0.3101 | 0.3051 | 0.9060 | 0.3073 | 0.9096 |
| | | | (0.0311) | (0.0302) | (0.0335) | (0.0365) | (0.0231) | (0.0232) | (0.0549) |
| | | GQL | 0.2944 | 0.2976 | 0.3026 | 0.3022 | 0.9017 | 0.3002 | 0.9025 |
| | | | (0.0215) | (0.0206) | (0.0239) | (0.0267) | (0.0130) | (0.0127) | |

**TABLE 3.3**

Estimates of the Parameters and Standard Errors under Stationary BINMA(1) Process for $\rho_{kk} = [\rho_{11} = 0.3, \rho_{22} = 0.3]$

| $\kappa_{12,1}$ | $T$ | Method | $\hat{\beta}_1^{[1]}$ | $\hat{\beta}_2^{[1]}$ | $\hat{\beta}_1^{[2]}$ | $\hat{\beta}_2^{[2]}$ | $\hat{\rho}_{11}$ | $\hat{\rho}_{22}$ | $\hat{\kappa}_{12,1}$ |
|---|---|---|---|---|---|---|---|---|---|
| 0.1 | 60 | FGLS | 0.2722 | 0.2784 | 0.3223 | 0.3276 | 0.3214 | 0.3211 | 0.1298 |
| | | | (0.1104) | (0.1152) | (0.1175) | (0.1172) | (0.1083) | (0.1016) | (0.0546) |
| | | GQL | 0.2855 | 0.2813 | 0.3119 | 0.3109 | 0.3163 | 0.3149 | 0.1177 |
| | | | (0.0945) | (0.0997) | (0.0952) | (0.0993) | (0.0832) | (0.0899) | |
| | 100 | FGLS | 0.2862 | 0.3156 | 0.3112 | 0.2894 | 0.3151 | 0.3161 | 0.1272 |
| | | | (0.0981) | (0.0913) | (0.0906) | (0.0948) | (0.0811) | (0.0848) | (0.0422) |
| | | GQL | 0.2972 | 0.3066 | 0.3021 | 0.2923 | 0.3073 | 0.3016 | 0.1116 |
| | | | (0.0740) | (0.0771) | (0.0760) | (0.0733) | (0.0699) | (0.0659) | |
| | 600 | FGLS | 0.2914 | 0.3063 | 0.3041 | 0.2916 | 0.2918 | 0.3028 | 0.1109 |
| | | | (0.0396) | (0.0380) | (0.0311) | (0.0373) | (0.0241) | (0.0247) | (0.0144) |
| | | GQL | 0.2991 | 0.3017 | 0.3011 | 0.2949 | 0.2987 | 0.3011 | 0.1011 |
| | | | (0.0209) | (0.0207) | (0.0205) | (0.0202) | (0.0115) | (0.0110) | |
| 0.3 | 60 | FGLS | 0.2813 | 0.3268 | 0.2785 | 0.2752 | 0.3257 | 0.3266 | 0.3293 |
| | | | (0.1167) | (0.1117) | (0.1110) | (0.1153) | (0.1032) | (0.1072) | (0.1226) |
| | | GQL | 0.2845 | 0.3150 | 0.2813 | 0.2822 | 0.3113 | 0.3186 | 0.3156 |
| | | | (0.0975) | (0.0947) | (0.0944) | (0.0961) | (0.0852) | (0.0840) | |
| | 100 | FGLS | 0.3127 | 0.3120 | 0.2850 | 0.2847 | 0.3180 | 0.2831 | 0.3123 |
| | | | (0.0972) | (0.0908) | (0.0975) | (0.0974) | (0.0814) | (0.0885) | (0.1082) |
| | | GQL | 0.3032 | 0.3096 | 0.2927 | 0.2921 | 0.3088 | 0.2988 | 0.2911 |
| | | | (0.0750) | (0.0741) | (0.0745) | (0.0737) | (0.0688) | (0.0670) | |
| | 600 | FGLS | 0.3058 | 0.3119 | 0.2899 | 0.2915 | 0.2890 | 0.2904 | 0.3089 |
| | | | (0.0324) | (0.0308) | (0.0316) | (0.0316) | (0.0277) | (0.0275) | (0.0578) |
| | | GQL | 0.3008 | 0.3088 | 0.2932 | 0.2938 | 0.2975 | 0.2989 | 0.3029 |
| | | | (0.0255) | (0.0205) | (0.0210) | (0.0206) | (0.0144) | (0.0134) | |
| 0.5 | 60 | FGLS | 0.2818 | 0.3219 | 0.2780 | 0.3271 | 0.3250 | 0.2834 | 0.5290 |
| | | | (0.1129) | (0.1134) | (0.1136) | (0.1196) | (0.1021) | (0.1083) | (0.1260) |
| | | GQL | 0.2892 | 0.3142 | 0.2824 | 0.3148 | 0.3173 | 0.2885 | 0.5188 |
| | | | (0.0956) | (0.0993) | (0.0946) | (0.0999) | (0.0880) | (0.0841) | |
| | 100 | FGLS | 0.3124 | 0.3145 | 0.2909 | 0.2913 | 0.3170 | 0.3114 | 0.5164 |
| | | | (0.0985) | (0.0919) | (0.0936) | (0.0968) | (0.0891) | (0.0890) | (0.1027) |
| | | GQL | 0.3031 | 0.3094 | 0.2901 | 0.2942 | 0.3034 | 0.3025 | 0.5051 |
| | | | (0.0740) | (0.0741) | (0.0784) | (0.0735) | (0.0693) | (0.0650) | |
| | 600 | FGLS | 0.3066 | 0.2953 | 0.2910 | 0.3071 | 0.3047 | 0.2923 | 0.5105 |
| | | | (0.0335) | (0.0314) | (0.0317) | (0.0338) | (0.0222) | (0.0223) | (0.0526) |
| | | GQL | 0.3018 | 0.2973 | 0.2917 | 0.3014 | 0.3025 | 0.2976 | 0.5022 |
| | | | (0.0205) | (0.0212) | (0.0255) | (0.0256) | (0.0160) | (0.0124) | |
| 0.9 | 60 | FGLS | 0.2830 | 0.3254 | 0.2810 | 0.3267 | 0.3261 | 0.3295 | 0.9254 |
| | | | (0.1174) | (0.1105) | (0.1117) | (0.1189) | (0.1029) | (0.1044) | (0.1213) |
| | | GQL | 0.2861 | 0.3148 | 0.2882 | 0.3115 | 0.3189 | 0.3176 | 0.9106 |
| | | | (0.0954) | (0.0970) | (0.0955) | (0.0998) | (0.0844) | (0.0850) | |
| | 100 | FGLS | 0.3188 | 0.3160 | 0.3138 | 0.2918 | 0.3147 | 0.3102 | 0.9153 |
| | | | (0.0981) | (0.0952) | (0.0994) | (0.0974) | (0.0863) | (0.0851) | (0.1055) |
| | | GQL | 0.3096 | 0.3078 | 0.3069 | 0.2997 | 0.3086 | 0.3021 | 0.9044 |
| | | | (0.0799) | (0.0757) | (0.0780) | (0.0795) | (0.0661) | (0.0670) | |
| | 600 | FGLS | 0.2948 | 0.2944 | 0.2924 | 0.2940 | 0.3112 | 0.2945 | 0.9090 |
| | | | (0.0337) | (0.0371) | (0.0323) | (0.0366) | (0.0266) | (0.0281) | (0.0588) |
| | | GQL | 0.2975 | 0.2982 | 0.2966 | 0.2984 | 0.3040 | 0.2989 | 0.9025 |
| | | | (0.0215) | (0.0207) | (0.0249) | (0.0279) | (0.0115) | (0.0114) | |

**TABLE 3.4**
Number of Non-Convergent Simulations under GQL and FGLS for the Different Combinations

| $\hat{\rho}_{11}$ | $\hat{\rho}_{22}$ | $\hat{\kappa}_{12,1}$ | T | GQL | FGLS | $\hat{\rho}_{11}$ | $\hat{\rho}_{22}$ | $\hat{\kappa}_{12,1}$ | T | GQL | FGLS |
|---|---|---|---|---|---|---|---|---|---|---|---|
| 0.9 | 0.9 | 0.1 | 60 | 160 | 320 | 0.9 | 0.3 | 0.5 | 60 | 170 | 350 |
| | | | 100 | 130 | 280 | | | | 100 | 150 | 320 |
| | | | 600 | 50 | 230 | | | | 600 | 75 | 260 |
| 0.9 | 0.9 | 0.3 | 60 | 170 | 350 | 0.9 | 0.3 | 0.9 | 60 | 180 | 390 |
| | | | 100 | 135 | 320 | | | | 100 | 140 | 360 |
| | | | 600 | 70 | 250 | | | | 600 | 80 | 300 |
| 0.9 | 0.9 | 0.5 | 60 | 175 | 380 | 0.3 | 0.3 | 0.1 | 60 | 150 | 275 |
| | | | 100 | 150 | 350 | | | | 100 | 110 | 240 |
| | | | 600 | 75 | 280 | | | | 600 | 45 | 200 |
| 0.9 | 0.9 | 0.9 | 60 | 190 | 400 | 0.3 | 0.3 | 0.3 | 60 | 160 | 290 |
| | | | 100 | 160 | 375 | | | | 100 | 120 | 260 |
| | | | 600 | 90 | 300 | | | | 600 | 65 | 210 |
| 0.9 | 0.3 | 0.1 | 60 | 140 | 300 | 0.3 | 0.3 | 0.5 | 60 | 170 | 320 |
| | | | 100 | 110 | 270 | | | | 100 | 140 | 290 |
| | | | 600 | 40 | 210 | | | | 600 | 70 | 240 |
| 0.9 | 0.3 | 0.3 | 60 | 150 | 330 | 0.3 | 0.3 | 0.9 | 60 | 180 | 340 |
| | | | 100 | 115 | 300 | | | | 100 | 140 | 315 |
| | | | 600 | 60 | 240 | | | | 600 | 75 | 270 |

From Table 3.4, we conclude that GQL yields considerably lesser non-convergent simulations than FGLS. Hence, based on the standard errors and number of non-convergent simulations, it is remarked here as well that GQL is a more optimal method of estimation and is therefore extended in the next subsection to analyze the non-stationary BINMA(1) series.

### 3.3.2 Non-Stationary Conditions

In this subsection, the focus is on developing the GQL equation under non-stationarity and presenting the estimation procedure.

#### 3.3.2.1 Generalized Quasi-Likelihood (GQL)

The GQL equation to estimate $\boldsymbol{\beta}^{[k]}$ and $\rho_{kk}$ is specified as

$$\boldsymbol{D}'_{\boldsymbol{\beta},\boldsymbol{\rho}}\boldsymbol{\Sigma}^{-1}_{\boldsymbol{\beta},\boldsymbol{\rho}}(\boldsymbol{y}-\boldsymbol{\mu}) = 0, \tag{3.24}$$

where

$$\boldsymbol{y} = \begin{pmatrix} \boldsymbol{y}^{[1]} \\ \boldsymbol{y}^{[2]} \end{pmatrix}_{2T\times 1}, \boldsymbol{y}^{[1]} = [y_1^{[1]}, y_2^{[1]}, \dots, y_T^{[1]}]', \boldsymbol{y}^{[2]} = [y_1^{[2]}, y_2^{[2]}, \dots, y_T^{[2]}]' \tag{3.25}$$

and

$$\boldsymbol{\mu} = \begin{pmatrix} \boldsymbol{\mu}^{[1]} \\ \boldsymbol{\mu}^{[2]} \end{pmatrix}_{2T\times 1}, \boldsymbol{\mu}^{[1]} = [\mu_1^{[1]}, \mu_2^{[1]}, \dots, \mu_T^{[1]}]', \boldsymbol{\mu}^{[2]} = [\mu_1^{[2]}, \mu_2^{[2]}, \dots, \mu_T^{[2]}]' \tag{3.26}$$

while $\boldsymbol{\Sigma}_{\boldsymbol{\beta},\boldsymbol{\rho}}$ is similar to covariance matrix (3.20), with the entries computed using Equations (3.3)–(3.9).
As for the derivative matrix, $\boldsymbol{D}_{\beta,\rho} = [\boldsymbol{D}_1, \boldsymbol{D}_2, \ldots, \boldsymbol{D}_t, \ldots, \boldsymbol{D}_T]'$ with

$$\boldsymbol{D}_t = \begin{pmatrix} \frac{\partial \mu_t^{[1]}}{\partial \boldsymbol{\beta}^{[1]}} & 0 \\ 0 & \frac{\partial \mu_t^{[2]}}{\partial \boldsymbol{\beta}^{[2]}} \\ \frac{\partial \mu_t^{[1]}}{\partial \rho_{11}} & 0 \\ 0 & \frac{\partial \mu_t^{[2]}}{\partial \rho_{22}} \end{pmatrix}_{(2p+2)\times 2} \tag{3.27}$$

where $\mu_t^{[k]} = E(Y_t^{[k]}) = \rho_{kk}\lambda_{t-1}^{[k]} + \lambda_t^{[k]}$
The entries of the derivative matrix are:

1. $\frac{\partial \mu_t^{[k]}}{\partial \beta_j^{[k]}} = \rho_{kk}\lambda_{t-1}^{[k]} x_{t-1,j} + \lambda_t^{[k]} x_{tj}.$

2. $\frac{\partial \mu_t^{[k]}}{\partial \rho_{kk}} = \lambda_{t-1}^{[k]}.$

The cross-correlation parameter $\kappa_{12,t}$ is estimated iteratively using Equation (3.6) as follows:

$$\hat{\kappa}_{12,t} = \begin{cases} \frac{\tilde{\text{Cov}}(Y_t^{[1]}, Y_t^{[2]})}{(1+\hat{\rho}_{11}\hat{\rho}_{22})\sqrt{\hat{\lambda}_t^{[1]}}\sqrt{\hat{\lambda}_t^{[2]}}} & t = 1, \\ \frac{\tilde{\text{Cov}}(Y_t^{[1]}, Y_t^{[2]}) - \hat{\rho}_{11}\hat{\rho}_{22}\hat{\kappa}_{12,t-1}\sqrt{\hat{\lambda}_{t-1}^{[1]}}\sqrt{\hat{\lambda}_{t-1}^{[2]}}}{\sqrt{\hat{\lambda}_t^{[1]}}\sqrt{\hat{\lambda}_t^{[2]}}} & t = 2, \ldots, T \end{cases} \tag{3.28}$$

where $\tilde{\text{Cov}}(Y_t^{[1]}, Y_t^{[2]}) = \frac{1}{T}\sum_{t=1}^{T}(y_t^{[1]} - \hat{\mu}_t^{[1]})(y_t^{[2]} - \hat{\mu}_t^{[2]})$, $\hat{\kappa}_{12,0} = \hat{\kappa}_{12,1}$ $\hat{\lambda}_0^{[k]} = \hat{\lambda}_1^{[k]}$.

The GQL Equation (3.24) is solved using the Newton-Raphson iterative scheme as follows:

$$(\hat{\boldsymbol{\psi}}_{r+1}) = (\hat{\boldsymbol{\psi}}_r) + [\boldsymbol{D}_{\boldsymbol{\psi}}{}' \boldsymbol{\Sigma}_{\boldsymbol{\psi}}{}^{-1} \boldsymbol{D}_{\boldsymbol{\psi}}]_r^{-1} [\boldsymbol{D}_{\boldsymbol{\psi}}{}' \boldsymbol{\Sigma}_{\boldsymbol{\psi}}{}^{-1} (\boldsymbol{y} - \boldsymbol{\mu})]_r \tag{3.29}$$

where $\hat{\boldsymbol{\psi}} = [\hat{\beta}_1^{[1]}, \hat{\beta}_2^{[1]}, \hat{\beta}_1^{[2]}, \hat{\beta}_2^{[2]}, \hat{\rho}_{11}, \hat{\rho}_{22}]$ and $\hat{\boldsymbol{\psi}}_r$ are the estimates at the $r^{th}$ iteration and $[.]_r$ are the values of the expression at the $r^{th}$ iteration.

The working of this algorithm is as follows: For an initial $\hat{\boldsymbol{\psi}}$, $\kappa_{12,t}$ are obtained using Equation (3.28). These estimates are then used to calculate the score and Hessian components and replaced in the Equation (3.29) to obtain an updated $\hat{\boldsymbol{\psi}}$. These updated estimates are then used to re-calculate $\kappa_{12,t}$, the score and Hessian components. This cycle of iteration continues until convergence. Under mild regulatory conditions, the regression and dependence parameters are consistent and $(\hat{\boldsymbol{\psi}} - \boldsymbol{\psi})'$ follows an asymptotic normal distribution with mean $\mathbf{0}$ and covariance matrix $[\boldsymbol{D}_{\psi}' \boldsymbol{\Sigma}_{\psi}^{-1} \boldsymbol{D}_{\psi}]^{-1}$ (Refer to Subsection 2.3.1.2).

Using the converged estimates, we now derive the forecasting equations of the constrained BINMA(1) model. Following Brannas & Hall (2001), Quoreshi (2006a, 2008) and Brannas & Quoreshi (2010), the conditional expectation and variance of the one-step ahead forecast $Y_{t+1}^{[k]}$ given $r_t^{[k]}$ are as follows:

$$\begin{aligned} E(Y_{t+1}^{[k]}|r_t^{[k]}) &= E(\rho_{kk} \circ r_t^{[k]}|r_t^{[k]}) + E(R_{t+1}^{[k]}) \\ &= \rho_{kk} r_t^{[k]} + \lambda_{t+1}^{[k]} \end{aligned} \tag{3.30}$$

and

$$\begin{aligned} \text{Var}(Y_{t+1}^{[k]}|r_t^{[k]}) &= \text{Var}(\rho_{kk} \circ r_t^{[k]}|r_t^{[k]}) + \text{Var}(R_{t+1}^{[k]}) \\ &= \rho_{kk}(1-\rho_{kk}) r_t^{[k]} + \lambda_{t+1}^{[k]}, \end{aligned} \tag{3.31}$$

where $r_t^{[k]}$ are approximated by their respective means.

#### 3.3.2.2 Simulation

In this section, we generate non-stationary BINMA(1) time series using the OD model (3.1) for T=60, 100 and 600. The $(2 \times 1)$ vector of covariates is given by

$$x_{t1} = \begin{cases} -1+t & (t = 1, \ldots, T/4), \\ rnorm(1,0,1) & (t = (T/4)+1, \ldots, 3T/4), \\ 1+t & (t = (3T/4)+1, \ldots, T), \end{cases}$$

where *rnorm* generates random numbers from normal distribution and $x_{t2}$ is a fixed Poisson random value generated with mean parameter 2. For this simulation process, the pair of innovation terms is generated using the R codes developed in Subsection 2.3.1.3, where $R = matrix(c(1, \kappa_{12,1}, \kappa_{12,1}, 1), 2, 2)$ and $lambda = matrix((\lambda_t^{[1]}, \lambda_t^{[2]}), 2, 1)$.

$\kappa_{12,1}$ takes values $0.1, 0.3, 0.5, 0.9$, $\beta_1^{[k]} = \beta_2^{[k]} = 0.3$ and $[\rho_{11}, \rho_{22}]$=[0.9,0.9], [0.9,0.3], [0.3,0.3]. For each time point, 5000 simulations are run and the mean estimates of the regression and correlation parameters are shown in Tables 3.5–3.7.

From the results displayed in Tables 3.5–3.7, it is remarked that similar to the stationary case, these estimates are close to the true values. The correlation for each time point are estimated but due to the high-dimensional number of time points, various lagged serial-correlations and the cross-correlations at each time point are not displayed, but have been reliable estimates between 0 and 1, as expected. It is also remarked that as T increases, the standard errors decrease. However, due to ill-conditioned covariance matrix and singular double-derivative Hessian for some cluster size, some non-convergent simulations are noted. For $\kappa_{12,1} = 0.1$ and $\rho_{11} = \rho_{22} = 0.9$ and $T = 60$, the number of non-convergent simulations is around 180. As $T$ increases to 100 and 600, the number of non-convergent simulations is 150 and 75, respectively.

**TABLE 3.5**

Estimates of the Parameters and Standard Errors under Non-Stationary BINMA(1) Process for $\rho_{kk} = [\rho_{11} = 0.9, \rho_{22} = 0.9]$

| $\kappa_{12,1}$ | $T$ | $\hat{\beta}_1^{[1]}$ | $\hat{\beta}_2^{[1]}$ | $\hat{\beta}_1^{[2]}$ | $\hat{\beta}_2^{[2]}$ | $\hat{\rho}_{11}$ | $\hat{\rho}_{22}$ | $\hat{\kappa}_{12,1}$ |
|---|---|---|---|---|---|---|---|---|
| 0.1 | 60 | 0.3115 | 0.3121 | 0.3187 | 0.3195 | 0.9171 | 0.9195 | 0.1172 |
| | | (0.0811) | (0.0808) | (0.0851) | (0.0832) | (0.0748) | (0.0719) | |
| | 100 | 0.3083 | 0.3084 | 0.3089 | 0.3098 | 0.8992 | 0.9029 | 0.1078 |
| | | (0.0602) | (0.0658) | (0.0612) | (0.0610) | (0.0517) | (0.0547) | |
| | 600 | 0.3055 | 0.3080 | 0.3084 | 0.3095 | 0.8998 | 0.9025 | 0.1019 |
| | | (0.0131) | (0.0101) | (0.0116) | (0.0128) | (0.0012) | (0.0071) | |
| 0.3 | 60 | 0.3164 | 0.3171 | 0.3125 | 0.3108 | 0.8865 | 0.8861 | 0.3159 |
| | | (0.0804) | (0.0815) | (0.0825) | (0.0811) | (0.0797) | (0.0775) | |
| | 100 | 0.3029 | 0.3025 | 0.3014 | 0.3038 | 0.9052 | 0.9032 | 0.3040 |
| | | (0.0619) | (0.0723) | (0.0702) | (0.0617) | (0.0676) | (0.0596) | |
| | 600 | 0.3007 | 0.3098 | 0.3012 | 0.3004 | 0.9052 | 0.9024 | 0.3011 |
| | | (0.0149) | (0.0295) | (0.0239) | (0.0221) | (0.0179) | (0.0028) | |
| 0.5 | 60 | 0.3191 | 0.3169 | 0.3116 | 0.3127 | 0.9178 | 0.9187 | 0.5111 |
| | | (0.0939) | (0.0918) | (0.1047) | (0.0815) | (0.0719) | (0.0830) | |
| | 100 | 0.3020 | 0.3023 | 0.3045 | 0.3061 | 0.9065 | 0.8958 | 0.4912 |
| | | (0.0624) | (0.0711) | (0.0725) | (0.0677) | (0.0579) | (0.0651) | |
| | 600 | 0.3012 | 0.3010 | 0.3010 | 0.3011 | 0.8957 | 0.9032 | 0.5032 |
| | | (0.0242) | (0.0317) | (0.0151) | (0.0318) | (0.0125) | (0.0174) | |
| 0.9 | 60 | 0.3199 | 0.3134 | 0.3190 | 0.3181 | 0.8815 | 0.9110 | 0.8872 |
| | | (0.1003) | (0.1114) | (0.0843) | (0.0916) | (0.0915) | (0.0807) | |
| | 100 | 0.3019 | 0.3078 | 0.3098 | 0.3015 | 0.9035 | 0.8962 | 0.8919 |
| | | (0.0702) | (0.0607) | (0.0812) | (0.0836) | (0.0727) | (0.0525) | |
| | 600 | 0.3011 | 0.3042 | 0.3087 | 0.3005 | 0.8983 | 0.8970 | 0.8991 |
| | | (0.0339) | (0.0295) | (0.0341) | (0.0111) | (0.0288) | (0.0220) | |

**TABLE 3.6**

Estimates of the Parameters and Standard Errors under Non-Stationary BINMA(1) Process for $\rho_{kk} = [\rho_{11} = 0.9, \rho_{22} = 0.3]$

| $\kappa_{12,1}$ | $T$ | $\hat{\beta}_1^{[1]}$ | $\hat{\beta}_2^{[1]}$ | $\hat{\beta}_1^{[2]}$ | $\hat{\beta}_2^{[2]}$ | $\hat{\rho}_{11}$ | $\hat{\rho}_{22}$ | $\hat{\kappa}_{12,1}$ |
|---|---|---|---|---|---|---|---|---|
| 0.1 | 60 | 0.3122 | 0.3194 | 0.3111 | 0.3120 | 0.8889 | 0.3118 | 0.1119 |
| | | (0.0820) | (0.0839) | (0.0815) | (0.0821) | (0.0715) | (0.0707) | |
| | 100 | 0.3037 | 0.3039 | 0.3029 | 0.3031 | 0.8911 | 0.2962 | 0.1089 |
| | | (0.0603) | (0.0606) | (0.0601) | (0.0617) | (0.0514) | (0.0516) | |
| | 600 | 0.3028 | 0.3029 | 0.3011 | 0.3021 | 0.9012 | 0.3015 | 0.1055 |
| | | (0.0112) | (0.0115) | (0.0117) | (0.0153) | (0.0038) | (0.0081) | |
| 0.3 | 60 | 0.3121 | 0.3133 | 0.3122 | 0.3198 | 0.8891 | 0.3175 | 0.3178 |
| | | (0.0922) | (0.0941) | (0.0845) | (0.0811) | (0.0835) | (0.0747) | |
| | 100 | 0.3041 | 0.3024 | 0.3098 | 0.3036 | 0.9099 | 0.3019 | 0.3020 |
| | | (0.0711) | (0.0612) | (0.0608) | (0.0768) | (0.0517) | (0.0622) | |
| | 600 | 0.3023 | 0.3016 | 0.3069 | 0.3020 | 0.9061 | 0.3011 | 0.3010 |
| | | (0.0212) | (0.0216) | (0.0115) | (0.0235) | (0.0167) | (0.0003) | |

*(Continued)*

**TABLE 3.6**
*(Continued)*

| $\kappa_{12,1}$ | $T$ | $\hat{\beta}_1^{[1]}$ | $\hat{\beta}_2^{[1]}$ | $\hat{\beta}_1^{[2]}$ | $\hat{\beta}_2^{[2]}$ | $\hat{\rho}_{11}$ | $\hat{\rho}_{22}$ | $\hat{\kappa}_{12,1}$ |
|---|---|---|---|---|---|---|---|---|
| 0.5 | 60 | 0.3154 | 0.3162 | 0.3113 | 0.3125 | 0.9129 | 0.3112 | 0.5191 |
| | | (0.1011) | (0.1112) | (0.0913) | (0.0821) | (0.0957) | (0.0703) | |
| | 100 | 0.3012 | 0.3036 | 0.3081 | 0.3035 | 0.8902 | 0.3072 | 0.5011 |
| | | (0.0718) | (0.0656) | (0.0751) | (0.0812) | (0.0613) | (0.0552) | |
| | 600 | 0.3010 | 0.3009 | 0.3002 | 0.3007 | 0.9017 | 0.3066 | 0.5004 |
| | | (0.0394) | (0.0212) | (0.0121) | (0.0311) | (0.0286) | (0.0095) | |
| 0.9 | 60 | 0.3191 | 0.3146 | 0.3181 | 0.3184 | 0.9171 | 0.2867 | 0.8878 |
| | | (0.1218) | (0.1113) | (0.0922) | (0.1042) | (0.0924) | (0.0729) | |
| | 100 | 0.3050 | 0.3031 | 0.3086 | 0.3019 | 0.8915 | 0.3015 | 0.8909 |
| | | (0.0729) | (0.0813) | (0.0612) | (0.0815) | (0.0523) | (0.0769) | |
| | 600 | 0.3015 | 0.3010 | 0.3007 | 0.3008 | 0.9021 | 0.3009 | 0.9021 |
| | | (0.0301) | (0.0316) | (0.0127) | (0.0346) | (0.0224) | (0.0111) | |

**TABLE 3.7**
Estimates of the Parameters and Standard Errors under Non-Stationary BINMA(1) Process for $\rho_{kk} = [\rho_{11} = 0.3, \rho_{22} = 0.3]$

| $\kappa_{12,1}$ | $T$ | $\hat{\beta}_1^{[1]}$ | $\hat{\beta}_2^{[1]}$ | $\hat{\beta}_1^{[2]}$ | $\hat{\beta}_2^{[2]}$ | $\hat{\rho}_{11}$ | $\hat{\rho}_{22}$ | $\hat{\kappa}_{12,1}$ |
|---|---|---|---|---|---|---|---|---|
| 0.1 | 60 | 0.3138 | 0.3198 | 0.3193 | 0.3152 | 0.2811 | 0.3123 | 0.1111 |
| | | (0.0811) | (0.0823) | (0.0817) | (0.0860) | (0.0771) | (0.0756) | |
| | 100 | 0.3018 | 0.3081 | 0.3028 | 0.3019 | 0.3041 | 0.2996 | 0.1085 |
| | | (0.0601) | (0.0604) | (0.0615) | (0.0613) | (0.0570) | (0.0540) | |
| | 600 | 0.3012 | 0.3011 | 0.3013 | 0.3010 | 0.2972 | 0.3001 | 0.1010 |
| | | (0.0138) | (0.0111) | (0.0122) | (0.0114) | (0.0017) | (0.0029) | |
| 0.3 | 60 | 0.3114 | 0.3113 | 0.3111 | 0.3116 | 0.3133 | 0.3143 | 0.3114 |
| | | (0.0931) | (0.0816) | (0.0911) | (0.0815) | (0.0855) | (0.0734) | |
| | 100 | 0.3011 | 0.3012 | 0.3025 | 0.3021 | 0.3099 | 0.3088 | 0.3092 |
| | | (0.0618) | (0.0713) | (0.0617) | (0.0711) | (0.0631) | (0.0509) | |
| | 600 | 0.3003 | 0.3008 | 0.3019 | 0.3015 | 0.2994 | 0.2956 | 0.3078 |
| | | (0.0145) | (0.0210) | (0.0137) | (0.0226) | (0.0131) | (0.0084) | |
| 0.5 | 60 | 0.3113 | 0.3112 | 0.3114 | 0.3111 | 0.3135 | 0.2858 | 0.5187 |
| | | (0.1017) | (0.0813) | (0.1014) | (0.0975) | (0.0815) | (0.0709) | |
| | 100 | 0.3012 | 0.3010 | 0.3011 | 0.3016 | 0.3094 | 0.3091 | 0.5089 |
| | | (0.0719) | (0.0615) | (0.0761) | (0.0812) | (0.0706) | (0.0545) | |
| | 600 | 0.3002 | 0.3005 | 0.3010 | 0.3011 | 0.3051 | 0.2971 | 0.5019 |
| | | (0.0226) | (0.0153) | (0.0248) | (0.0128) | (0.0178) | (0.0045) | |
| 0.9 | 60 | 0.3122 | 0.3135 | 0.3120 | 0.3131 | 0.3148 | 0.3192 | 0.8885 |
| | | (0.1111) | (0.1222) | (0.0912) | (0.0935) | (0.0812) | (0.0715) | |
| | 100 | 0.3011 | 0.3017 | 0.3013 | 0.3016 | 0.3088 | 0.3096 | 0.8901 |
| | | (0.0882) | (0.0797) | (0.0655) | (0.0636) | (0.0535) | (0.0621) | |
| | 600 | 0.3009 | 0.3007 | 0.3011 | 0.3010 | 0.3025 | 0.2948 | 0.8995 |
| | | (0.0311) | (0.0145) | (0.0213) | (0.0252) | (0.0145) | (0.0251) | |

For same values of $\rho_{11}, \rho_{22}$, as $\kappa_{12,1}$ increases, a slight increase in the number of non-convergent simulations is remarked. It is important to note that for non-stationary bivariate time series process, the non-stationarity of the cross-covariance terms between the two series should be modeled appropriately.

## 3.4 Summary

In this chapter, a novel constrained BINMA(1) model with Poisson innovations under both stationary and non-stationary moment conditions has been developed. The FGLS has been used to estimate the parameters for the stationary BINMA(1) process. In addition, we have formulated the GQL approach and have compared its performance with FGLS where it was noticed that GQL has shown to yield estimates with significant lower standard errors and fewer non-convergent simulations. The GQL has thus been extended to analyze the non-stationary BINMA(1) series with Poisson innovations. The simulation results in the non-stationary BINMA(1) series have also illustrated that the estimates under the extended GQL are consistent. Based on the results of Chapter 2 and 3, we conclude that the GQL is a more feasible, practical and optimal estimation approach than CMLE and FGLS. For the subsequent chapters, and especially for the over-dispersed and unconstrained models, it is of interest to explore some extended approaches of GQL to estimate the different sets of parameters under the non-stationary conditions. The next chapter extends the constrained bivariate model to an unconstrained bivariate model by allowing a full correlation structure.

# 4

# *Unconstrained BINAR(1) Model with Poisson Innovations*

## 4.1 Introduction

The constrained BINAR(1) model introduced in Chapter 2 may not be sufficient to model some real-life situations. It may be useful to consider the relationship between the lag responses between the series and hence the need to extend the constrained BINAR(1) model to an unconstrained BINAR(1) model. In terms of inferential procedure, different estimation methods to estimate the unknown model parameters of the BINAR(1), CMLE, FGLS and GQL, have been developed and compared in Chapter 2. Based on the Monte-Carlo simulation results, it has been illustrated that GQL yields estimators of considerably lower bias than FGLS and asymptotically equally efficient estimates as CMLE. In addition, CMLE has been found to be extremely time-consuming and resulted in huge number of non-convergent simulations compared to GQL.

Thus, in this chapter we develop the unconstrained BINAR(1) model with Poisson innovations under cross-correlation and alongside implement a bivariate GQL-based inference procedure. However, the specification of the GQL components will not have similar structure as in the constrained BINAR(1) model developed in Chapter 2, since the derivatives and auto-covariance components appear in a difference mode.

Hence the chapter is organized as follows: In Section 4.2, the extension to unconstrained BINAR(1) model with Poisson innovations that incorporates the local cross-correlation is presented. This section also considers the derivation of the marginal and joint moments and discusses on the possible limitations of deriving the joint pgf under non-stationarity by making an analogy with the special case of stationarity. In Section 4.3, the GQL approach is formulated along side the asymptotic normality properties and forecasting equations. This chapter ends with the conclusion.

DOI: 10.1201/9781003600435-4

## 4.2 The Model Construction

In this section, we introduce the BINAR(1) model $\{[Y_t^{[1]}, Y_t^{[2]}]', \ t \in \mathbb{Z}\}$ where

$$Y_t^{[1]} = \rho_{11} \circ Y_{t-1}^{[1]} + \rho_{12} \circ Y_{t-1}^{[2]} + R_t^{[1]} \tag{4.1}$$

$$Y_t^{[2]} = \rho_{21} \circ Y_{t-1}^{[1]} + \rho_{22} \circ Y_{t-1}^{[2]} + R_t^{[2]} \tag{4.2}$$

based on the following conditions:

(a) $R_t^{[k]}$ following the Poisson distribution such that $R_t^{[k]} \sim \text{Poisson}(\lambda_t^{[k]})$, for $k \in \{1, 2\}$ and $\text{Corr}(R_t^{[1]}, R_t^{[2]})$=0.

(b) '$\circ$' is the usual binomial thinning operator (Steutel & Van Harn (1979), Wei$\beta$ (2008b)) where

$$\{\rho_{jk} \circ Y_{t-1}^{[k]}\} = \begin{cases} \sum_{l=1}^{Y_{t-1}^{[k]}} b_l(\rho_{jk}), & Y_{t-1}^{[k]} > 0, \\ 0 & Y_{t-1}^{[k]} = 0, \end{cases}$$

where the counting series $\{b_l(\rho_{jk})\}$ is a sequence of independent and identically distributed Bernoulli random variable with $\text{P}\{b_l(\rho_{jk})\} = 1 - \text{P}\{b_l(\rho_{jk})\} = \rho_{jk}$, or alternatively, $\{\rho_{jk} \circ Y_{t-1}^{[k]} | Y_{t-1}^{[k]}\} \sim$ Binomial $(Y_{t-1}^{[k]}, \rho_{jk})$. It follows that $\text{Var}(\rho_{jk} \circ Y_{t-1}^{[k]} | Y_{t-1}^{[k]}) = \rho_{jk}(1 - \rho_{jk}) Y_{t-1}^{[k]}$ and hence $\text{Var}(\rho_{jk} \circ Y_{t-1}^{[k]}) = \rho_{jk}(1 - \rho_{jk}) E(Y_{t-1}^{[k]}) + \rho_{jk}^2 \text{Var}(Y_{t-1}^{[k]})$ for $\rho_{jk} \in [0, 1]$ and $j, k \in \{1, 2\}$.

(c) The random variables $\rho_{ik} \circ Y_{t-1}^{[k]}$ and $\rho_{jk'} \circ Y_{t-1}^{[k']}$, where $k \neq k'$, for given $Y_{t-1}^{[k]}$ and $Y_{t-1}^{[k']}$ are dependent binomial random variables with parameters $Y_{t-1}^{[k]}$ and $\rho_{ik}$, and $Y_{t-1}^{[k']}$ and $\rho_{jk'}$ respectively, where $i, j, k, k' \in \{1, 2\}$.

Equations (4.1) and (4.2) may be combined in vector form as follows:

$$\boldsymbol{Y_t} = \boldsymbol{A} \circ \boldsymbol{Y_{t-1}} + \boldsymbol{R_t} \tag{4.3}$$

where $\boldsymbol{Y_t} = [Y_t^{[1]}, Y_t^{[2]}]'$, $\boldsymbol{R_t} = [R_t^{[1]}, R_t^{[2]}]'$ and $\boldsymbol{A} = \begin{pmatrix} \rho_{11} & \rho_{12} \\ \rho_{21} & \rho_{22} \end{pmatrix}$.

**Lemma 4.1.** (a) $E(\boldsymbol{Y_t}) = \boldsymbol{A} E(\boldsymbol{Y_{t-1}}) + E(\boldsymbol{R_t})$.

(b) $\text{Var}(\boldsymbol{Y_t}) = \boldsymbol{A} \text{Var}(\boldsymbol{Y_{t-1}}) \boldsymbol{A}' + \text{Var}(\boldsymbol{A} \circ \boldsymbol{Y_{t-1}}) + \text{Var}(\boldsymbol{R_t})$

*Proof:*

(a)

$$\begin{aligned} E(\boldsymbol{Y_t}) &= E(\boldsymbol{A} \circ \boldsymbol{Y_{t-1}}) + E(\boldsymbol{R_t}) \\ &= E_{\boldsymbol{Y_{t-1}}}[E(\boldsymbol{A} \circ \boldsymbol{Y_{t-1}}|\boldsymbol{Y_{t-1}})] + E(\boldsymbol{R_t}) \\ &= E_{\boldsymbol{Y_{t-1}}}(\boldsymbol{A}\boldsymbol{Y_{t-1}}) + E(\boldsymbol{R_t}) \\ &= \boldsymbol{A}E(\boldsymbol{Y_{t-1}}) + E(\boldsymbol{R_t}) \end{aligned}$$

(b)

$$\begin{aligned} \mathrm{Var}(\boldsymbol{Y_t}) &= \mathrm{Var}[E(\boldsymbol{A} \circ \boldsymbol{Y_{t-1}} + \boldsymbol{R_t}|\boldsymbol{Y_{t-1}})] + E[\mathrm{Var}(\boldsymbol{A} \circ \boldsymbol{Y_{t-1}} + \boldsymbol{R_t}|\boldsymbol{Y_{t-1}})] \\ &= \mathrm{Var}[\boldsymbol{A}\boldsymbol{Y_{t-1}} + E(\boldsymbol{R_t})] + E[\mathrm{Var}(\boldsymbol{A} \circ \boldsymbol{Y_{t-1}}) + \mathrm{Var}(\boldsymbol{R_t})] \\ &= \boldsymbol{A}\mathrm{Var}(\boldsymbol{Y_{t-1}})\boldsymbol{A}' + \mathrm{Var}(\boldsymbol{A} \circ \boldsymbol{Y_{t-1}}) + \mathrm{Var}(\boldsymbol{R_t}) \end{aligned}$$

(See Franke & Rao (1995), Latour (1997), Pedeli & Karlis (2013a) for more details).

Under Lemma 4.1 and assuming $\mathrm{E}(\boldsymbol{Y_t}) = \boldsymbol{\mu_t}$, where $\boldsymbol{\mu_t} = [\mu_t^{[1]}, \mu_t^{[2]}]'$, we obtain

$$\begin{aligned} E(\boldsymbol{Y_t}) &= \boldsymbol{A}E(\boldsymbol{Y_{t-1}}) + E(\boldsymbol{R_t}) \\ \boldsymbol{\mu_t} &= \boldsymbol{A}\boldsymbol{\mu_{t-1}} + \boldsymbol{\lambda_t}, \end{aligned} \tag{4.4}$$

where $\boldsymbol{\lambda_t} = [\lambda_t^{[1]}, \lambda_t^{[2]}]'$.
Hence, by expanding Equation (4.4), we obtain

$$\begin{aligned} \mu_t^{[1]} &\equiv E(Y_t^{[1]}) = \rho_{11}\mu_{t-1}^{[1]} + \rho_{12}\mu_{t-1}^{[2]} + \lambda_t^{[1]} \quad \text{and} \\ \mu_t^{[2]} &\equiv E(Y_t^{[2]}) = \rho_{21}\mu_{t-1}^{[1]} + \rho_{22}\mu_{t-1}^{[2]} + \lambda_t^{[2]}. \end{aligned} \tag{4.5}$$

Denoting $\boldsymbol{\Sigma}_{0,t} = \begin{bmatrix} \mathrm{Var}(Y_t^{[1]}) & \mathrm{Cov}(Y_t^{[1]}, Y_t^{[2]}) \\ \mathrm{Cov}(Y_t^{[2]}, Y_t^{[1]}) & \mathrm{Var}(Y_t^{[2]}) \end{bmatrix}$, we have

$$\boldsymbol{\Sigma}_{0,t} = \boldsymbol{A}\mathrm{Var}(\boldsymbol{Y_{t-1}})\boldsymbol{A}' + \mathrm{diag}(\boldsymbol{B}\boldsymbol{\mu_{t-1}}) + \mathrm{Var}(\boldsymbol{R_t}) \tag{4.6}$$

where

$$\boldsymbol{B} = \begin{pmatrix} \rho_{11}(1-\rho_{11}) & \rho_{12}(1-\rho_{12}) \\ \rho_{21}(1-\rho_{21}) & \rho_{22}(1-\rho_{22}) \end{pmatrix}.$$

Hence, by expanding (4.6), we obtain

$$\begin{aligned} \sigma_t^{2[1]} = \mathrm{Var}(Y_t^{[1]}) &= \rho_{11}(1-\rho_{11})\mu_{t-1}^{[1]} + \rho_{11}^2\mathrm{Var}(Y_{t-1}^{[1]}) \\ &+ \rho_{12}(1-\rho_{12})\mu_{t-1}^{[2]} + \rho_{12}^2\mathrm{Var}(Y_{t-1}^{[2]}) \\ &+ 2\rho_{11}\rho_{12}\mathrm{Cov}(Y_{t-1}^{[1]}, Y_{t-1}^{[2]}) + \lambda_t^{[1]} \\ &= \mu_t^{[1]} + \rho_{11}^2[\mathrm{Var}(Y_{t-1}^{[1]}) - \mu_{t-1}^{[1]}] + \rho_{12}^2[\mathrm{Var}(Y_{t-1}^{[2]}) - \mu_{t-1}^{[2]}] \\ &+ 2\rho_{11}\rho_{12}\mathrm{Cov}(Y_{t-1}^{[1]}, Y_{t-1}^{[2]}), \end{aligned} \tag{4.7}$$

$$\begin{aligned}\sigma_t^{2[2]} = \mathrm{Var}(Y_t^{[2]}) &= \rho_{21}(1-\rho_{21})\mu_{t-1}^{[1]} + \rho_{21}^2\mathrm{Var}(Y_{t-1}^{[1]}) + \rho_{22}(1-\rho_{22})\mu_{t-1}^{[2]} \\ &+ \rho_{22}^2\mathrm{Var}(Y_{t-1}^{[2]}) + 2\rho_{21}\rho_{22}\mathrm{Cov}(Y_{t-1}^{[1]}, Y_{t-1}^{[2]}) + \lambda_t^{[2]} \\ &= \mu_t^{[2]} + \rho_{21}^2[\mathrm{Var}(Y_{t-1}^{[1]}) - \mu_{t-1}^{[1]}] + \rho_{22}^2[\mathrm{Var}(Y_{t-1}^{[2]}) - \mu_{t-1}^{[2]}] \\ &+ 2\rho_{21}\rho_{22}\mathrm{Cov}(Y_{t-1}^{[1]}, Y_{t-1}^{[2]}) \end{aligned} \tag{4.8}$$

and

$$\begin{aligned}\mathrm{Cov}(Y_t^{[1]}, Y_t^{[2]}) &= (\rho_{11}\rho_{22} + \rho_{12}\rho_{21})\mathrm{Cov}(Y_{t-1}^{[1]}, Y_{t-1}^{[2]}) + \rho_{11}\rho_{21}\mathrm{Var}(Y_{t-1}^{[1]}) \\ &+ \rho_{22}\rho_{12}\mathrm{Var}(Y_{t-1}^{[2]}). \end{aligned} \tag{4.9}$$

**Remark 4.1.** Under stationary moment condition, Equations (4.5) and (4.7)–(4.9) can be simplified to

$$\mu_1^{[1]} = \frac{\rho_{12}\mu_1^{[2]} + \lambda_1^{[1]}}{1-\rho_{11}}, \quad \mu_1^{[2]} = \frac{\rho_{21}\mu_1^{[1]} + \lambda_1^{[2]}}{1-\rho_{22}}, \tag{4.10}$$

$$\begin{aligned}\mathrm{Var}(Y_t^{[1]}) &= \left(\frac{1}{1-\rho_{11}^2}\right)\{\rho_{11}(1-\rho_{11})\mu_1^{[1]} + \rho_{12}(1-\rho_{12})\mu_1^{[2]} + \rho_{12}^2\mathrm{Var}(Y_t^{[2]}) \\ &+ 2\rho_{11}\rho_{12}\mathrm{Cov}(Y_t^{[1]}, Y_t^{[2]}) + \lambda_1^{[1]}\}, \end{aligned} \tag{4.11}$$

$$\begin{aligned}\mathrm{Var}(Y_t^{[2]}) &= \left(\frac{1}{1-\rho_{22}^2}\right)\{\rho_{21}(1-\rho_{21})\mu_1^{[1]} + \rho_{21}^2\mathrm{Var}(Y_t^{[1]}) + \rho_{22}(1-\rho_{22})\mu_1^{[2]} \\ &+ 2\rho_{21}\rho_{22}\mathrm{Cov}(Y_t^{[1]}, Y_t^{[2]}) + \lambda_1^{[2]}\}, \end{aligned} \tag{4.12}$$

$$\mathrm{Cov}(Y_t^{[1]}, Y_t^{[2]}) = \frac{\rho_{11}\rho_{21}\mathrm{Var}(Y_t^{[1]}) + \rho_{22}\rho_{12}\mathrm{Var}(Y_t^{[2]})}{1-\rho_{11}\rho_{22}-\rho_{12}\rho_{21}} \tag{4.13}$$

and solving simultaneously, we obtain

$$\mu_1^{[1]} = \frac{(1-\rho_{22})\lambda_1^{[1]} + \rho_{12}\lambda_1^{[2]}}{(1-\rho_{11})(1-\rho_{22}) - \rho_{12}\rho_{21}}, \quad \mu_1^{[2]} = \frac{(1-\rho_{11})\lambda_1^{[2]} + \rho_{21}\lambda_1^{[1]}}{(1-\rho_{11})(1-\rho_{22}) - \rho_{12}\rho_{21}}, \tag{4.14}$$

$$\begin{aligned}\mathrm{Var}(Y_t^{[1]}) &= \{[(1-\rho_{11}^2)(1-\rho_{22}^2)(1-\rho_{11}\rho_{22}) - \rho_{12}\rho_{21}(1-\rho_{11}^2)(1-\rho_{22}^2) \\ &- \rho_{12}^2\rho_{21}^2(1-\rho_{11}\rho_{22}-\rho_{12}\rho_{21})]\mu_1^{[1]} + 2\rho_{12}(\rho_{11}+\rho_{12}\rho_{21}\rho_{22}-\rho_{11}\rho_{22}^2) \\ &\times (\rho_{12}\rho_{22}\mu_1^{[2]})\}/(\delta), \end{aligned} \tag{4.15}$$

$$\begin{aligned}\mathrm{Var}(Y_t^{[2]}) &= \{[(1-\rho_{11}^2)(1-\rho_{22}^2)(1-\rho_{11}\rho_{22}) - \rho_{12}\rho_{21}(1-\rho_{11}^2)(1-\rho_{22}^2) \\ &- \rho_{12}^2\rho_{21}^2(1-\rho_{11}\rho_{22}-\rho_{12}\rho_{21})]\mu_1^{[2]} + 2\rho_{21}(\rho_{22}+\rho_{11}\rho_{12}\rho_{21}-\rho_{11}^2\rho_{22}) \\ &\times (\rho_{11}\rho_{21}\mu_1^{[1]})\}/(\delta), \end{aligned} \tag{4.16}$$

$$\mathrm{Cov}(Y_t^{[1]}, Y_t^{[2]}) = [(1-\rho_{11}^2)(1-\rho_{22}^2) - \rho_{12}^2\rho_{21}^2](\rho_{11}\rho_{21}\mu_1^{[1]} + \rho_{12}\rho_{22}\mu_1^{[2]})/(\delta) \tag{4.17}$$

where $\delta = (1-\rho_{11}^2)(1-\rho_{22}^2)(1-\rho_{11}\rho_{22}) - \rho_{12}\rho_{21}(1+\rho_{11}^2+\rho_{22}^2-3\rho_{11}^2\rho_{22}^2) - \rho_{11}^2\rho_{21}^2(1-\rho_{12}\rho_{21}+3\rho_{11}\rho_{22})$.

**Remark 4.2.** These moment expressions are analogous to Equations (5), (6), (18) and (19) from Pedeli & Karlis (2013a) and hence under the stationary condition, from Equation (4.3), $G_{\boldsymbol{Y}_t}(\boldsymbol{s}) = G_{\boldsymbol{Y}_t}(\boldsymbol{A}'\boldsymbol{s})G_{\boldsymbol{R}_t}(\boldsymbol{s})$ which yields the joint pgf expressions (9) and (10) from the stationary BINAR(1) model of Pedeli & Karlis (2013a). However, it is remarked that, though under stationarity, the derivation of the joint probability and marginal distribution of $[Y_t^{[1]}, Y_t^{[2]}]'$ compose of tedious algebra. In fact, Pedeli & Karlis (2013a) managed to derive the marginal distribution of $Y_t^{[k]}$ which is proved to be Hermite under stationarity, for $k \in \{1,2\}$. Under the specific case of non-stationarity as described in Equations (4.5), (4.7) and (4.8) in the above, the same conditions pertaining to the joint pgf derivation do not hold and besides the marginal moments indicate clearly that the distribution of $Y_t^{[1]}$ and $Y_t^{[2]}$ is unknown. In fact, for non-stationary INAR(1) process, Nastic et al. (2016a) suggested using $G_{\boldsymbol{Y}_t}(\boldsymbol{s}) = G_{\boldsymbol{Y}_{t-1}}(\boldsymbol{A}'\boldsymbol{s})G_{\boldsymbol{R}_t}(\boldsymbol{s})$, where $G_{\boldsymbol{Y}_t}$ is the pgf of the random variable $\boldsymbol{Y}_t$, but argued that the pgf derivation under the single series becomes hard and hence it is increasingly difficult for bivariate non-stationary series. Thus, it is clear that the likelihood specification or probability distribution specification becomes difficult.

**Note.** Under the assumption that there exists some covariates that commonly influence $Y_t^{[1]}$ and $Y_t^{[2]}$, that is, $\boldsymbol{x}_t = [x_{t1}, x_{t2}, \ldots, x_{tj}, \ldots, x_{tp}]'$ with $p$ explanatory effects, then $\lambda_t^{[k]}$ is represented as $\lambda_t^{[k]} = \exp(\boldsymbol{x}_t'\boldsymbol{\beta}^{[k]})$ where $\boldsymbol{\beta}^{[k]} = [\beta_1^{[k]}, \beta_2^{[k]}, \ldots, \beta_j^{[k]}, \ldots, \beta_p^{[k]}]'$ for $k \in \{1,2\}$. Hence, the unknown parameters of the BINAR(1) model constitute the vector of regression parameters $[\boldsymbol{\beta}^{[1]}, \boldsymbol{\beta}^{[2]}]'$, the set of serial and cross thinning coefficients $[\rho_{11}, \rho_{12}, \rho_{21}, \rho_{22}]'$. The next section presents the estimation method to be employed for obtaining the parameter estimates described above.

To facilitate the computation and in view of exploring the BINAR(1) structure, we denote

$$\boldsymbol{\Sigma}_{h,t} = \begin{bmatrix} \mathrm{Cov}(Y_t^{[1]}, Y_{t+h}^{[1]}) & \mathrm{Cov}(Y_{t+h}^{[1]}, Y_t^{[2]}) \\ \mathrm{Cov}(Y_t^{[1]}, Y_{t+h}^{[2]}) & \mathrm{Cov}(Y_t^{[2]}, Y_{t+h}^{[2]}) \end{bmatrix} = \begin{bmatrix} \Sigma_{11,h,t} & \Sigma_{12,h,t} \\ \Sigma_{21,h,t} & \Sigma_{22,h,t} \end{bmatrix} \tag{4.18}$$

as described in Pedeli & Karlis (2013a) and Nastic et al. (2016b). Under this representation, further to the autoregressive structure,
when $h = 1$,

$$\begin{aligned}\mathrm{Cov}(Y_t^{[1]}, Y_{t+1}^{[1]}) &= \mathrm{Cov}[Y_t^{[1]}, (\rho_{11} \circ Y_t^{[1]} + \rho_{12} \circ Y_t^{[2]} + R_{t+1}^{[1]})] \\ &= \rho_{11}\mathrm{Var}(Y_t^{[1]}) + \rho_{12}\mathrm{Cov}(Y_t^{[1]}, Y_t^{[2]}).\end{aligned}$$

Similarly,

$$\mathrm{Cov}(Y_t^{[2]}, Y_{t+1}^{[2]}) = \rho_{22}\mathrm{Var}(Y_t^{[2]}) + \rho_{21}\mathrm{Cov}(Y_t^{[2]}, Y_t^{[1]}).$$

Thus,
$\Sigma_{11,1,t} = \rho_{11}\Sigma_{11,0,t} + \rho_{12}\Sigma_{21,0,t}$
$\Sigma_{22,1,t} = \rho_{22}\Sigma_{22,0,t} + \rho_{21}\Sigma_{12,0,t}$

$$\begin{aligned}\mathrm{Cov}(Y_t^{[1]}, Y_{t+1}^{[2]}) &= \mathrm{Cov}[Y_t^{[1]}, (\rho_{21} \circ Y_t^{[1]} + \rho_{22} \circ Y_t^{[2]} + R_{t+1}^{[2]})] \\ &= \rho_{21}\mathrm{Var}(Y_t^{[1]}) + \rho_{22}\mathrm{Cov}(Y_t^{[1]}, Y_t^{[2]}).\end{aligned}$$

Similarly,

$$\mathrm{Cov}(Y_t^{[2]}, Y_{t+1}^{[1]}) = \rho_{12}\mathrm{Var}(Y_t^{[2]}) + \rho_{11}\mathrm{Cov}(Y_t^{[2]}, Y_t^{[1]}).$$

Thus,
$\Sigma_{21,1,t} = \rho_{21}\Sigma_{11,0,t} + \rho_{22}\Sigma_{21,0,t}$
$\Sigma_{12,1,t} = \rho_{12}\Sigma_{22,0,t} + \rho_{11}\Sigma_{12,0,t}$

$$\boldsymbol{\Sigma}_{1,t} = \begin{bmatrix} \Sigma_{11,1,t} & \Sigma_{12,1,t} \\ \Sigma_{21,1,t} & \Sigma_{22,1,t} \end{bmatrix} = \begin{bmatrix} \rho_{11} & \rho_{12} \\ \rho_{21} & \rho_{22} \end{bmatrix} \begin{bmatrix} \Sigma_{11,0,t} & \Sigma_{12,0,t} \\ \Sigma_{21,0,t} & \Sigma_{22,0,t} \end{bmatrix}.$$

Hence, $\boldsymbol{\Sigma}_{1,t} = \boldsymbol{A}\boldsymbol{\Sigma}_{0,t}$.
When $h = 2$,
$\Sigma_{11,2,t} = \rho_{11}\Sigma_{11,1,t} + \rho_{12}\Sigma_{21,1,t}$
$\Sigma_{22,2,t} = \rho_{22}\Sigma_{22,1,t} + \rho_{21}\Sigma_{12,1,t}$
and
$\Sigma_{21,2,t} = \rho_{21}\Sigma_{11,1,t} + \rho_{22}\Sigma_{21,1,t}$
$\Sigma_{12,2,t} = \rho_{12}\Sigma_{22,1,t} + \rho_{11}\Sigma_{12,1,t}$
Hence, $\boldsymbol{\Sigma}_{2,t} = \boldsymbol{A}\boldsymbol{\Sigma}_{1,t} = \boldsymbol{A}^2\boldsymbol{\Sigma}_{0,t}$.
In general,
$\Sigma_{11,h,t} = \rho_{11}\Sigma_{11,h-1,t} + \rho_{12}\Sigma_{21,h-1,t}$
$\Sigma_{22,h,t} = \rho_{22}\Sigma_{22,h-1,t} + \rho_{21}\Sigma_{12,h-1,t}$
and
$\Sigma_{21,h,t} = \rho_{21}\Sigma_{11,h-1,t} + \rho_{22}\Sigma_{21,h-1,t}$
$\Sigma_{12,h,t} = \rho_{12}\Sigma_{22,h-1,t} + \rho_{11}\Sigma_{12,h-1,t}$
Hence,

$$\boldsymbol{\Sigma}_{h,t} = \boldsymbol{A}\boldsymbol{\Sigma}_{h-1,t} = \boldsymbol{A}^2\boldsymbol{\Sigma}_{h-2,t} = \ldots = \boldsymbol{A}^h\boldsymbol{\Sigma}_{0,t}. \tag{4.19}$$

**Remark 4.3.** Under $\rho_{12} = \rho_{21} = 0$, $Y_{t-1}^{[k]} \sim \mathrm{Poisson}(\mu_{t-1}^{[k]})$ and $\lambda_t^{[k]} = \mu_t^{[k]} - \rho_{kk}\mu_{t-1}^{[k]}$, for $k \in \{1, 2\}$, the marginal moments reduce to

$$E(Y_t^{[k]}) = \mathrm{Var}(Y_t^{[k]}) = \mu_t^{[k]} \tag{4.20}$$

similar to Equations (2.5) and (2.6). Hence, the constrained BINAR(1) model with Poisson innovations introduced in Chapter 2 is a special case of the unconstrained BINAR(1) model with Poisson innovations introduced in this chapter.

## 4.3 Estimation of Parameters

It is clear from Section 4.2 that the likelihood equation specification is very cumbersome and difficult to express since the joint pgf under non-stationarity condition is complex (See Nastic et al. (2016a)). This section alternatively proposes the GQL method that estimates the unknown regression and dependence effects. In the proposed GQL, the auto-covariance function of the scores is constructed from the autoregressive assumption as in Equation (4.19). Since under non-stationarity the moments vary, the derivation of the scores in the GQL approach is conducted in different scheme. The asymptotic and consistency properties of the GQL estimates are established. The following subsections provide details on the GQL estimation procedure.

The GQL equation to estimate the regression and dependence parameters is given by

$$\boldsymbol{D_{\beta,\rho}}' \boldsymbol{\Sigma_{\beta,\rho}}^{-1}(\boldsymbol{f} - \boldsymbol{\mu}) = 0. \tag{4.21}$$

In the above, $\boldsymbol{f} = [\boldsymbol{f}_1, \boldsymbol{f}_2, \ldots, \boldsymbol{f}_t, \ldots, \boldsymbol{f}_{t+h}, \ldots, \boldsymbol{f}_T]$, where $\boldsymbol{f}_t = [Y_t^{[1]}, Y_t^{[2]}]'$, with corresponding mean $\boldsymbol{\mu}_t = [\mu_t^{[1]}, \mu_t^{[2]}]'$. The other components of the GQL Equation (4.21) are explained as follows.

### 4.3.1 The Derivative Structure

The derivative matrix $\boldsymbol{D_{\beta,\rho}}$ is denoted by $\boldsymbol{D_{\beta,\rho}} = [\boldsymbol{D}_1, \boldsymbol{D}_2, \ldots, \boldsymbol{D}_t, \ldots, \boldsymbol{D}_T]'$ with

$$\boldsymbol{D}_t = \begin{pmatrix} \frac{\partial \mu_t^{[1]}}{\partial \boldsymbol{\beta}^{[1]}} & \frac{\partial \mu_t^{[2]}}{\partial \boldsymbol{\beta}^{[1]}} \\ \frac{\partial \mu_t^{[1]}}{\partial \boldsymbol{\beta}^{[2]}} & \frac{\partial \mu_t^{[2]}}{\partial \boldsymbol{\beta}^{[2]}} \\ \frac{\partial \mu_t^{[1]}}{\partial \rho_{11}} & \frac{\partial \mu_t^{[2]}}{\partial \rho_{11}} \\ \frac{\partial \mu_t^{[1]}}{\partial \rho_{12}} & \frac{\partial \mu_t^{[2]}}{\partial \rho_{12}} \\ \frac{\partial \mu_t^{[1]}}{\partial \rho_{21}} & \frac{\partial \mu_t^{[2]}}{\partial \rho_{21}} \\ \frac{\partial \mu_t^{[1]}}{\partial \rho_{22}} & \frac{\partial \mu_t^{[2]}}{\partial \rho_{22}} \end{pmatrix}_{(2p+4)\times 2}, \tag{4.22}$$

where

$$\frac{\partial \mu_t^{[k]}}{\partial \boldsymbol{\beta}^{[l]}} = \left[ \frac{\partial \mu_t^{[k]}}{\partial \beta_1^{[l]}}, \frac{\partial \mu_t^{[k]}}{\partial \beta_2^{[l]}}, \ldots, \frac{\partial \mu_t^{[k]}}{\partial \beta_j^{[l]}}, \ldots, \frac{\partial \mu_t^{[k]}}{\partial \beta_p^{[l]}} \right]', \quad k, l \in \{1, 2\}.$$

The above derivative expressions are specified in different form as follows:

$$\frac{\partial\mu_t^{[1]}}{\partial\beta_j^{[1]}} = \rho_{11}\frac{\partial\mu_{t-1}^{[1]}}{\partial\beta_j^{[1]}} + \rho_{12}\frac{\partial\mu_{t-1}^{[2]}}{\partial\beta_j^{[1]}} + \lambda_t^{[1]}x_{tj}, \qquad \frac{\partial\mu_t^{[2]}}{\partial\beta_j^{[1]}} = \rho_{21}\frac{\partial\mu_{t-1}^{[1]}}{\partial\beta_j^{[1]}} + \rho_{22}\frac{\partial\mu_{t-1}^{[2]}}{\partial\beta_j^{[1]}},$$
$$\frac{\partial\mu_t^{[1]}}{\partial\beta_j^{[2]}} = \rho_{11}\frac{\partial\mu_{t-1}^{[1]}}{\partial\beta_j^{[2]}} + \rho_{12}\frac{\partial\mu_{t-1}^{[2]}}{\partial\beta_j^{[2]}}, \qquad \frac{\partial\mu_t^{[2]}}{\partial\beta_j^{[2]}} = \rho_{21}\frac{\partial\mu_{t-1}^{[1]}}{\partial\beta_j^{[2]}} + \rho_{22}\frac{\partial\mu_{t-1}^{[2]}}{\partial\beta_j^{[2]}} + \lambda_t^{[2]}x_{tj}. \tag{4.23}$$

Equation (4.23) is computed iteratively for $t = 2, \ldots, T$, using the following initial derivatives from Equation (4.14):

$$\frac{\partial\mu_1^{[1]}}{\partial\boldsymbol{\beta}_j^{[1]}} = \frac{(1-\rho_{22})\lambda_1^{[1]}x_{1j}}{(1-\rho_{11})(1-\rho_{22})-\rho_{12}\rho_{21}}, \quad \frac{\partial\mu_1^{[2]}}{\partial\boldsymbol{\beta}_j^{[1]}} = \frac{\rho_{21}\lambda_1^{[1]}x_{1j}}{(1-\rho_{11})(1-\rho_{22})-\rho_{12}\rho_{21}},$$
$$\frac{\partial\mu_1^{[1]}}{\partial\boldsymbol{\beta}_j^{[2]}} = \frac{\rho_{12}\lambda_1^{[2]}x_{1j}}{(1-\rho_{11})(1-\rho_{22})-\rho_{12}\rho_{21}}, \quad \frac{\partial\mu_1^{[2]}}{\partial\boldsymbol{\beta}_j^{[2]}} = \frac{(1-\rho_{11})\lambda_1^{[2]}x_{1j}}{(1-\rho_{11})(1-\rho_{22})-\rho_{12}\rho_{21}}. \tag{4.24}$$

Similarly, the other derivatives with respect to the dependence parameters $\boldsymbol{\rho}$ for $t = 2, \ldots, T$ are given by

$$\frac{\partial\mu_t^{[1]}}{\partial\rho_{11}} = (1+\rho_{11})\frac{\partial\mu_{t-1}^{[1]}}{\partial\rho_{11}} + \rho_{12}\frac{\partial\mu_{t-1}^{[2]}}{\partial\rho_{11}}, \qquad \frac{\partial\mu_t^{[2]}}{\partial\rho_{11}} = \rho_{21}\frac{\partial\mu_{t-1}^{[1]}}{\partial\rho_{11}} + \rho_{22}\frac{\partial\mu_{t-1}^{[2]}}{\partial\rho_{11}},$$
$$\frac{\partial\mu_t^{[1]}}{\partial\rho_{12}} = \rho_{11}\frac{\partial\mu_{t-1}^{[1]}}{\partial\rho_{12}} + (1+\rho_{12})\frac{\partial\mu_{t-1}^{[2]}}{\partial\rho_{12}}, \qquad \frac{\partial\mu_t^{[2]}}{\partial\rho_{12}} = \rho_{21}\frac{\partial\mu_{t-1}^{[1]}}{\partial\rho_{12}} + \rho_{22}\frac{\partial\mu_{t-1}^{[2]}}{\partial\rho_{12}},$$
$$\frac{\partial\mu_t^{[1]}}{\partial\rho_{21}} = \rho_{11}\frac{\partial\mu_{t-1}^{[1]}}{\partial\rho_{21}} + \rho_{12}\frac{\partial\mu_{t-1}^{[2]}}{\partial\rho_{21}}, \qquad \frac{\partial\mu_t^{[2]}}{\partial\rho_{21}} = (1+\rho_{21})\frac{\partial\mu_{t-1}^{[1]}}{\partial\rho_{21}} + \rho_{22}\frac{\partial\mu_{t-1}^{[2]}}{\partial\rho_{21}},$$
$$\frac{\partial\mu_t^{[1]}}{\partial\rho_{22}} = \rho_{11}\frac{\partial\mu_{t-1}^{[1]}}{\partial\rho_{22}} + \rho_{12}\frac{\partial\mu_{t-1}^{[2]}}{\partial\rho_{22}}, \qquad \frac{\partial\mu_t^{[2]}}{\partial\rho_{22}} = \rho_{21}\frac{\partial\mu_{t-1}^{[1]}}{\partial\rho_{22}} + (1+\rho_{22})\frac{\partial\mu_{t-1}^{[2]}}{\partial\rho_{22}}, \tag{4.25}$$

where the initial derivatives are given by

$$\frac{\partial\mu_1^{[1]}}{\partial\rho_{11}} = \frac{[(1-\rho_{22})\lambda_1^{[1]}+\rho_{12}\lambda_1^{[2]}][1-\rho_{22}]}{\theta_1^2}, \qquad \frac{\partial\mu_1^{[2]}}{\partial\rho_{11}} = \frac{[(1-\rho_{11})\lambda_1^{[2]}+\rho_{21}\lambda_1^{[1]}][1-\rho_{22}]-\lambda_1^{[2]}\theta_1}{\theta_1^2},$$
$$\frac{\partial\mu_1^{[1]}}{\partial\rho_{12}} = \frac{\lambda_1^{[2]}\theta_1+[(1-\rho_{22})\lambda_1^{[1]}+\rho_{12}\lambda_1^{[2]}][\rho_{21}]}{\theta_1^2}, \qquad \frac{\partial\mu_1^{[2]}}{\partial\rho_{12}} = \frac{[(1-\rho_{11})\lambda_1^{[2]}+\rho_{21}\lambda_1^{[1]}][\rho_{21}]}{\theta_1^2},$$
$$\frac{\partial\mu_1^{[1]}}{\partial\rho_{21}} = \frac{[(1-\rho_{22})\lambda_1^{[1]}+\rho_{12}\lambda_1^{[2]}][\rho_{12}]}{\theta_1^2}, \qquad \frac{\partial\mu_1^{[2]}}{\partial\rho_{21}} = \frac{\lambda_1^{[1]}\theta_1+[(1-\rho_{11})\lambda_1^{[2]}+\rho_{21}\lambda_1^{[1]}][\rho_{12}]}{\theta_1^2},$$
$$\frac{\partial\mu_1^{[1]}}{\partial\rho_{22}} = \frac{[(1-\rho_{22})\lambda_1^{[1]}+\rho_{12}\lambda_1^{[2]}][1-\rho_{11}]-\lambda_1^{[1]}\theta_1}{\theta_1^2}, \qquad \frac{\partial\mu_1^{[2]}}{\partial\rho_{22}} = \frac{[(1-\rho_{11})\lambda_1^{[2]}+\rho_{21}\lambda_1^{[1]}][1-\rho_{11}]}{\theta_1^2}, \tag{4.26}$$

where $\theta_1 = (1-\rho_{11})(1-\rho_{22}) - \rho_{12}\rho_{21}$.

### 4.3.2 The Construction of the Covariance Structure $\Sigma_{\beta,\rho}$

The covariance matrix $\boldsymbol{\Sigma}_{\boldsymbol{\beta},\boldsymbol{\rho}}$ is specified as:

$$
\begin{array}{c}
\\
\boldsymbol{f}_1 \\ \boldsymbol{f}_2 \\ \boldsymbol{f}_3 \\ \vdots \\ \boldsymbol{f}_t \\ \vdots \\ \boldsymbol{f}_{t+h} \\ \vdots \\ \boldsymbol{f}_T
\end{array}
\begin{array}{c}
\begin{array}{ccccccccc}
\boldsymbol{f}_1 & \boldsymbol{f}_2 & \boldsymbol{f}_3 & \cdots & \boldsymbol{f}_t & \cdots & \boldsymbol{f}_{t+h} & \cdots & \boldsymbol{f}_T
\end{array} \\
\begin{pmatrix}
\boldsymbol{\Sigma}_{0,1} & \boldsymbol{\Sigma}'_{1,1} & \boldsymbol{\Sigma}'_{2,1} & \cdots & \boldsymbol{\Sigma}'_{t-1,1} & \cdots & \boldsymbol{\Sigma}'_{t+h-1,1} & \cdots & \boldsymbol{\Sigma}'_{T-1,1} \\
\boldsymbol{\Sigma}_{1,1} & \boldsymbol{\Sigma}_{0,2} & \boldsymbol{\Sigma}'_{1,2} & \cdots & \boldsymbol{\Sigma}'_{t-2,2} & \cdots & \boldsymbol{\Sigma}'_{t+h-2,2} & \cdots & \boldsymbol{\Sigma}'_{T-2,2} \\
\boldsymbol{\Sigma}_{2,1} & \boldsymbol{\Sigma}_{1,2} & \boldsymbol{\Sigma}_{0,3} & \cdots & \boldsymbol{\Sigma}'_{t-3,3} & \cdots & \boldsymbol{\Sigma}'_{t+h-3,3} & \cdots & \boldsymbol{\Sigma}'_{T-3,3} \\
\vdots & \vdots & \vdots & \ddots & \vdots & \vdots & \vdots & \vdots & \vdots \\
\boldsymbol{\Sigma}_{t-1,1} & \boldsymbol{\Sigma}_{t-2,2} & \boldsymbol{\Sigma}_{t-3,3} & \cdots & \boldsymbol{\Sigma}_{0,t} & \cdots & \boldsymbol{\Sigma}'_{t+h-t,t} & \cdots & \boldsymbol{\Sigma}'_{T-t,t} \\
\vdots & \vdots & \vdots & \vdots & \vdots & \ddots & \vdots & \vdots & \vdots \\
\boldsymbol{\Sigma}_{t+h-1,1} & \boldsymbol{\Sigma}_{t+h-2,2} & \boldsymbol{\Sigma}_{t+h-3,3} & \cdots & \boldsymbol{\Sigma}_{t+h-t,t} & \cdots & \boldsymbol{\Sigma}_{0,t+h} & \cdots & \boldsymbol{\Sigma}'_{T-t,t} \\
\vdots & \vdots & \vdots & \vdots & \vdots & \vdots & \vdots & \ddots & \vdots \\
\boldsymbol{\Sigma}_{T-1,1} & \boldsymbol{\Sigma}_{T-2,2} & \boldsymbol{\Sigma}_{T-3,3} & \cdots & \boldsymbol{\Sigma}_{T-t,t} & \cdots & \boldsymbol{\Sigma}_{T-t+h,t} & \cdots & \boldsymbol{\Sigma}_{0,T}
\end{pmatrix}_{(2T\times 2T)}
\end{array}
\tag{4.27}
$$

where the above entries are computed iteratively for $t = 2, \ldots, T$ using Equation (4.19), with initial variances and covariance given by Equations (4.15)–(4.17).

The Newton-Raphson iterative technique is applied on the GQL (4.21) to yield an iterative Equation of the form

$$(\ \hat{\boldsymbol{\psi}}_{r+1}\ ) = (\ \hat{\boldsymbol{\psi}}_r\ ) + [\boldsymbol{D}'\boldsymbol{\Sigma}^{-1}\boldsymbol{D}]_r^{-1}[\boldsymbol{D}'\boldsymbol{\Sigma}^{-1}(\boldsymbol{f} - \boldsymbol{\mu})]_r \tag{4.28}$$

where $\hat{\boldsymbol{\psi}} = [\hat{\beta}_1^{[1]}, \hat{\beta}_2^{[1]}, \hat{\beta}_1^{[2]}, \hat{\beta}_2^{[2]}, \hat{\rho}_{11}, \hat{\rho}_{12}, \hat{\rho}_{21}, \hat{\rho}_{22}]$ and $\hat{\boldsymbol{\psi}}_r$ are the estimates at the $r^{th}$ iteration and $[.]_r$ are the values of the expression at the $r^{th}$ iteration.

The overall algorithm works as follows: For an initial value of $\hat{\boldsymbol{\psi}}$, $[\beta_1^{[1]} = \beta_2^{[1]} = 0.01, \beta_1^{[2]} = \beta_2^{[2]} = 0.02, \rho_{11} = \rho_{22} = 0.1, \rho_{12} = \rho_{12} = 0.15]$, the covariance structure is computed and these values are used in the iterative Equation (4.28) to obtain an updated estimate of the regression and dependence parameters. These updated regression and dependence values are in turn used to re-calculate the covariance structure and used again in the iterative Equation (4.28) to obtain another updated set of values. This cycle of iteration continues until convergence. Under mild regulatory conditions, the parameters obtained from the iterative Equation (4.28) are consistent and $(\hat{\boldsymbol{\psi}} - \boldsymbol{\psi})'$ follows an asymptotic normal distribution with mean $\mathbf{0}$ and covariance matrix $[\boldsymbol{D}'\boldsymbol{\Sigma}^{-1}\boldsymbol{D}]^{-1}$, as discussed in Subsection 2.3.1.2.

### 4.3.3 Forecasting Equations

Using the converged estimates, we derive the forecasting equations of the unconstrained BINAR(1) model with Poisson innovations. The conditional expectation and variance of the one-step ahead forecast $Y_{t+1}^{[k]}$ given $y_t^{[k]}$ are expressed as follows:

$$\begin{aligned} E(Y_{t+1}^{[1]}|y_t^{[1]}, y_t^{[2]}) &= E(\rho_{11} \circ y_t^{[1]}|y_t^{[1]}) + E(\rho_{12} \circ y_t^{[2]}|y_t^{[2]}) + E(R_{t+1}^{[1]}) \\ &= \rho_{11} y_t^{[1]} + \rho_{12} y_t^{[2]} + \lambda_{t+1}^{[1]}, \end{aligned} \tag{4.29}$$

$$\begin{aligned} E(Y_{t+1}^{[2]}|y_t^{[1]}, y_t^{[2]}) &= E(\rho_{21} \circ y_t^{[1]}|y_t^{[1]}) + E(\rho_{22} \circ y_t^{[2]}|y_t^{[2]}) + E(R_{t+1}^{[2]}) \\ &= \rho_{21} y_t^{[1]} + \rho_{22} y_t^{[2]} + \lambda_{t+1}^{[2]}, \end{aligned} \tag{4.30}$$

$$\begin{aligned} \text{Var}(Y_{t+1}^{[1]}|y_t^{[1]}, y_t^{[2]}) &= \text{Var}(\rho_{11} \circ y_t^{[1]}|y_t^{[1]}) + \text{Var}(\rho_{12} \circ y_t^{[2]}|y_t^{[2]}) + \text{Var}(R_{t+1}^{[1]}) \\ &= \rho_{11}(1 - \rho_{11}) y_t^{[1]} + \rho_{12}(1 - \rho_{12}) y_t^{[2]} + \lambda_{t+1}^{[1]} \end{aligned} \tag{4.31}$$

and

$$\begin{aligned} \text{Var}(Y_{t+1}^{[2]}|y_t^{[1]}, y_t^{[2]}) &= \text{Var}(\rho_{21} \circ y_t^{[1]}|y_t^{[1]}) + \text{Var}(\rho_{22} \circ y_t^{[2]}|y_t^{[2]}) + \text{Var}(R_{t+1}^{[2]}) \\ &= \rho_{21}(1 - \rho_{21}) y_t^{[1]} + \rho_{22}(1 - \rho_{22}) y_t^{[2]} + \lambda_{t+1}^{[2]}. \end{aligned} \tag{4.32}$$

## 4.4 Summary

In this chapter, we have introduced a novel BINAR(1) model where the innovations are Poisson marginals and the cross-correlation between the series is induced by the relation between the current series observations and the previous-lagged observations of the other series. The marginal moments and joint covariance expressions have been derived. In terms of the inferential procedures, a GQL approach has been developed since likelihood-based approaches are unfeasible due to the cumbersomeness in deriving the likelihood equation. The properties of the GQL estimators have been derived. The next chapter presents an unconstrained BINMA(1) model with Poisson innovations by adopting a similar approach.

# 5

# *Unconstrained BINMA(1) Model with Poisson Innovations*

## 5.1 Introduction

The constrained BINMA(1) model introduced in Chapter 3 is based on the assumption of diagonality of the autocorrelation parameters with certain limitations, as explained in Subsection 1.2.2. Hence, we generalize the constrained BINMA(1) to an unconstrained BINMA(1) model and derive its marginal moments and covariance structure using the matrix multivariate binomial thinning mechanism under non-stationary moment conditions. As for the inference procedure, FGLS and GQL estimation methods have been developed in Chapter 3 and based on some simulation experiments, it has been illustrated that GQL yields more efficient estimates than FGLS. Hence, we propose to set-up the GQL estimating equation to estimate the regression, serial- and cross-correlation parameters.

The plan of this chapter is as follows: In Section 5.2, the unconstrained BINMA(1) model with Poisson innovations is developed and the marginal moments and covariance structure are derived using the matrix multivariate binomial thinning mechanism. In Section 5.3, the GQL equation is developed to estimate the regression and dependence parameters. In the same section, the one-step ahead forecasting equations are derived for the proposed model. In Section 5.4, some numerical evaluation wherein non-stationary BINMA(1) time series data are generated under different combinations of low and high serial- and cross-correlation parameters. This chapter ends with the conclusion.

## 5.2 The Unconstrained BINMA(1) Model

The unconstrained BINMA(1) model is specified as:

$$Y_t^{[1]} = \rho_{11} \circ R_{t-1}^{[1]} + \rho_{12} \circ R_{t-1}^{[2]} + R_t^{[1]} \tag{5.1}$$

$$Y_t^{[2]} = \rho_{21} \circ R_{t-1}^{[1]} + \rho_{22} \circ R_{t-1}^{[2]} + R_t^{[2]}. \tag{5.2}$$

 DOI: 10.1201/9781003600435-5

In the above, $Y_t^{[k]}$ is the random counting observation for the $k^{th}$ series at the $t^{th}$ time point where $R_t^{[k]}$ is its corresponding innovation term with $R_t^{[k]}$ marginally following a Poisson distribution with mean $\lambda_t^{[k]}$, where $\lambda_t^{[k]} = \exp(\boldsymbol{x}_t'\boldsymbol{\beta}^{[k]})$ with $\boldsymbol{x}_t = [x_{t1}, x_{t2}, \ldots, x_{tj}, \ldots, x_{tp}]'$ and $\boldsymbol{\beta}^{[k]} = [\beta_1^{[k]}, \beta_2^{[k]}, \ldots, \beta_j^{[k]}, \ldots, \beta_p^{[k]}]'$ for $k \in \{1,2\}$ and $t \in \mathbb{Z}$. As for the pair of innovations $(R_t^{[1]}, R_t^{[2]})$ in the above model, $\text{Corr}(R_t^{[1]}, R_t^{[2]})$ is assumed to be zero.

'$\circ$' is the binomial thinning operator such that $\{\rho_{jk} \circ R_{t-1}^{[k]} \mid R_{t-1}^{[k]}\} \sim \text{Binomial}(R_{t-1}^{[k]}, \rho_{jk})$, $j, k \in \{1,2\}$ with $\rho_{jk} \in [0,1]$. By applying some binomial thinning properties (Silva & Oliveira (2004), Steutel & Van Harn (1979)), we have

(a) $E(\rho_{jk} \circ R_{t-1}^{[k]}) = \rho_{jk} E(R_{t-1}^{[k]})$.

(b) $\text{Var}(\rho_{jk} \circ R_{t-1}^{[k]}) = \rho_{jk}(1-\rho_{jk})E(R_{t-1}^{[k]}) + \rho_{jk}^2 \text{Var}(R_{t-1}^{[k]})$.

(c) $\text{Cov}(\rho_{j1} \circ R_{t-1}^{[1]}, \rho_{j2} \circ R_{t-1}^{[2]}) = \rho_{j1}\rho_{j2}\text{Cov}(R_{t-1}^{[1]}, R_{t-1}^{[2]})$.

By expressing Equations (5.1) and (5.2) in vector form, we obtain

$$\boldsymbol{Y_t} = \boldsymbol{A} \circ \boldsymbol{R_{t-1}} + \boldsymbol{R_t}, \tag{5.3}$$

where $\boldsymbol{Y_t} = [Y_t^{[1]}, Y_t^{[2]}]'$, $\boldsymbol{R_t} = [R_t^{[1]}, R_t^{[2]}]'$ and $\boldsymbol{A} = \begin{pmatrix} \rho_{11} & \rho_{12} \\ \rho_{21} & \rho_{22} \end{pmatrix}$.

**Lemma 5.1.** (a) $E(\boldsymbol{Y_t}) = \boldsymbol{A}E(\boldsymbol{R_{t-1}}) + E(\boldsymbol{R_t})$.

(b) $\text{Var}(\boldsymbol{Y_t}) = \boldsymbol{A}\text{Var}(\boldsymbol{R_{t-1}})\boldsymbol{A}' + \text{Var}(\boldsymbol{A} \circ \boldsymbol{R_{t-1}}) + \text{Var}(\boldsymbol{R_t})$.

*Proof:*

(a)

$$\begin{aligned} E(\boldsymbol{Y_t}) &= E(\boldsymbol{A} \circ \boldsymbol{R_{t-1}}) + E(\boldsymbol{R_t}) \\ &= E_{\boldsymbol{R_{t-1}}}[E(\boldsymbol{A} \circ \boldsymbol{R_{t-1}} | \boldsymbol{R_{t-1}})] + E(\boldsymbol{R_t}) \\ &= E_{\boldsymbol{R_{t-1}}}(\boldsymbol{A}\boldsymbol{R_{t-1}}) + E(\boldsymbol{R_t}) \\ &= \boldsymbol{A}E(\boldsymbol{R_{t-1}}) + E(\boldsymbol{R_t}) \end{aligned}$$

(b)

$$\begin{aligned} \text{Var}(\boldsymbol{Y_t}) &= \text{Var}[E(\boldsymbol{A} \circ \boldsymbol{R_{t-1}} + \boldsymbol{R_t} | \boldsymbol{R_{t-1}})] \\ &\quad + E[\text{Var}(\boldsymbol{A} \circ \boldsymbol{R_{t-1}} + \boldsymbol{R_t} | \boldsymbol{R_{t-1}})] \\ &= \text{Var}[\boldsymbol{A}\boldsymbol{R_{t-1}} + E(\boldsymbol{R_t})] + E[\text{Var}(\boldsymbol{A} \circ \boldsymbol{R_{t-1}}) + \text{Var}(\boldsymbol{R_t})] \\ &= \boldsymbol{A}\text{Var}(\boldsymbol{R_{t-1}})\boldsymbol{A}' + \text{Var}(\boldsymbol{A} \circ \boldsymbol{R_{t-1}}) + \text{Var}(\boldsymbol{R_t}) \end{aligned}$$

Using Lemma 5.1 and assuming $\mathrm{E}(\boldsymbol{Y_t}) = \boldsymbol{\mu_t}$, where $\boldsymbol{\mu_t} = [\mu_t^{[1]}, \mu_t^{[2]}]'$, we have

$$\begin{aligned} E(\boldsymbol{Y_t}) &= \boldsymbol{A}E(\boldsymbol{R_{t-1}}) + E(\boldsymbol{R_t}) \\ \boldsymbol{\mu_t} &= \boldsymbol{A\lambda_{t-1}} + \boldsymbol{\lambda_t} \end{aligned} \tag{5.4}$$

where $\boldsymbol{\lambda_t} = [\lambda_t^{[1]}, \lambda_t^{[2]}]'$.

Hence, expanding Equation (5.4), we obtain

$$\begin{aligned} \mu_t^{[1]} &\equiv E(Y_t^{[1]}) = \rho_{11}\lambda_{t-1}^{[1]} + \rho_{12}\lambda_{t-1}^{[2]} + \lambda_t^{[1]}, \\ \mu_t^{[2]} &\equiv E(Y_t^{[2]}) = \rho_{21}\lambda_{t-1}^{[1]} + \rho_{22}\lambda_{t-1}^{[2]} + \lambda_t^{[2]}. \end{aligned} \tag{5.5}$$

Denoting $\boldsymbol{\Sigma}_{0,t} = \begin{bmatrix} \mathrm{Var}(Y_t^{[1]}) & \mathrm{Cov}(Y_t^{[1]}, Y_t^{[2]}) \\ \mathrm{Cov}(Y_t^{[2]}, Y_t^{[1]}) & \mathrm{Var}(Y_t^{[2]}) \end{bmatrix}$, we have

$$\boldsymbol{\Sigma}_{0,t} = \boldsymbol{A}\mathrm{Var}(\boldsymbol{R_{t-1}})\boldsymbol{A}' + \mathrm{diag}(\boldsymbol{B\lambda_{t-1}}) + \mathrm{Var}(\boldsymbol{R_t}) \tag{5.6}$$

where

$$\boldsymbol{B} = \begin{pmatrix} \rho_{11}(1-\rho_{11}) & \rho_{12}(1-\rho_{12}) \\ \rho_{21}(1-\rho_{21}) & \rho_{22}(1-\rho_{22}) \end{pmatrix}.$$

Expanding (5.6), we obtain

$$\mathrm{Var}(Y_t^{[1]}) = \rho_{11}\lambda_{t-1}^{[1]} + \rho_{12}\lambda_{t-1}^{[2]} + \lambda_t^{[1]}, \quad \mathrm{Var}(Y_t^{[2]}) = \rho_{21}\lambda_{t-1}^{[1]} + \rho_{22}\lambda_{t-1}^{[2]} + \lambda_t^{[2]} \tag{5.7}$$

and

$$\mathrm{Cov}(Y_t^{[1]}, Y_t^{[2]}) = \rho_{11}\rho_{21}\lambda_{t-1}^{[1]} + \rho_{22}\rho_{12}\lambda_{t-1}^{[2]}. \tag{5.8}$$

**Remark 5.1.** The marginal moments (5.5) and (5.7) indicate that the unconstrained BINMA(1) model follows marginally the Poisson distribution.

**Remark 5.2.** Under stationary moment conditions, Equations (5.5), (5.7) and (5.8) reduce to

$$E(Y_t^{[1]}) = \mathrm{Var}(Y_t^{[1]}) = (1+\rho_{11})\lambda_1^{[1]} + \rho_{12}\lambda_1^{[2]}, \tag{5.9}$$

$$E(Y_t^{[2]}) = \mathrm{Var}(Y_t^{[2]}) = \rho_{21}\lambda_1^{[1]} + (1+\rho_{22})\lambda_1^{[2]}, \tag{5.10}$$

and

$$\mathrm{Cov}(Y_t^{[1]}, Y_t^{[2]}) = \rho_{11}\rho_{21}\lambda_1^{[1]} + \rho_{22}\rho_{12}\lambda_1^{[2]}. \tag{5.11}$$

In general, the covariance matrix at lag-$h$ can be expressed as:

$$\boldsymbol{\Sigma}_{h,t} = \begin{bmatrix} \mathrm{Cov}(Y_t^{[1]}, Y_{t+h}^{[1]}) & \mathrm{Cov}(Y_{t+h}^{[1]}, Y_t^{[2]}) \\ \mathrm{Cov}(Y_t^{[1]}, Y_{t+h}^{[2]}) & \mathrm{Cov}(Y_t^{[2]}, Y_{t+h}^{[2]}) \end{bmatrix} = \begin{bmatrix} \Sigma_{11,h,t} & \Sigma_{12,h,t} \\ \Sigma_{21,h,t} & \Sigma_{22,h,t} \end{bmatrix}. \tag{5.12}$$

When $h = 1$,

$$\begin{aligned}\text{Cov}(Y_t^{[1]}, Y_{t+1}^{[1]}) &= \text{Cov}\left[Y_t^{[1]}, (\rho_{11} \circ R_t^{[1]} + \rho_{12} \circ R_t^{[2]} + R_{t+1}^{[1]})\right] \\ &= \rho_{11}\text{Var}(R_t^{[1]}).\end{aligned}$$

Similarly,

$$\text{Cov}(Y_t^{[2]}, Y_{t+1}^{[2]}) = \rho_{22}\text{Var}(R_t^{[2]}).$$

Thus,
$\Sigma_{11,1,t} = \rho_{11}\text{Var}(R_t^{[1]})$

$\Sigma_{22,1,t} = \rho_{22}\text{Var}(R_t^{[2]})$

$$\begin{aligned}\text{Cov}(Y_t^{[1]}, Y_{t+1}^{[2]}) &= \text{Cov}\left[Y_t^{[1]}, (\rho_{21} \circ R_t^{[1]} + \rho_{22} \circ R_t^{[2]} + R_{t+1}^{[2]})\right] \\ &= \rho_{21}\text{Var}(R_t^{[1]}).\end{aligned}$$

Similarly

$$\text{Cov}(Y_{t+1}^{[1]}, Y_t^{[2]}) = \rho_{12}\text{Var}(R_t^{[2]}).$$

Thus,
$\Sigma_{21,1,t} = \rho_{21}\text{Var}(R_t^{[1]})$

$\Sigma_{12,1,t} = \rho_{12}\text{Var}(R_t^{[2]})$
When $h = 2$,

$$\text{Cov}(Y_t^{[1]}, Y_{t+2}^{[1]}) = \text{Cov}\left[Y_t^{[1]}, (\rho_{11} \circ R_{t+1}^{[1]} + \rho_{12} \circ R_{t+1}^{[2]} + R_{t+2}^{[1]})\right] = 0.$$

Similarly,

$$\text{Cov}(Y_t^{[2]}, Y_{t+2}^{[2]}) = 0.$$

and same will hold in $\text{Cov}(Y_t^{[1]}, Y_{t+h}^{[1]})$ and $\text{Cov}(Y_t^{[2]}, Y_{t+h}^{[2]})$ for $h > 1$. Thus,
$\Sigma_{11,h,t} = 0$ for $h > 1$.
$\Sigma_{22,h,t} = 0$ for $h > 1$.

$$\text{Cov}(Y_t^{[1]}, Y_{t+2}^{[2]}) = \text{Cov}\left[Y_t^{[1]}, (\rho_{21} \circ R_{t+1}^{[1]} + \rho_{22} \circ R_{t+1}^{[2]} + R_{t+2}^{[2]})\right] = 0.$$

Similarly,

$$\text{Cov}(Y_t^{[2]}, Y_{t+2}^{[1]}) = 0.$$

and same will hold in $\text{Cov}(Y_t^{[1]}, Y_{t+h}^{[2]})$ and $\text{Cov}(Y_t^{[2]}, Y_{t+h}^{[1]})$ for $h > 1$.
$\Sigma_{21,h,t} = 0$ for $h > 1$.
$\Sigma_{12,h,t} = 0$ for $h > 1$.
Hence,

$$\boldsymbol{\Sigma}_{h,t} = \begin{cases} \boldsymbol{A}^h \begin{bmatrix} \text{Var}(R_t^{[1]}) & 0 \\ 0 & \text{Var}(R_t^{[2]}) \end{bmatrix}, & \text{for} \quad h = 1 \\ 0 & h > 1. \end{cases} \tag{5.13}$$

**Remark 5.3.** Under $\rho_{12} = \rho_{21} = 0$, the unconstrained BINMA(1) model (5.3) reduces to the constrained BINMA(1) model with Poisson innovations, where

$$E(Y_t^{[k]}) = \text{Var}(Y_t^{[k]}) = \rho_{kk}\lambda_{t-1}^{[k]} + \lambda_t^{[k]} \tag{5.14}$$

which is same as Equations (3.2) and (3.3). Thus, we conclude that the constrained BINMA(1) model with Poisson innovations introduced in Chapter 3 is a special case of the unconstrained BINMA(1) model with Poisson innovations introduced in this chapter.

## 5.3 Estimation of Parameters

The GQL equation to estimate the regression, serial- and cross-correlation coefficients is expressed as:

$$\boldsymbol{D_{\beta,\rho}}'\boldsymbol{\Sigma_{\beta,\rho}}^{-1}(\boldsymbol{f} - \boldsymbol{\mu}) = 0 \tag{5.15}$$

where the score vector $\boldsymbol{f} = [\boldsymbol{f}_1, \boldsymbol{f}_2, \ldots, \boldsymbol{f}_t, \ldots, \boldsymbol{f}_{t+h}, \ldots, \boldsymbol{f}_T]$, where $\boldsymbol{f}_t = [Y_t^{[1]}, Y_t^{[2]}]'$ and $\boldsymbol{\mu} = [\boldsymbol{\mu}_1, \boldsymbol{\mu}_2, \ldots, \boldsymbol{\mu}_t, \ldots, \boldsymbol{\mu}_T]$, where $\boldsymbol{\mu}_t = [\mu_t^{[1]}, \mu_t^{[2]}]'$ for $t = 1, 2, \ldots, T$.
As for the derivative matrix, $\boldsymbol{D_{\beta,\rho}} = [\boldsymbol{D}_1, \boldsymbol{D}_2, \ldots, \boldsymbol{D}_t, \ldots, \boldsymbol{D}_T]'$ with

$$\boldsymbol{D}_t = \begin{pmatrix} \frac{\partial \mu_t^{[1]}}{\partial \boldsymbol{\beta}^{[1]}} & \frac{\partial \mu_t^{[2]}}{\partial \boldsymbol{\beta}^{[1]}} \\ \frac{\partial \mu_t^{[1]}}{\partial \boldsymbol{\beta}^{[2]}} & \frac{\partial \mu_t^{[2]}}{\partial \boldsymbol{\beta}^{[2]}} \\ \frac{\partial \mu_t^{[1]}}{\partial \rho_{11}} & 0 \\ \frac{\partial \mu_t^{[1]}}{\partial \rho_{12}} & 0 \\ 0 & \frac{\partial \mu_t^{[2]}}{\partial \rho_{21}} \\ 0 & \frac{\partial \mu_t^{[2]}}{\partial \rho_{22}} \end{pmatrix}_{(2p+4)\times 2} \tag{5.16}$$

where

$$\frac{\partial \mu_t^{[k]}}{\partial \boldsymbol{\beta}^{[l]}} = \left[\frac{\partial \mu_t^{[k]}}{\partial \beta_1^{[l]}}, \frac{\partial \mu_t^{[k]}}{\partial \beta_2^{[l]}}, \ldots, \frac{\partial \mu_t^{[k]}}{\partial \beta_j^{[l]}}, \ldots, \frac{\partial \mu_t^{[k]}}{\partial \beta_p^{[l]}}\right]', \quad k, l \in \{1, 2\}.$$

The entries of the derivative matrix are:

1. $\frac{\partial \mu_t^{[k]}}{\partial \boldsymbol{\beta}^{[k']}} = [\frac{\partial \mu_t^{[k]}}{\partial \beta_j^{[k']}}]_{j=1:p}$ where $\frac{\partial \mu_t^{[k]}}{\partial \beta_j^{[k]}} = \rho_{kk}\lambda_{t-1}^{[k]} x_{t-1,j} + \lambda_t^{[k]} x_{tj}$ and $\frac{\partial \mu_t^{[k]}}{\partial \beta_j^{[k']}} = \rho_{kk'}\lambda_{t-1}^{[k']} x_{t-1,j}$.

2. $\frac{\partial \mu_t^{[1]}}{\partial \rho_{11}} = \frac{\partial \mu_t^{[2]}}{\partial \rho_{21}} = \lambda_{t-1}^{[1]}$.

3. $\frac{\partial \mu_t^{[1]}}{\partial \rho_{12}} = \frac{\partial \mu_t^{[2]}}{\partial \rho_{22}} = \lambda_{t-1}^{[2]}$.

The covariance matrix $\boldsymbol{\Sigma}_{\boldsymbol{\beta},\boldsymbol{\rho}}$ is expressed as

$$\boldsymbol{\Sigma}_{\boldsymbol{\beta},\boldsymbol{\rho}} = \begin{pmatrix} \boldsymbol{\Sigma}_{0,1} & \boldsymbol{\Sigma}'_{1,1} & 0 & 0 & \dots & 0 & 0 \\ \boldsymbol{\Sigma}_{1,1} & \boldsymbol{\Sigma}_{0,2} & \boldsymbol{\Sigma}'_{1,2} & 0 & \dots & 0 & 0 \\ 0 & \boldsymbol{\Sigma}_{1,2} & \boldsymbol{\Sigma}_{0,3} & \boldsymbol{\Sigma}'_{1,3} & \dots & 0 & 0 \\ \vdots & \vdots & \vdots & \vdots & \ddots & \vdots & \vdots \\ 0 & 0 & 0 & 0 & \dots & \boldsymbol{\Sigma}_{0,T-1} & \boldsymbol{\Sigma}'_{1,T-1} \\ 0 & 0 & 0 & 0 & \dots & \boldsymbol{\Sigma}_{1,T-1} & \boldsymbol{\Sigma}_{0,T} \end{pmatrix}_{2T\times 2T}.$$

The Newton-Raphson iterative scheme is used to solve Equation (5.15) which yields an iteration of the form

$$\left( \hat{\boldsymbol{\psi}}_{r+1} \right) = \left( \hat{\boldsymbol{\psi}}_r \right) + [\boldsymbol{D}'_{\psi}\boldsymbol{\Sigma}_{\psi}^{-1}\boldsymbol{D}_{\psi}]_r^{-1}[\boldsymbol{D}'_{\psi}\boldsymbol{\Sigma}_{\psi}^{-1}(\boldsymbol{f}-\boldsymbol{\mu})]_r, \tag{5.17}$$

where $\hat{\boldsymbol{\psi}} = [\hat{\beta}_1^{[1]}, \hat{\beta}_2^{[1]}, \hat{\beta}_1^{[2]}, \hat{\beta}_2^{[2]}, \hat{\rho}_{11}, \hat{\rho}_{12}, \hat{\rho}_{21}, \hat{\rho}_{22}]$ and $\hat{\boldsymbol{\psi}}_r$ are the estimates at the $r^{th}$ iteration and $[.]_r$ are the values of the expression at the $r^{th}$ iteration.

The working of this algorithm is as follows: For an initial $\hat{\boldsymbol{\psi}}$, $[\beta_1^{[1]} = \beta_2^{[1]} = 0.01, \beta_1^{[2]} = \beta_2^{[2]} = 0.02, \rho_{11} = \rho_{22} = 0.1, \rho_{12} = \rho_{12} = 0.15]$, the score and Hessian components are calculated and replaced in the iterative Equation (5.17) to obtain an updated $\hat{\boldsymbol{\psi}}$. These updated values are in turn used to re-calculate the score and Hessian components which are in turn replaced in the iterative Equation (5.17) to obtain another set of values. This cycle of iteration continues until convergence. Under mild regulatory conditions, the regression and dependence estimators are consistent and $(\hat{\boldsymbol{\psi}} - \boldsymbol{\psi})'$ follows an asymptotic normal distribution with mean $\mathbf{0}$ and covariance matrix $[\boldsymbol{D}'_{\psi}\boldsymbol{\Sigma}_{\boldsymbol{\psi}}{}^{-1}\boldsymbol{D}_{\psi}]^{-1}$ (Refer to Subsection 2.3.1.2).

Using the converged estimates, the forecasting equations of the unconstrained BINMA(1) model with Poisson innovations are derived. The expectation and variance of the one-step ahead forecast $Y_{t+1}^{[k]}$ given $r_t^{[1]}$ and $r_t^{[2]}$ are:

$$\begin{aligned} E(Y_{t+1}^{[1]}|r_t^{[1]}, r_t^{[2]}) &= E(\rho_{11} \circ r_t^{[1]}|r_t^{[1]}) + E(\rho_{12} \circ r_t^{[2]}|r_t^{[2]}) + E(R_{t+1}^{[1]}) \\ &= \rho_{11} r_t^{[1]} + \rho_{12} r_t^{[2]} + \lambda_{t+1}^{[1]}, \end{aligned} \tag{5.18}$$

$$\begin{aligned} E(Y_{t+1}^{[2]}|r_t^{[1]}, r_t^{[2]}) &= E(\rho_{21} \circ r_t^{[1]}|r_t^{[1]}) + E(\rho_{22} \circ r_t^{[2]}|r_t^{[2]}) + E(R_{t+1}^{[2]}) \\ &= \rho_{21} r_t^{[1]} + \rho_{22} r_t^{[2]} + \lambda_{t+1}^{[2]}, \end{aligned} \tag{5.19}$$

$$\begin{aligned} \text{Var}(Y_{t+1}^{[1]}|r_t^{[1]}, r_t^{[2]}) &= \text{Var}(\rho_{11} \circ r_t^{[1]}|r_t^{[1]}) + \text{Var}(\rho_{12} \circ r_t^{[2]}|r_t^{[2]}) + \text{Var}(R_{t+1}^{[1]}) \\ &= \rho_{11}(1-\rho_{11}) r_t^{[1]} + \rho_{12}(1-\rho_{12}) r_t^{[2]} + \lambda_{t+1}^{[1]} \end{aligned} \tag{5.20}$$

and

$$\begin{aligned}\text{Var}(Y_{t+1}^{[2]}|r_t^{[1]}, r_t^{[2]}) &= \text{Var}(\rho_{21} \circ r_t^{[1]}|r_t^{[1]}) + \text{Var}(\rho_{22} \circ r_t^{[2]}|r_t^{[2]}) + \text{Var}(R_{t+1}^{[2]}) \\ &= \rho_{21}(1-\rho_{21})r_t^{[1]} + \rho_{22}(1-\rho_{22})r_t^{[2]} + \lambda_{t+1}^{[2]}, \qquad (5.21)\end{aligned}$$

where $r_t^{[k]}$ are approximated by their respective means.

## 5.4 Simulation Study

In this section, we use the OD Equations (5.1) and (5.2) to generate non-stationary time series data, by simulating the innovation terms, $R_t^{[k]} \sim \text{Poisson}(\lambda_t^{[k]})$, for $T = 60, 100$ and 500. The population values of $[\beta_1^{[1]}, \beta_2^{[1]}, \beta_1^{[2]}, \beta_2^{[2]}] = [0.3, 0.3, 0.3, 0.3]$ while the population values for $\rho_{11}, \rho_{22}, \rho_{12}$, and $\rho_{21}$ are combinations of 0.2 and 0.6, highlighting the cases of low and high serial- and cross-correlation coefficients. A $(2 \times 1)$ covariate structure of the following form is assumed:

$$x_{t1} = \begin{cases} \cos(2\pi t/6) + 0.01 & (t = 1, \ldots, T/5), \\ \cos(2\pi t/6) - 0.01 & (t = (T/5)+1, \ldots, 2T/5), \\ 2\pi t & (t = (2T/5)+1, \ldots, 3T/5), \\ -2\pi t & (t = (3T/5)+1, \ldots, 4T/5), \\ \sin(2\pi t/6) + 0.05 & (T = (4T/5)+1, \ldots, T), \end{cases}$$

and $x_{t2}$ is a random standard normal variable generated from the R function $rnorm(0,1)$ to compute $\lambda_t^{[k]} = \exp(\boldsymbol{x}_t' \boldsymbol{\beta}^{[k]})$. The number of Monte Carlo replication for each combination of the serial- and cross-correlation is 5000. Tables 5.1–5.2 show the simulated mean estimates of the different parameters by applying GQL.

As displayed in Tables 5.1–5.2, the estimates are close to the population parameters under both low and high correlation and the standard errors for all the parameters decrease with increasing time points. However, some non-convergence problems are noted for all cluster sizes under low and high correlation. For example, when $\rho_{11} = \rho_{22} = 0.6$ and $\rho_{12} = \rho_{21} = 0.2$, the number of non-convergent simulations is 290 for T=60. As $T$ increases to 100 and 500, the number of non-convergent simulations becomes 275 and 180, respectively. For lower values of serial-correlation but higher values of cross-correlation, that is, $\rho_{11} = \rho_{22} = 0.2$ and $\rho_{12} = \rho_{21} = 0.6$, GQL did not converge in around 310 simulations for $T = 60$, 280 for $T = 100$ and 235 for $T = 500$. When $\rho_{11} = 0.2, \rho_{22} = 0.6$, $\rho_{12} = 0.2$ and $\rho_{21} = 0.6$, the number of failures is 275 for $T = 60$, 240 for $T = 100$ and 170 for $T = 500$. When $\rho_{11} = 0.6, \rho_{22} = 0.2$, $\rho_{12} = 0.6$ and $\rho_{21} = 0.2$, GQL did not converge in 285 simulations for $T = 60$,

**TABLE 5.1**
Estimates of the Parameters and Standard Errors under the Unconstrained Non-Stationary BINMA(1) Process

| $T$ | $\hat{\beta}_1^{[1]}$ | $\hat{\beta}_2^{[1]}$ | $\hat{\beta}_1^{[2]}$ | $\hat{\beta}_2^{[2]}$ | $\hat{\rho}_{11}$=0.6 | $\hat{\rho}_{22}$=0.6 | $\hat{\rho}_{12}=0.6$ | $\hat{\rho}_{21}=0.6$ |
|---|---|---|---|---|---|---|---|---|
| 60 | 0.3120 | 0.3106 | 0.3158 | 0.3191 | 0.6190 | 0.6108 | 0.6195 | 0.6198 |
| | (0.0832) | (0.0990) | (0.0820) | (0.0999) | (0.0716) | (0.0893) | (0.0745) | (0.0981) |
| 100 | 0.3071 | 0.3056 | 0.3067 | 0.3077 | 0.6072 | 0.5912 | 0.6071 | 0.6077 |
| | (0.0791) | (0.0671) | (0.0858) | (0.0678) | (0.0589) | (0.0553) | (0.0667) | (0.0592) |
| 500 | 0.3024 | 0.3016 | 0.3021 | 0.3008 | 0.6012 | 0.6013 | 0.6018 | 0.6011 |
| | (0.0330) | (0.0239) | (0.0123) | (0.0240) | (0.0133) | (0.0247) | (0.0035) | (0.0215) |
| $T$ | $\hat{\beta}_1^{[1]}$ | $\hat{\beta}_2^{[1]}$ | $\hat{\beta}_1^{[2]}$ | $\hat{\beta}_2^{[2]}$ | $\hat{\rho}_{11}$=0.6 | $\hat{\rho}_{22}$=0.6 | $\hat{\rho}_{12}=0.2$ | $\hat{\rho}_{21}=0.2$ |
| 60 | 0.3131 | 0.3116 | 0.3142 | 0.3127 | 0.5854 | 0.5816 | 0.2191 | 0.2193 |
| | (0.1008) | (0.1131) | (0.0971) | (0.0823) | (0.0906) | (0.0998) | (0.0711) | (0.0961) |
| 100 | 0.3084 | 0.3085 | 0.3080 | 0.3096 | 0.6059 | 0.6091 | 0.2070 | 0.2075 |
| | (0.0847) | (0.0633) | (0.0711) | (0.0624) | (0.0663) | (0.0582) | (0.0691) | (0.0569) |
| 500 | 0.3053 | 0.3057 | 0.3044 | 0.3040 | 0.6025 | 0.6079 | 0.2017 | 0.2012 |
| | (0.0219) | (0.0117) | (0.0351) | (0.0324) | (0.0231) | (0.0021) | (0.0213) | (0.0125) |
| $T$ | $\hat{\beta}_1^{[1]}$ | $\hat{\beta}_2^{[1]}$ | $\hat{\beta}_1^{[2]}$ | $\hat{\beta}_2^{[2]}$ | $\hat{\rho}_{11}$=0.2 | $\hat{\rho}_{22}$=0.2 | $\hat{\rho}_{12}=0.2$ | $\hat{\rho}_{21}=0.2$ |
| 60 | 0.3122 | 0.3118 | 0.3144 | 0.3183 | 0.1839 | 0.2123 | 0.2196 | 0.2197 |
| | (0.0910) | (0.0898) | (0.0925) | (0.0816) | (0.0845) | (0.0785) | (0.0878) | (0.0918) |
| 100 | 0.3041 | 0.3030 | 0.3025 | 0.3088 | 0.2072 | 0.2030 | 0.2060 | 0.2059 |
| | (0.0753) | (0.0723) | (0.0830) | (0.0617) | (0.0720) | (0.0525) | (0.0613) | (0.0598) |
| 500 | 0.3016 | 0.3019 | 0.3003 | 0.3062 | 0.2019 | 0.1956 | 0.2013 | 0.2015 |
| | (0.0325) | (0.0214) | (0.0160) | (0.0126) | (0.0251) | (0.0023) | (0.0235) | (0.0097) |
| $T$ | $\hat{\beta}_1^{[1]}$ | $\hat{\beta}_2^{[1]}$ | $\hat{\beta}_1^{[2]}$ | $\hat{\beta}_2^{[2]}$ | $\hat{\rho}_{11}$=0.2 | $\hat{\rho}_{22}$=0.2 | $\hat{\rho}_{12}=0.6$ | $\hat{\rho}_{21}=0.6$ |
| 60 | 0.3110 | 0.3171 | 0.3199 | 0.3143 | 0.2191 | 0.1855 | 0.6197 | 0.6195 |
| | (0.1228) | (0.0933) | (0.1197) | (0.0826) | (0.0940) | (0.0909) | (0.0756) | (0.0813) |
| 100 | 0.3061 | 0.3066 | 0.3053 | 0.3044 | 0.2088 | 0.2041 | 0.6079 | 0.6069 |
| | (0.0853) | (0.0679) | (0.0996) | (0.0671) | (0.0774) | (0.0583) | (0.0770) | (0.0572) |
| 500 | 0.3024 | 0.3051 | 0.3016 | 0.3031 | 0.2045 | 0.2021 | 0.6020 | 0.6019 |
| | (0.0140) | (0.0110) | (0.0124) | (0.0105) | (0.0012) | (0.0045) | (0.0027) | (0.0033) |
| T | $\hat{\beta}_1^{[1]}$ | $\hat{\beta}_2^{[1]}$ | $\hat{\beta}_1^{[2]}$ | $\hat{\beta}_2^{[2]}$ | $\hat{\rho}_{11}$=0.2 | $\hat{\rho}_{22}$=0.6 | $\hat{\rho}_{12}=0.2$ | $\hat{\rho}_{21}=0.6$ |
| 60 | 0.3137 | 0.3195 | 0.3124 | 0.3157 | 0.2153 | 0.6169 | 0.2161 | 0.6183 |
| | (0.0985) | (0.0853) | (0.0964) | (0.1016) | (0.0788) | (0.0823) | (0.0926) | (0.0713) |
| 100 | 0.3093 | 0.3073 | 0.3021 | 0.3083 | 0.2074 | 0.6053 | 0.2060 | 0.6071 |
| | (0.0761) | (0.0652) | (0.0780) | (0.0697) | (0.0660) | (0.0595) | (0.0692) | (0.0582) |
| 500 | 0.3081 | 0.3010 | 0.3008 | 0.3059 | 0.2034 | 0.6021 | 0.2024 | 0.6020 |
| | (0.0204) | (0.0120) | (0.0217) | (0.0124) | (0.0114) | (0.0021) | (0.0026) | (0.0131) |
| $T$ | $\hat{\beta}_1^{[1]}$ | $\hat{\beta}_2^{[1]}$ | $\hat{\beta}_1^{[2]}$ | $\hat{\beta}_2^{[2]}$ | $\hat{\rho}_{11}$=0.6 | $\hat{\rho}_{22}$=0.2 | $\hat{\rho}_{12}=0.6$ | $\hat{\rho}_{21}=0.2$ |
| 60 | 0.3115 | 0.3150 | 0.3121 | 0.3181 | 0.6171 | 0.2114 | 0.6193 | 0.2188 |
| | (0.1235) | (0.0911) | (0.1178) | (0.1102) | (0.0964) | (0.0877) | (0.0710) | (0.0990) |
| 100 | 0.3069 | 0.3053 | 0.3032 | 0.3057 | 0.6022 | 0.2043 | 0.6081 | 0.2076 |
| | (0.0876) | (0.0792) | (0.0961) | (0.0659) | (0.0786) | (0.0571) | (0.0667) | (0.0593) |
| 500 | 0.3019 | 0.3040 | 0.3004 | 0.3015 | 0.6001 | 0.2016 | 0.6021 | 0.2019 |
| | (0.0342) | (0.0248) | (0.0113) | (0.0247) | (0.0217) | (0.0015) | (0.0246) | (0.0131) |

**TABLE 5.2**
Estimates of the Parameters and Standard Errors under the Unconstrained Non-Stationary BINMA(1) Process

| T | $\hat{\beta}_1^{[1]}$ | $\hat{\beta}_2^{[1]}$ | $\hat{\beta}_1^{[2]}$ | $\hat{\beta}_2^{[2]}$ | $\hat{\rho}_{11}$=0.2 | $\hat{\rho}_{22}$=0.2 | $\hat{\rho}_{12}=0.2$ | $\hat{\rho}_{21}=0.6$ |
|---|---|---|---|---|---|---|---|---|
| 60 | 0.3155 | 0.3169 | 0.3106 | 0.3164 | 0.2151 | 0.2103 | 0.2192 | 0.6194 |
| | (0.0921) | (0.0804) | (0.0913) | (0.0876) | (0.0821) | (0.0727) | (0.0777) | (0.0863) |
| 100 | 0.3057 | 0.3084 | 0.3086 | 0.3071 | 0.2052 | 0.2081 | 0.2068 | 0.6072 |
| | (0.0625) | (0.0740) | (0.0655) | (0.0714) | (0.0673) | (0.0710) | (0.0525) | (0.0542) |
| 500 | 0.3032 | 0.3051 | 0.3014 | 0.3095 | 0.2040 | 0.2012 | 0.2020 | 0.6031 |
| | (0.0217) | (0.0125) | (0.0230) | (0.0107) | (0.0145) | (0.0002) | (0.0111) | (0.0019) |
| $T$ | $\hat{\beta}_1^{[1]}$ | $\hat{\beta}_2^{[1]}$ | $\hat{\beta}_1^{[2]}$ | $\hat{\beta}_2^{[2]}$ | $\hat{\rho}_{11}$=0.6 | $\hat{\rho}_{22}$=0.6 | $\hat{\rho}_{12}=0.6$ | $\hat{\rho}_{21}=0.2$ |
| 60 | 0.3192 | 0.3161 | 0.3111 | 0.3153 | 0.6147 | 0.6184 | 0.6194 | 0.2186 |
| | (0.0888) | (0.0883) | (0.1151) | (0.1286) | (0.0768) | (0.0934) | (0.0775) | (0.0961) |
| 100 | 0.3076 | 0.3043 | 0.3091 | 0.3050 | 0.6083 | 0.6016 | 0.6075 | 0.2066 |
| | (0.0716) | (0.0627) | (0.0919) | (0.0623) | (0.0610) | (0.0726) | (0.0588) | (0.0738) |
| 500 | 0.3070 | 0.3013 | 0.3076 | 0.3028 | 0.6041 | 0.6003 | 0.6011 | 0.2022 |
| | (0.0307) | (0.0112) | (0.0368) | (0.0108) | (0.0202) | (0.0009) | (0.0063) | (0.0220) |

250 for $T = 100$ and 185 for $T = 500$. Note that these non-convergent simulations were due to the auto-covariance function and the Hessian structure in the iterative Equation (5.17) were close to singularity.

## 5.5 Summary

In this chapter, we have introduced the unconstrained BINMA(1) model with Poisson innovations by assuming that the counting series may be influenced by some time-varying covariates. Similar to the constrained BINMA(1) model introduced in Chapter 3, the counting series of the unconstrained BINMA(1) model in this chapter follow marginally the Poisson distribution. Under this set-up, the main parameters of interest in this model have been estimated using the GQL approach. The simulation results demonstrate that the GQL approach yields consistent parameter estimates with few non-convergent simulations. In the next chapter, we introduce a constrained BINAR(1) model with NB innovations that can analyze data which exhibit low, moderate or high over-dispersion.

# 6

# *Constrained BINAR(1) Model with Correlated NB Innovations*

## 6.1 Introduction

So far, the unconstrained BINAR(1) model introduced in Chapter 4 could only model series that exhibit mild over-dispersion though at different levels. As for significant over-dispersed series, we could only refer to the over-dispersed BINAR(1) model of Pedeli & Karlis (2009, 2011) where the innovations were pairwise NB under the BVNB model of Marshall & Olkin (1990). However, this time series model can only cater for series subject to same level of over-dispersion. Above all, this BINAR(1) model has only been designed under stationary condition. Thus, this chapter overcomes these limitations by developing a novel constrained BINAR(1) model under the non-stationary moment condition with paired NB innovations under different levels of over-dispersion (Ng et al. (2010)). As for the inferential procedures, a GQL estimation approach is developed to estimate the model parameters.

The organization of this chapter is as follows: In the next section, a novel BINAR(1) model with paired NB innovations under Ng et al. (2010) bivariate model is developed. In Section 6.3, the GQL inference procedure is developed to estimate the different model parameters. In this process, some new lemmas are proposed to compute the covariance expressions exactly. This section is followed by some Monte Carlo simulation experiments and the performance of the GQL approach is investigated. The conclusion is presented in the last section.

## 6.2 The Constrained BINAR(1) Model

We suppose the BINAR(1) model $\{[Y_t^{[1]}, Y_t^{[2]}]',\ t \in \mathbb{Z}\}$ is defined as

$$\begin{pmatrix} Y_t^{[1]} \\ Y_t^{[2]} \end{pmatrix} = \begin{pmatrix} \alpha_{11,t} & 0 \\ 0 & \alpha_{22,t} \end{pmatrix} \circ \begin{pmatrix} Y_{t-1}^{[1]} \\ Y_{t-1}^{[2]} \end{pmatrix} + \begin{pmatrix} R_t^{[1]} \\ R_t^{[2]} \end{pmatrix}, \tag{6.1}$$

DOI: 10.1201/9781003600435-6

where $Y_t^{[k]}$ is the random count observation for the $k^{th}$ series at the $t^{th}$ time point with $R_t^{[k]}$ representing its corresponding innovation terms for $k \in \{1,2\}$.

The pair of innovations $(R_t^{[1]}, R_t^{[2]})$ is allowed to follow the BVNB model proposed by Ng et al. (2010) where $R_t^{[1]}$ and $R_t^{[2]}$ exhibit different levels of over-dispersion. From Ng et al. (2010), the pgf of the BVNB under a re-parametrization of the basic parameters is derived as follows.
From Ng et al. (2010), the BVNB distribution of $(R_t^{[1]}, R_t^{[2]})$ has joint pgf of the form

$$G_{(R_t^{[1]},R_t^{[2]})}(u,v) = \left(\frac{q_1}{1-p_1u}\right)^{\alpha_1}\left(\frac{q_2}{1-p_2v}\right)^{\alpha_2} \times \left(\frac{\theta}{1-\theta_1u-\theta_2v}\right)^{\nu} \tag{6.2}$$

where $P_i(i=1,2)$ represents the probability and $\alpha_i(i=1,2)$ and $\nu$ are dispersion parameters such that $p_1 = \left(\frac{\theta_1}{1-\theta_2}\right)$, $p_2 = \left(\frac{\theta_2}{1-\theta_1}\right)$, $q_k = 1-p_k$ ($k$=1,2), $\theta = 1-\theta_1-\theta_2$, where $\theta_1$ and $\theta_2$ are to be determined. From Equation (6.2), it is shown in Ng et al. (2010) that the marginal pgf of $R_t^{[1]}$ and $R_t^{[2]}$ are given by $G_{R_t^{[1]}}(u) = \left(\frac{q_1}{1-p_1u}\right)^{\alpha_1+\nu}$ and $G_{R_t^{[2]}}(v) = \left(\frac{q_2}{1-p_2v}\right)^{\alpha_2+\nu}$, which implies that $R_t^{[k]}$ is marginally NB with parameters $(\alpha_k+\nu)$ and $p_k$ and $E(R_t^{[k]}) = \frac{(\alpha_k+\nu)p_k}{(1-p_k)}$.

In the present context, we let $\alpha_k+\nu = \frac{1}{c_k}$ and $\mathrm{E}(R_t^{[k]}) = \lambda_t^{[k]} = \frac{(\alpha_k+\nu)p_k}{(1-p_k)}$, such that by substituting $p_k$ and $\alpha_k$, we obtain in terms of $\theta_1$ and $\theta_2$,

$$\lambda_t^{[k]} = \frac{\theta_k}{c_k(1-\theta_1-\theta_2)}, \quad \text{for} \quad k=1,2. \tag{6.3}$$

Thus, the parameters $\theta_1$ and $\theta_2$ are obtained from Equation (6.3) by solving simultaneously, which yields

$$\theta_1 = \frac{c_1\lambda_t^{[1]}}{1+c_1\lambda_t^{[1]}+c_2\lambda_t^{[2]}}, \quad \theta_2 = \frac{c_2\lambda_t^{[2]}}{1+c_1\lambda_t^{[1]}+c_2\lambda_t^{[2]}} \tag{6.4}$$

which by substituting into (6.2) gives the pgf, which is of the form

$$G_{(R_t^{[1]},R_t^{[2]})}(u,v) = \left(\frac{1}{1+c_1\lambda_t^{[1]}-c_1\lambda_t^{[1]}u}\right)^{{c_1}^{-1}-\nu}\left(\frac{1}{1+c_2\lambda_t^{[2]}-c_2\lambda_t^{[2]}v}\right)^{{c_2}^{-1}-\nu} \times \left(\frac{1}{1+c_1\lambda_t^{[1]}+c_2\lambda_t^{[2]}-c_1\lambda_t^{[1]}u-c_2\lambda_t^{[2]}v}\right)^{\nu} \tag{6.5}$$

where $c_1, c_2, \lambda_t^{[1]}, \lambda_t^{[2]} > 0$ and $0 < \nu < \min(1/c_1, 1/c_2)$. We suppose that the random vectors $(R_t^{[1]}, R_t^{[2]})$ and $(R_u^{[1]}, R_u^{[2]})$ are independent for all $t \neq u$ and hence the marginal distribution of $R_t^{[k]}$ is given by

$$P_{R_t^{[k]}}(r_k) = \frac{\Gamma({c_k}^{-1} + r_k)}{\Gamma({c_k}^{-1}) r_k!} \left(\frac{1}{1 + c_k \lambda_t^{[k]}}\right)^{{c_k}^{-1}} \left(\frac{c_k \lambda_t^{[k]}}{1 + c_k \lambda_t^{[k]}}\right)^{r_k},$$
$$c_k, \lambda_t^{[k]} > 0, \quad r_k \geq 0, \tag{6.6}$$

with $E(R_t^{[k]}) = \lambda_t^{[k]}$ and $\mathrm{Var}(R_t^{[k]}) = \lambda_t^{[k]}(1 + c_k \lambda_t^{[k]})$, while

$$\mathrm{Cov}(R_t^{[1]}, R_{t+h}^{[2]}) = \begin{cases} c_1 c_2 \nu \lambda_t^{[1]} \lambda_t^{[2]}, & h = 0 \\ 0, & h \neq 0. \end{cases} \tag{6.7}$$

From Equation (6.6), we can simply write $R_t^{[k]} \sim \mathrm{NB}(\frac{1}{c_k}, c_k \lambda_t^{[k]})$, where $\lambda_t^{[k]} = \exp(\boldsymbol{x}_t' \boldsymbol{\beta}^{[k]})$, with $\boldsymbol{x}_t = [x_{t1}, x_{t2}, \ldots, x_{tj}, \ldots, x_{tp}]'$ and $\boldsymbol{\beta}^{[k]} = [\beta_1^{[k]}, \beta_2^{[k]}, \ldots, \beta_j^{[k]}, \ldots, \beta_p^{[k]}]'$ for $k \in \{1, 2\}$.

The other conditions based on the above BINAR(1) model (6.1) include:

(a) The sequences of random variables $\{\alpha_{11,t}\}$ and $\{\alpha_{22,t}\}$ are mutually independent sequences of independent and identically distributed random variables with $\alpha_{kk,t} \sim \mathrm{Beta}(\frac{\rho_{kk}}{c_k}, \frac{1-\rho_{kk}}{c_k})$, where $\rho_{kk} \in [0, 1]$.

(b) '$\circ$' is the binomial thinning operator indicating $\{\alpha_{kk,t} \circ Y_{t-1}^{[k]} \mid Y_{t-1}^{[k]}, \alpha_{kk,t}\}$ follows the binomial model with parameters $Y_{t-1}^{[k]}$ and $\alpha_{kk,t}$. From (a) and (b), we have that $E(\alpha_{kk,t} \circ Y_{t-1}^{[k]}) = E(\alpha_{kk,t}) E(Y_{t-1}^{[k]})$.

(c) The random variables $\alpha_{kk,t}$ and $R_u^{[k]}$ are independent random variables for $k \in \{1, 2\}$ and all $t, u \in \mathbb{Z}$. Also, under same notation, the random variables $\alpha_{kk,t}$ and $Y_u^{[k]}$ are independent random variables for $k \in \{1, 2\}$ and all $t, u \in \mathbb{Z}$.

(d) The random variables $Y_t^{[k]}$ and $R_u^{[l]}$ are independent for all $k, l \in \{1, 2\}$, $t, u \in \mathbb{Z}$ and $t \neq u$. Then it follows that $\mathrm{Cov}(\alpha_{kk,t} \circ Y_{t-1}^{[k]}, R_{t+h}^{[l]}) = 0$ for any $h \geq 0$ and any $k, l \in \{1, 2\}$.

Under condition (a) we have that $\mathrm{E}(\alpha_{kk,t}) = \rho_{kk}$ and $\mathrm{Var}(\alpha_{kk,t}) = \frac{\rho_{kk}(1-\rho_{kk})c_k}{1+c_k}$.

The following lemma contains auxiliary results that we will use later in this chapter.

**Lemma 6.1.** *Suppose that the above conditions hold. Then,*

(i) (a) $\mathrm{Var}(\alpha_{kk,t} \circ Y_{t-1}^{[k]} | Y_{t-1}^{[k]}, \alpha_{kk,t}) = \alpha_{kk,t}(1 - \alpha_{kk,t}) Y_{t-1}^{[k]}, k \in \{1, 2\}$.

(b)

$$\mathrm{Var}(\alpha_{kk,t} \circ Y_{t-1}^{[k]}) = \frac{\rho_{kk}(c_k + \rho_{kk})}{1 + c_k} E({Y_{t-1}^{[k]}}^2) - \rho_{kk}^2 \left(E(Y_{t-1}^{[k]})\right)^2 + \frac{\rho_{kk}(1 - \rho_{kk})}{1 + c_k} E(Y_{t-1}^{[k]}), \quad k \in \{1, 2\}.$$

(ii) (a) $\text{Cov}(Y_t^{[k]}, \alpha_{jj,t+h+1} \circ Y_{t+h}^{[j]}) = \rho_{jj}\text{Cov}(Y_t^{[k]}, Y_{t+h}^{[j]})$ *for* $h \geq 0$ *and* $k, j \in \{1, 2\}$.

(b) $\text{Cov}(\alpha_{kk,t+1} \circ Y_t^{[k]}, \alpha_{jj,t+h+1} \circ Y_{t+h}^{[j]}) = \rho_{jj}\rho_{kk}\text{Cov}(Y_t^{[k]}, Y_{t+h}^{[j]})$ *for* $h \geq 0$ *and* $k, j \in \{1, 2\}$.

(iii) (a) $\text{Cov}(Y_t^{[k]}, R_t^{[k]}) = \text{Var}(R_t^{[k]}), k \in \{1, 2\}$.

(b)

$$\text{Cov}(Y_t^{[k]}, R_{t+h}^{[j]}) = \begin{cases} \text{Cov}(R_t^{[k]}, R_t^{[j]}), & h = 0 \\ 0, & h \neq 0, j, k \in \{1, 2\}, j \neq k. \end{cases}$$

*Proof:*

(i) (a) *The proof follows from the fact that the random variable* $\{\alpha_{kk,t} \circ Y_{t-1}^{[k]}\}$ *for given* $Y_{t-1}^{[k]}$ *and* $\alpha_{kk,t}$ *has the binomial distribution with parameters* $Y_{t-1}^{[k]}$ *and* $\alpha_{kk,t}$.

(b) *Using the law of total variance and the independency of the corresponding random variables, we obtain that*

$$\begin{aligned}
\text{Var}(\alpha_{kk,t} \circ Y_{t-1}^{[k]}) &= \text{Var}[E(\alpha_{kk,t} \circ Y_{t-1}^{[k]} | Y_{t-1}^{[k]}, \alpha_{kk,t})] \\
&\quad + E[\text{Var}(\alpha_{kk,t} \circ Y_{t-1}^{[k]} | Y_{t-1}^{[k]}, \alpha_{kk,t})] \\
&= \text{Var}[\alpha_{kk,t} Y_{t-1}^{[k]}] + E[\alpha_{kk,t}(1 - \alpha_{kk,t}) Y_{t-1}^{[k]}] \\
&= \frac{\rho_{kk}(c_k + \rho_{kk})}{1 + c_k} E({Y_{t-1}^{[k]}}^2) - \rho_{kk}^2 \left(E(Y_{t-1}^{[k]})\right)^2 \\
&\quad + \frac{\rho_{kk}(1 - \rho_{kk})}{1 + c_k} E(Y_{t-1}^{[k]}).
\end{aligned}$$

(ii) (a) *The proof follows from the two properties of the binomial thinning operator* $E(\alpha_{jj,t+h+1} \circ Y_{t+h}^{[j]}) = \rho_{jj} E(Y_{t+h}^{[j]})$ *and* $E(Y_t^{[k]}(\alpha_{jj,t+h+1} \circ Y_{t+h}^{[j]})) = \rho_{jj} E(Y_t^{[k]} Y_{t+h}^{[j]})$.

(b) *Similarly, the proof follows from the independency of the random variables* $\alpha_{kk,t+1}$ *and* $\alpha_{jj,t+h+1}$ *and the property of the binomial thinning operator*

$$\begin{aligned}
&E((\alpha_{kk,t+1} \circ Y_t^{[k]})(\alpha_{jj,t+h+1} \circ Y_{t+h}^{[j]})) \\
&\quad = E(\alpha_{kk,t+1}) E(\alpha_{jj,t+h+1}) E(Y_t^{[k]} Y_{t+h}^{[j]}) \\
&\quad = \rho_{kk}\rho_{jj} E(Y_t^{[k]} Y_{t+h}^{[j]}).
\end{aligned}$$

(iii) (a)

$$\begin{aligned}
\mathrm{Cov}(Y_t^{[k]}, R_t^{[k]}) &= \mathrm{Cov}(\alpha_{kk,t} \circ Y_{t-1}^{[k]} + R_t^{[k]}, R_t^{[k]}) \\
&= \mathrm{Cov}(\alpha_{kk,t} \circ Y_{t-1}^{[k]}, R_t^{[k]}) + \mathrm{Cov}(R_t^{[k]}, R_t^{[k]}) \\
&= \rho_{kk}\mathrm{Cov}(Y_{t-1}^{[k]}, R_t^{[k]}) + \mathrm{Var}(R_t^{[k]}) \\
&= \mathrm{Var}(R_t^{[k]}).
\end{aligned}$$

(b)

$$\begin{aligned}
\mathrm{Cov}(Y_t^{[k]}, R_{t+h}^{[j]}) &= \mathrm{Cov}(\alpha_{kk,t} \circ Y_{t-1}^{[k]} + R_t^{[k]}, R_{t+h}^{[j]}) \\
&= \mathrm{Cov}(\alpha_{kk,t} \circ Y_{t-1}^{[k]}, R_{t+h}^{[j]}) + \mathrm{Cov}(R_t^{[k]}, R_{t+h}^{[j]}) \\
&= \rho_{kk}\mathrm{Cov}(Y_{t-1}^{[k]}, R_{t+h}^{[j]}) + \mathrm{Cov}(R_t^{[k]}, R_{t+h}^{[j]})
\end{aligned}$$

$$= \begin{cases} \mathrm{Cov}(R_t^{[k]}, R_t^{[j]}), & h = 0 \\ 0, & h \neq 0, j, k \in \{1, 2\}, j \neq k. \end{cases}$$

By assuming $Y_{t-1}^{[k]} \sim \mathrm{NB}(\frac{1}{c_k^*}, c_k^* \mu_{t-1}^{[k]})$, where $c_k^*$ is to be determined, we show, using Lemma 6.1, that

$$\begin{aligned}
E(Y_t^{[k]}) \equiv \mu_t^{[k]} &= E(\alpha_{kk,t} \circ Y_{t-1}^{[k]}) + E(R_t^{[k]}) \\
&= E_{\alpha_{kk,t}} E_{Y_{t-1}^{[k]}} E(\alpha_{kk,t} \circ Y_{t-1}^{[k]} | Y_{t-1}^{[k]}, \alpha_{kk,t}) + E(R_t^{[k]}) \\
&= \rho_{kk}\mu_{t-1}^{[k]} + \lambda_t^{[k]} \qquad (6.8)
\end{aligned}$$

and

$$\begin{aligned}
\mathrm{Var}(Y_t^{[k]}) &= \mathrm{Var}(\alpha_{kk,t} \circ Y_{t-1}^{[k]}) + \mathrm{Var}(R_t^{[k]}) \\
&= \frac{\rho_{kk}(c_k + \rho_{kk})}{1 + c_k} E({Y_{t-1}^{[k]}}^2) - \rho_{kk}^2 \left(E(Y_{t-1}^{[k]})\right)^2 \\
&+ \frac{\rho_{kk}(1 - \rho_{kk})}{1 + c_k} E(Y_{t-1}^{[k]}) + \mathrm{Var}(R_t^{[k]}) \\
&= \frac{\rho_{kk}(c_k + \rho_{kk})}{1 + c_k} \left[\mu_{t-1}^{[k]} + c_k {\mu_{t-1}^{[k]}}^2 + {\mu_{t-1}^{[k]}}^2\right] \\
&- \rho_{kk}^2 {\mu_{t-1}^{[k]}}^2 + \frac{\rho_{kk}(1 - \rho_{kk})}{1 + c_k} \mu_{t-1}^{[k]} \\
&+ \lambda_t^{[k]}(1 + c_k \lambda_t^{[k]})
\end{aligned}$$

$$
\begin{aligned}
&= \frac{\rho_{kk}(c_k+\rho_{kk})}{1+c_k} c_k^* {\mu_{t-1}^{[k]}}^2 + \left[\frac{\rho_{kk}(c_k+\rho_{kk})}{1+c_k} - \rho_{kk}^2\right] {\mu_{t-1}^{[k]}}^2 \\
&+ \left[\frac{\rho_{kk}(c_k+\rho_{kk})}{1+c_k} + \frac{\rho_{kk}(1-\rho_{kk})}{1+c_k}\right] \mu_{t-1}^{[k]} + \lambda_t^{[k]}(1+c_k\lambda_t^{[k]}) \\
&= \frac{\rho_{kk}(c_k+\rho_{kk})}{1+c_k} c_k^* {\mu_{t-1}^{[k]}}^2 + \frac{\rho_{kk}c_k(1-\rho_{kk})}{1+c_k} {\mu_{t-1}^{[k]}}^2 \\
&+ \rho_{kk}\mu_{t-1}^{[k]} + \lambda_t^{[k]}(1+c_k\lambda_t^{[k]}).
\end{aligned} \tag{6.9}
$$

Hence, from Jowaheer & Sutradhar (2002) and Mallick & Sutradhar (2008), $Y_t^{[k]} \sim \text{NB}(\frac{1}{c_k^*}, c_k^*\mu_t^{[k]})$ with

$$
c_k^* = \frac{c_k\left[\rho_{kk}(1-\rho_{kk}){\mu_{t-1}^{[k]}}^2 + {\lambda_t^{[k]}}^2(1+c_k)\right]}{({\lambda_t^{[k]}}^2 + 2\rho_{kk}\mu_{t-1}^{[k]}\lambda_t^{[k]})(1+c_k) + \rho_{kk}(\rho_{kk}-c_k){\mu_{t-1}^{[k]}}^2}. \tag{6.10}
$$

Using (6.9) and the representation of $c_k^*$ given by (6.10), the variance of $Y_t^{[k]}$ is expressed as

$$
\text{Var}(Y_t^{[k]}) = (\rho_{kk}\mu_{t-1}^{[k]} + \lambda_t^{[k]})\left[1 + c_k^*(\rho_{kk}\mu_{t-1}^{[k]} + \lambda_t^{[k]})\right]. \tag{6.11}
$$

The covariance between the random variables $Y_t^{[k]}$ and $Y_{t+h}^{[k]}$, $h \geq 1$, is derived as

$$
\begin{aligned}
\text{Cov}(Y_t^{[k]}, Y_{t+h}^{[k]}) &= \text{Cov}(Y_t^{[k]}, \alpha_{kk,t+h} \circ Y_{t+h-1}^{[k]} + R_{t+h}^{[k]}) \\
&= \rho_{kk}\text{Cov}(Y_t^{[k]}, Y_{t+h-1}^{[k]}) + \text{Cov}(Y_t^{[k]}, R_{t+h}^{[k]}) \\
&= \rho_{kk}\text{Cov}(Y_t^{[k]}, Y_{t+h-1}^{[k]}) \\
&= \rho_{kk}\text{Cov}(Y_t^{[k]}, \alpha_{kk,t+h-1} \circ Y_{t+h-2}^{[k]} + R_{t+h-1}^{[k]}) \\
&= \rho_{kk}^2\text{Cov}(Y_t^{[k]}, Y_{t+h-2}^{[k]}) + \rho_{kk}\text{Cov}(Y_t^{[k]}, R_{t+h-1}^{[k]}) \\
&= \rho_{kk}^2 Cov(Y_t^{[k]}, Y_{t+h-2}^{[k]}) \\
&\ \ \vdots \\
&= \rho_{kk}^h\text{Cov}(Y_t^{[k]}, Y_t^{[k]}) \\
&= \rho_{kk}^h\mu_t^{[k]}(1 + c_k^*\mu_t^{[k]})
\end{aligned} \tag{6.12}
$$

and hence the lag-$h$ serial correlation is derived as

$$\mathrm{Corr}(Y_t^{[k]}, Y_{t+h}^{[k]}) = \rho_{kk}^h \frac{\sqrt{\mu_t^{[k]}(1 + c_k^* \mu_t^{[k]})}}{\sqrt{\mu_{t+h}^{[k]}(1 + c_k^* \mu_{t+h}^{[k]})}}. \tag{6.13}$$

The cross-covariance between the two series is given by

$$\begin{aligned}
\mathrm{Cov}(Y_t^{[1]}, Y_t^{[2]}) &= \mathrm{Cov}(\alpha_{11,t} \circ Y_{t-1}^{[1]} + R_t^{[1]}, \alpha_{22,t} \circ Y_{t-1}^{[2]} + R_t^{[2]}) \\
&= \mathrm{Cov}(\alpha_{11,t} \circ Y_{t-1}^{[1]}, \alpha_{22,t} \circ Y_{t-1}^{[2]}) + \mathrm{Cov}(R_t^{[1]}, R_t^{[2]}) \\
&= \rho_{11}\rho_{22}\mathrm{Cov}(Y_{t-1}^{[1]}, Y_{t-1}^{[2]}) + \mathrm{Cov}(R_t^{[1]}, R_t^{[2]}) \\
&= \rho_{11}\rho_{22}\mathrm{Cov}(Y_{t-1}^{[1]}, Y_{t-1}^{[2]}) + c_1 c_2 \nu \lambda_t^{[1]} \lambda_t^{[2]}.
\end{aligned} \tag{6.14}$$

We now derive the covariances $\mathrm{Cov}(Y_t^{[1]}, Y_{t+h}^{[2]})$ and $\mathrm{Cov}(Y_t^{[2]}, Y_{t+h}^{[1]})$. The first one is derived as

$$\begin{aligned}
\mathrm{Cov}(Y_t^{[1]}, Y_{t+h}^{[2]}) &= \mathrm{Cov}(Y_t^{[1]}, \alpha_{22,t+h} \circ Y_{t+h-1}^{[2]} + R_{t+h}^{[2]}) \\
&= \mathrm{Cov}(Y_t^{[1]}, \alpha_{22,t+h} \circ Y_{t+h-1}^{[2]}) + \mathrm{Cov}(Y_t^{[1]}, R_{t+h}^{[2]}) \\
&= \rho_{22}\mathrm{Cov}(Y_t^{[1]}, Y_{t+h-1}^{[2]}) \\
&= \rho_{22}\mathrm{Cov}(Y_t^{[1]}, \alpha_{22,t+h-1} \circ Y_{t+h-2}^{[2]} + R_{t+h-1}^{[2]}) \\
&= \rho_{22}\mathrm{Cov}(Y_t^{[1]}, \alpha_{22,t+h-1} \circ Y_{t+h-2}^{[2]}) + \mathrm{Cov}(Y_t^{[1]}, R_{t+h-1}^{[2]}) \\
&= \rho_{22}^2\mathrm{Cov}(Y_t^{[1]}, Y_{t+h-2}^{[2]}) \\
&\vdots \\
&= \rho_{22}^h\mathrm{Cov}(Y_t^{[1]} Y_t^{[2]}).
\end{aligned} \tag{6.15}$$

Similarly, we obtain

$$\mathrm{Cov}(Y_{t+h}^{[1]}, Y_t^{[2]}) = \rho_{11}^h \mathrm{Cov}(Y_t^{[1]}, Y_t^{[2]}). \tag{6.16}$$

**Remark 6.1.** Since $c_k = 0 \Rightarrow c_k^* = 0$, we obtain from Equations (6.8) and (6.11)

$$E(Y_t^{[k]}) = \mathrm{Var}(Y_t^{[k]}) = \rho_{kk}\mu_{t-1}^{[k]} + \lambda_t^{[k]}. \tag{6.17}$$

Hence, $Y_t^{[k]} \sim \mathrm{Poisson}(\mu_t^{[k]})$, where $\mu_t^{[k]} = \rho_{kk}\mu_{t-1}^{[k]} + \lambda_t^{[k]}$.

**Remark 6.2.** Under stationary moment condition,

$$E(Y_t^{[k]}) \equiv \mu_1^{[k]} = \frac{\lambda_1^{[k]}}{1-\rho_{kk}}, \tag{6.18}$$

$$\mathrm{Var}(Y_t^{[k]}) = (\rho_{kk}\mu_1^{[k]} + \lambda_1^{[k]})\left[1 + c_k^*(\rho_{kk}\mu_1^{[k]} + \lambda_1^{[k]})\right] \tag{6.19}$$

and

$$\mathrm{Cov}(Y_t^{[1]}, Y_t^{[2]}) = \frac{c_1 c_2 \nu \lambda_1^{[1]} \lambda_1^{[2]}}{1-\rho_{11}\rho_{22}} \tag{6.20}$$

## 6.3 Estimation of Parameters

This section develops two GQL equations to estimate the different model parameters.

### 6.3.1 GQL-I

The first GQL equation estimates the parameters $\boldsymbol{\beta}^{[k]} = [\beta_1^{[1]}, \beta_2^{[1]}, \beta_1^{[2]}, \beta_2^{[2]}]'$ and $\boldsymbol{\rho} = [\rho_{11}, \rho_{22}]'$ and is specified as

$$\boldsymbol{D}'_{\boldsymbol{\beta},\boldsymbol{\rho}} \boldsymbol{\Sigma}_{\boldsymbol{\beta},\boldsymbol{\rho}}{}^{-1}(\boldsymbol{y} - \boldsymbol{\mu}) = 0, \tag{6.21}$$

where

$$\boldsymbol{y} = \begin{pmatrix} \boldsymbol{y}^{[1]} \\ \boldsymbol{y}^{[2]} \end{pmatrix}_{2T\times 1}, \boldsymbol{y}^{[1]} = [y_1^{[1]}, y_2^{[1]}, \ldots, y_T^{[1]}]', \boldsymbol{y}^{[2]} = [y_1^{[2]}, y_2^{[2]}, \ldots, y_T^{[2]}]' \tag{6.22}$$

and

$$\boldsymbol{\mu} = \begin{pmatrix} \boldsymbol{\mu}^{[1]} \\ \boldsymbol{\mu}^{[2]} \end{pmatrix}_{2T\times 1}, \boldsymbol{\mu}^{[1]} = [\mu_1^{[1]}, \mu_2^{[1]}, \ldots, \mu_T^{[1]}]', \boldsymbol{\mu}^{[2]} = [\mu_1^{[2]}, \mu_2^{[2]}, \ldots, \mu_T^{[2]}]'. \tag{6.23}$$

The derivative matrix $\boldsymbol{D}_{\beta,\rho} = [\boldsymbol{D}_1, \boldsymbol{D}_2, \ldots, \boldsymbol{D}_T]'$ is a block diagonal $(2T \times (2p+2))$ matrix, with

$$\boldsymbol{D}_t = \begin{pmatrix} \frac{\partial \mu_t^{[1]}}{\partial \boldsymbol{\beta}^{[1]}} & 0 \\ 0 & \frac{\partial \mu_t^{[2]}}{\partial \boldsymbol{\beta}^{[2]}} \\ \frac{\partial \mu_t^{[1]}}{\partial \rho_{11}} & 0 \\ 0 & \frac{\partial \mu_t^{[2]}}{\partial \rho_{22}} \end{pmatrix}_{(2p+2)\times 2} \tag{6.24}$$

where

$$\frac{\partial \mu_t^{[k]}}{\partial \boldsymbol{\beta}^{[l]}} = \left[\frac{\partial \mu_t^{[k]}}{\partial \beta_1^{[l]}}, \frac{\partial \mu_t^{[k]}}{\partial \beta_2^{[l]}}, \ldots, \frac{\partial \mu_t^{[k]}}{\partial \beta_j^{[l]}}, \ldots, \frac{\partial \mu_t^{[k]}}{\partial \beta_p^{[l]}}\right]', \quad k, k', l \in \{1,2\}, k \neq k'.$$

The expressions for the above derivatives with respect to the regression vector $\{\boldsymbol{\beta}^{[k]}\}$ are given in different form as follows:

$$\frac{\partial \mu_t^{[k]}}{\partial \beta_j^{[k]}} = \rho_{kk} \frac{\partial \mu_{t-1}^{[k]}}{\partial \beta_j^{[k]}} + \lambda_t^{[k]} x_{tj} \tag{6.25}$$

where for $t = 1$, the initial derivative is obtained from Equation (6.18) as follows: $\frac{\partial \mu_1^{[k]}}{\partial \beta_j^{[k]}} = \frac{\lambda_1^{[k]} x_{1j}}{1-\rho_{kk}}$, while for $t = 2, \ldots, T$, the other derivatives in Equation (6.25) are obtained iteratively.

Similarly, the derivative with respect to the dependence parameters $\{\boldsymbol{\rho}\}$ for $t = 2, \ldots, T$ is given by

$$\frac{\partial \mu_t^{[k]}}{\partial \rho_{kk}} = \mu_{t-1}^{[k]} + \rho_{kk} \frac{\partial \mu_{t-1}^{[k]}}{\partial \rho_{kk}} \tag{6.26}$$

based on the initial condition, $\frac{\partial \mu_1^{[k]}}{\partial \rho_{kk}} = \frac{\lambda_1^{[k]}}{(1-\rho_{kk})^2}$ for $k \in \{1, 2\}$.

The covariance structure $\boldsymbol{\Sigma}_{\boldsymbol{\beta},\boldsymbol{\rho}}$ is specified as

$$
\begin{array}{c|ccccccccc}
 & Y_1^{[1]} & Y_2^{[1]} & \cdots & Y_T^{[1]} & & Y_1^{[2]} & Y_2^{[2]} & \cdots & Y_T^{[2]} \\
\hline
Y_1^{[1]} & V(Y_1^{[1]}) & Cov(Y_1^{[1]}, Y_2^{[1]}) & \cdots & Cov(Y_1^{[1]}, Y_T^{[1]}) & \vdots & Cov(Y_1^{[1]}, Y_1^{[2]}) & Cov(Y_1^{[1]}, Y_2^{[2]}) & \cdots & Cov(Y_1^{[1]}, Y_T^{[2]}) \\
Y_2^{[1]} & Cov(Y_2^{[1]}, Y_1^{[1]}) & V(Y_2^{[1]}) & \cdots & Cov(Y_2^{[1]}, Y_T^{[1]}) & \vdots & Cov(Y_2^{[1]}, Y_1^{[2]}) & Cov(Y_2^{[1]}, Y_2^{[2]}) & \cdots & Cov(Y_2^{[1]}, Y_T^{[2]}) \\
\vdots & \vdots & \vdots & \ddots & \vdots & \vdots & \vdots & \vdots & \ddots & \vdots \\
Y_T^{[1]} & Cov(Y_T^{[1]}, Y_1^{[1]}) & Cov(Y_T^{[1]}, Y_2^{[1]}) & \cdots & V(Y_T^{[1]}) & \vdots & Cov(Y_T^{[1]}, Y_1^{[2]}) & Cov(Y_T^{[1]}, Y_2^{[2]}) & \cdots & Cov(Y_T^{[1]}, Y_T^{[2]}) \\
\cdots & \cdots & \cdots & \cdots & \cdots & \vdots & \cdots & \cdots & \cdots & \cdots \\
Y_1^{[2]} & Cov(Y_1^{[2]}, Y_1^{[1]}) & Cov(Y_1^{[2]}, Y_2^{[1]}) & \cdots & Cov(Y_1^{[2]}, Y_T^{[1]}) & \vdots & V(Y_1^{[2]}) & Cov(Y_1^{[2]}, Y_2^{[2]}) & \cdots & Cov(Y_1^{[2]}, Y_T^{[2]}) \\
Y_2^{[2]} & Cov(Y_2^{[2]}, Y_1^{[1]}) & Cov(Y_2^{[2]}, Y_2^{[1]}) & \cdots & Cov(Y_2^{[2]}, Y_T^{[1]}) & \vdots & Cov(Y_2^{[2]}, Y_1^{[2]}) & V(Y_2^{[2]}) & \cdots & Cov(Y_2^{[2]}, Y_T^{[2]}) \\
\vdots & \vdots & \vdots & \ddots & \vdots & \vdots & \vdots & \vdots & \ddots & \vdots \\
Y_T^{[2]} & Cov(Y_T^{[2]}, Y_1^{[1]}) & Cov(Y_T^{[2]}, Y_2^{[1]}) & \cdots & Cov(Y_T^{[2]}, Y_T^{[1]}) & \vdots & Cov(Y_T^{[2]}, Y_1^{[2]}) & Cov(Y_T^{[2]}, Y_2^{[2]}) & \cdots & V(Y_T^{[2]})
\end{array},
\tag{6.27}
$$

where the entries are computed using Equations (6.11)–(6.16). As for $\text{Cov}(Y_t^{[1]}, Y_t^{[2]})$, we compute $\text{Cov}(Y_1^{[1]}, Y_1^{[2]})$ using Equation (6.20), while for $t = 2, \ldots, T$, the cross-covariances are computed iteratively using Equation (6.14).

The Newton-Raphson technique is then used to solve the first GQL Equation (6.21) yielding an iterative scheme of the form:

$$\left(\hat{\boldsymbol{\Theta}}_{r+1}\right) = \left(\hat{\boldsymbol{\Theta}}_r\right) + [\boldsymbol{D}_{\boldsymbol{\Theta}}' \boldsymbol{\Sigma}_{\boldsymbol{\Theta}}^{-1} \boldsymbol{D}_{\boldsymbol{\Theta}}]_r^{-1} [\boldsymbol{D}_{\boldsymbol{\Theta}}' \boldsymbol{\Sigma}_{\boldsymbol{\Theta}}^{-1} (\boldsymbol{y} - \boldsymbol{\mu})]_r, \tag{6.28}$$

where $\hat{\boldsymbol{\Theta}}=[\hat{\beta}_1^{[1]}, \hat{\beta}_2^{[1]}, \hat{\beta}_1^{[2]}, \hat{\beta}_2^{[2]}, \hat{\rho_{11}}, \hat{\rho_{22}}]$ and $\hat{\boldsymbol{\Theta}}_r$ are the estimates at the $r^{th}$ iteration and $[.]_r$ are the values of the expression at the $r^{th}$ iteration.

These estimates are consistent and under mild regularity conditions, it is shown that $(\hat{\boldsymbol{\Theta}} - \boldsymbol{\Theta})'$ is asymptotically normal with mean $\boldsymbol{0}$ and covariance matrix $[\boldsymbol{D}_{\boldsymbol{\Theta}}' \boldsymbol{\Sigma}_{\boldsymbol{\Theta}}^{-1} \boldsymbol{D}_{\boldsymbol{\Theta}}]^{-1}$. The asymptotic properties are derived in a similar way as in Subsection 2.3.1.2.

### 6.3.2 GQL-II

The second GQL equation is specified as:

$$\boldsymbol{D}_{\boldsymbol{c},\boldsymbol{\nu}}' \boldsymbol{\Sigma}_{\boldsymbol{c},\boldsymbol{\nu}}^{-1} (\boldsymbol{Z} - \boldsymbol{m}) = 0. \tag{6.29}$$

In the above equation $\boldsymbol{Z} = [Z_1, Z_2, \ldots, Z_T]'_{T\times 1}$ with $Z_t = Y_t^{[1]} Y_t^{[2]}$ and $\boldsymbol{m} = [m_1, m_2, \ldots, m_T]'_{T\times 1}$ with $m_t = E(Z_t | y_{t-1}^{[1]} y_{t-1}^{[2]}) = c_1 c_2 \nu \lambda_t^{[1]} \lambda_t^{[2]} + (\rho_{11} Y_{t-1}^{[1]} + \lambda_t^{[1]})(\rho_{22} Y_{t-1}^{[2]} + \lambda_t^{[2]})$.

Based on the above defined random variable $Z_t$, the entries of the conditional covariance matrix $\boldsymbol{\Sigma}_{\boldsymbol{c},\nu}$ consist of the conditional variances $\text{Var}(Z_t | y_{t-1}^{[1]}, y_{t-1}^{[2]})$ and the conditional covariances $\text{Cov}(Z_t, Z_{t+h} | y_{t-1}^{[1]}, y_{t-1}^{[2]}, y_{t+h-1}^{[1]}, y_{t+h-1}^{[2]})$, that is

$$\begin{pmatrix} \text{Var}(Z_1 | y_0^{[1]}, y_0^{[2]}) & \ldots & \text{Cov}(Z_1, Z_T | y_0^{[1]}, y_0^{[2]}, y_{T-1}^{[1]}, y_{T-1}^{[2]}) \\ \vdots & \ddots & \vdots \\ \text{Cov}(Z_1, Z_T | y_0^{[1]}, y_0^{[2]}, y_{T-1}^{[1]}, y_{T-1}^{[2]}) & \ldots & \text{Var}(Z_T | y_{T-1}^{[1]}, y_{T-1}^{[2]}) \end{pmatrix} \tag{6.30}$$

The derivation of these entries make use of the following lemmas. By denoting $W_{kk,t} = \alpha_{kk,t} \circ Y_{t-1}^{[k]}$ for $k \in \{1, 2\}$, we prove the following lemmas:

**Lemma 6.2.** *The conditional covariances* $\text{Cov}(Z_t, Z_{t+h} | y_{t-1}^{[1]}, y_{t-1}^{[2]}, y_{t+h-1}^{[1]}, y_{t+h-1}^{[2]})$ *are* 0 *for* $h > 0$.

*Proof: For given* $Y_{t-1}^{[1]}$, $Y_{t-1}^{[2]}$, $Y_{t+h-1}^{[1]}$ *and* $Y_{t+h-1}^{[2]}$, *the random variables* $Z_t$ *and* $Z_{t+h}$ *are independent random variables which proves the lemma.*

Thus, we only need to derive the conditional variances $\mathrm{Var}(Z_t|y_{t-1}^{[1]}, y_{t-1}^{[2]})$. They can be derived by using the results presented in the following lemmas.

**Lemma 6.3.**

(a) $\mathrm{Cov}(W_{11,t}W_{22,t}, R_t^{[1]}R_t^{[2]}|y_{t-1}^{[1]}, y_{t-1}^{[2]}) = 0$.

(b) $\mathrm{Cov}(W_{kk,t}R_t^{[j]}, W_{jj,t}R_t^{[k]}|y_{t-1}^{[1]}, y_{t-1}^{[2]}) = \nu\prod_{i=1}^{2}\rho_{ii}c_i\lambda_t^{[i]}Y_{t-1}^{[i]}$, for $j,k \in \{1,2\}$ and $j \neq k$.

(c) $\mathrm{Cov}(W_{kk,t}R_t^{[j]}, R_t^{[1]}R_t^{[2]}|y_{t-1}^{[1]}, y_{t-1}^{[2]}) = \rho_{kk}Y_{t-1}^{[k]}\lambda_t^{[1]}\lambda_t^{[2]}(1+\nu c_1c_2+c_j\lambda_t^{[j]}+\nu c_1c_2\lambda_t^{[j]}+2\nu c_j^2c_k\lambda_t^{[j]})$ for $j,k \in \{1,2\}$ and $j \neq k$.

(d) $\mathrm{Cov}(W_{11,t}W_{22,t}, W_{kk,t}R_t^{[j]}|y_{t-1}^{[1]}, y_{t-1}^{[2]}) = \frac{\rho_{jj}\rho_{kk}(1-\rho_{kk})}{1+c_k}\lambda_t^{[j]}Y_{t-1}^{[1]}Y_{t-1}^{[2]}(1+c_kY_{t-1}^{[k]})$ for $j,k \in \{1,2\}$ and $j \neq k$.

*Proof:*

(a) *For given* $Y_{t-1}^{[1]}$ *and* $Y_{t-1}^{[2]}$, *the product* $W_{11,t}W_{22,t}$ *is independent of the product* $R_t^{[1]}R_t^{[2]}$, *which implies that the corresponding covariance is 0.*

(b) *For given* $Y_{t-1}^{[1]}$ *and* $Y_{t-1}^{[2]}$, *the random variables* $W_{kk,t}$ *and* $W_{jj,t}$ *are mutually independent random variables independent of the random variables* $R_t^{[1]}$ *and* $R_t^{[2]}$. *Using this and the property of the binomial thinning operator that* $E(\alpha_{kk,t} \circ Y_{t-1}^{[k]}|Y_{t-1}^{[k]}) = \rho_{kk}Y_{t-1}^{[k]}$, *we have that*

$$\begin{aligned}\mathrm{Cov}(W_{kk,t}R_t^{[j]}, W_{jj,t}R_t^{[k]}|y_{t-1}^{[1]}, y_{t-1}^{[2]}) &= \prod_{i=1}^{2}E(W_{ii,t}|y_{t-1}^{[i]})\mathrm{Cov}(R_t^{[1]}, R_t^{[2]}) \\ &= \nu\prod_{i=1}^{2}\rho_{ii}c_i\lambda_t^{[i]}Y_{t-1}^{[i]}.\end{aligned}$$

(c) *The proof follows from the conditional independency of the random variables* $W_{kk,t}$ *and* $R_t^{[1]}$ *and* $R_t^{[2]}$ *for given* $Y_{t-1}^{[1]}$ *and* $Y_{t-1}^{[2]}$, *and the fact that the covariance between the random variables* $R_t^{[j]}$ *and* $R_t^{[1]}R_t^{[2]}$ *is given by*

$$\mathrm{Cov}(R_t^{[j]}, R_t^{[1]}R_t^{[2]}) = \lambda_t^{[1]}\lambda_t^{[2]}(1+\nu c_1c_2+c_j\lambda_t^{[j]}+\nu c_1c_2\lambda_t^{[j]}+2\nu c_j^2c_k\lambda_t^{[j]}).$$

(d) *From the conditional independency of the random variables* $W_{11,t}$, $W_{22,t}$ *and* $R_t^{[j]}$ *for given* $Y_{t-1}^{[1]}$ *and* $Y_{t-1}^{[2]}$, *we have first that*

$$\begin{aligned}&\mathrm{Cov}(W_{kk,t}R_t^{[j]}, W_{jj,t}R_t^{[k]}|y_{t-1}^{[1]}, y_{t-1}^{[2]}) \\ &= E(W_{jj,t}|y_{t-1}^{[j]})E(R_t^{[j]})\mathrm{Var}(W_{kk,t}|y_{t-1}^{[k]}) \\ &= \rho_{jj}\lambda_t^{[j]}Y_{t-1}^{[j]}\mathrm{Var}(W_{kk,t}|Y_{t-1}^{[k]}).\end{aligned}$$

*To derive the conditional variance* $\text{Var}(W_{kk,t}|y_{t-1}^{[k]})$, *we first derive the conditional second-order moment* $E(W_{kk,t}^2|y_{t-1}^{[k]})$. *We have that*

$$\begin{aligned} E(W_{kk,t}^2|y_{t-1}^{[k]}) &= E(E(W_{kk,t}^2|\alpha_{kk,t}, Y_{t-1}^{[k]})|Y_{t-1}^{[k]}) \\ &= E((\alpha_{kk,t} - \alpha_{kk,t}^2)Y_{t-1}^{[k]} + \alpha_{kk,t}^2 {Y_{t-1}^{[k]}}^2 |Y_{t-1}^{[k]}) \\ &= \frac{\rho_{kk}(1-\rho_{kk})}{1+c_k} Y_{t-1}^{[k]} + \frac{\rho_{kk}(c_k+\rho_{kk})}{1+c_k} {Y_{t-1}^{[k]}}^2. \end{aligned}$$

*Using this result we obtain that the conditional variance is given by* $\text{Var}(W_{kk,t}|y_{t-1}^{[k]}) = \frac{\rho_{kk}(1-\rho_{kk})}{1+c_k} Y_{t-1}^{[k]}(1 + c_k Y_{t-1}^{[k]})$. *Replacing this result in the above expression for the conditional covariance we prove this last statement.*

**Lemma 6.4.**

(a) $\text{Var}(W_{kk,t}R_t^{[j]}|y_{t-1}^{[k]}) = \lambda_t^{[j]}(1+\lambda_t^{[j]}+c_j\lambda_t^{[j]})[\frac{\rho_{kk}(1-\rho_{kk})}{1+c_k} Y_{t-1}^{[k]} + \frac{\rho_{kk}(c_k+\rho_{kk})}{1+c_k} {Y_{t-1}^{[k]}}^2] - \rho_{kk}^2 {\lambda_t^{[j]}}^2 {Y_{t-1}^{[k]}}^2$ *for* $k, j \in \{1,2\}$ *and* $k \neq j$.

(b) $\text{Var}(W_{11,t}W_{22,t}|y_{t-1}^{[1]}, y_{t-1}^{[2]}) = \prod_{i=1}^2 [\frac{\rho_{ii}(1-\rho_{ii})}{1+c_i} Y_{t-1}^{[i]} + \frac{\rho_{ii}(c_i+\rho_{ii})}{1+c_i} {Y_{t-1}^{[i]}}^2] - \prod_{i=1}^2 \rho_{ii}^2 {Y_{t-1}^{[i]}}^2$.

(c)

$$\begin{aligned} \text{Var}(R_t^{[1]}R_t^{[2]}) = (&c_1 + c_2 + c_1c_2 + 2c_1c_2\nu + 4\nu c_1^2 c_2 + 4\nu c_1 c_2^2 \\ &+ 6\nu c_1^2 c_2^2 + c_1^2 c_2^2 \nu^2) {\lambda_t^{[1]}}^2 {\lambda_t^{[2]}}^2 + 2(1 + 2\nu c_1 c_2) \\ &\times \lambda_t^{[1]}\lambda_t^{[2]} \prod_{i=1}^2 (1+c_i)\lambda_t^{[i]} + 2(1+\nu c_1 c_2)\lambda_t^{[1]}\lambda_t^{[2]}. \end{aligned}$$

*Proof:*

(a) *Using the results from the previous lemma we have that*

$$\begin{aligned} \text{Var}(W_{kk,t}R_t^{[j]}|y_{t-1}^{[k]}) &= E(W_{kk,t}^2|y_{t-1}^{[k]})E({R_t^{[j]}}^2) \\ &\quad - \left(E(W_{kk,t}|y_{t-1}^{[k]})E(R_t^{[j]})\right)^2 \\ &= \lambda_t^{[j]}(1+\lambda_t^{[j]}+c_j\lambda_t^{[j]}) \\ &\quad \times \left[\frac{\rho_{kk}(1-\rho_{kk})}{1+c_k} Y_{t-1}^{[k]} + \frac{\rho_{kk}(c_k+\rho_{kk})}{1+c_k} {Y_{t-1}^{[k]}}^2\right] \\ &\quad - \rho_{kk}^2 {\lambda_t^{[j]}}^2 {Y_{t-1}^{[k]}}^2. \end{aligned}$$

(b) *This statement follows from the conditional independency of the random variables* $W_{11,t}$ *and* $W_{22,t}$ *for given* $Y_{t-1}^{[1]}$ *and* $Y_{t-1}^{[2]}$.

(c) *Follows from the definition of the BVNB distribution of* Ng et al. (2010), *where the pgf* (6.5) *is differentiated repeatedly to obtain* $\mathrm{E}({R_t^{[1]}}^2, {R_t^{[2]}}^2)$.

Now we are able to derive the conditional variance of the random variable $Z_t$ for given $Y_{t-1}^{[1]}$ and $Y_{t-1}^{[2]}$. Using the results given in Lemmas 6.2, 6.3 and 6.4, the conditional variance is given by

$$
\begin{aligned}
&\mathrm{Var}(Z_t|y_{t-1}^{[1]}, y_{t-1}^{[2]})\\
&= \prod_{i=1}^{2} \frac{\rho_{ii}}{1+c_i}\left[c_1c_2 + \sum_{j=1}^{2} c_j\rho_{j^*j^*} - (c_1+c_2+c_1c_2)\rho_{11}\rho_{22}\right] {Y_{t-1}^{[1]}}^2 {Y_{t-1}^{[2]}}^2\\
&+ \prod_{i=1}^{2} \frac{\rho_{ii}}{1+c_i}\sum_{j=1}^{2}\Big[c_j + \rho_{jj} + 2c_j\lambda_t^{[j^*]} + 2c_1c_2\lambda_t^{[j^*]} - 2c_j\rho_{jj}\lambda_t^{[j^*]}\\
&- 2c_1c_2\rho_{jj}\lambda_t^{[j^*]} - c_j\rho_{j^*j^*} - \rho_{11}\rho_{22}\Big] Y_{t-1}^{[1]}Y_{t-1}^{[2]}Y_{t-1}^{[j]}\\
&+ \prod_{i=1}^{2} \frac{\rho_{ii}}{1+c_i}\Bigg[1 + 2\sum_{j=1}^{2}\lambda_t^{[j]}(1+c_j)(1-\rho_{j^*j^*})\\
&+ 2\nu c_1c_2\lambda_t^{[1]}\lambda_t^{[2]}(1+c_1)(1+c_2) - \sum_{j=1}^{2}\rho_{jj} + \rho_{11}\rho_{22}\Bigg] Y_{t-1}^{[1]}Y_{t-1}^{[2]}\\
&+ \sum_{j=1}^{2}\frac{\rho_{jj}\lambda_t^{[j^*]}}{1+c_j}\left[c_j(1+\lambda_t^{[j^*]} + c_{j^*}\lambda_t^{[j^*]}) + \rho_{jj}(1+(c_{j*}-c_j)\lambda_t^{[j^*]})\right] {Y_{t-1}^{[j]}}^2\\
&+ \sum_{j=1}^{2}\frac{\rho_{jj}\lambda_t^{[j^*]}}{1+c_j}[1 + 2(1+c_j)\lambda_t^{[j]} + \lambda_t^{[j^*]}(1+c_j\lambda_t^{[j^*]})\\
&- \rho_{jj}(1+\lambda_t^{[j^*]} + c_{j*}\lambda_t^{[j^*]}) + 2\nu c_1c_2\lambda_t^{[j]}(1+c_j) + 2c_j(1+c_j){\lambda_t^{[j]}}^2\\
&+ 2\nu c_1c_2{\lambda_t^{[j]}}^2(1+3c_j+4c_j^2)]Y_{t-1}^{[j]} + \mathrm{Var}(R_t^{[1]}R_t^{[1]}).
\end{aligned}
$$

The derivative matrix $\boldsymbol{D}'_{c,\nu}$ has the following components:

1. $\frac{\partial E(Y_t^{[1]}Y_t^{[2]}|Y_{t-1}^{[1]},Y_{t-1}^{[2]})}{\partial \nu} = c_1c_2\lambda_t^{[1]}\lambda_t^{[2]}.$

2. $\frac{\partial E(Y_t^{[1]}Y_t^{[2]}|Y_{t-1}^{[1]},Y_{t-1}^{[2]})}{\partial c_1} = \nu c_2\lambda_t^{[1]}\lambda_t^{[2]}.$

3. $\frac{\partial E(Y_t^{[1]}Y_t^{[2]}|Y_{t-1}^{[1]},Y_{t-1}^{[2]})}{\partial c_2} = \nu c_1\lambda_t^{[1]}\lambda_t^{[2]}.$

The Newton-Raphson iteration yields

$$
\left(\hat{\boldsymbol{\psi}}_{r+1}\right) = \left(\hat{\boldsymbol{\psi}}_{r}\right) + [\boldsymbol{D_\psi}'\boldsymbol{\Sigma_\psi}^{-1}\boldsymbol{D_\psi}]_r^{-1}[\boldsymbol{D_\psi}'\boldsymbol{\Sigma_\psi}^{-1}(\boldsymbol{Z}-\boldsymbol{m})]_r, \quad (6.31)
$$

where $\hat{\boldsymbol{\psi}} = [c_1, c_2, \nu]$ and $\hat{\boldsymbol{\psi}}_r$ are the estimates at the $r^{th}$ iteration and $[.]_r$ are the values of the expression at the $r^{th}$ iteration.

The estimation algorithm works as follows: For an initial estimate of the regression vector, dependence and over-dispersion parameters, the first GQL Equation (6.21) is solved iteratively until convergence to obtain updated regression vectors and dependence parameters. Using the converged estimates of the regression and dependence parameters from the first GQL, the second GQL Equation (6.29) is then solved iteratively to obtain new values of the over-dispersion coefficients using the regression and dependence estimates obtained from the first GQL equation. After obtaining the updated over-dispersion values, these are in turn substituted again in the first GQL Equation (6.21) to obtain another updated regression vectors and dependence parameters and thereafter the cycle of iteration continues until convergence of all sets of parameters with their corresponding standard errors obtained from the respective Hessian structures. These estimates are consistent and under mild regularity conditions, it is shown that $(\hat{\boldsymbol{\psi}} - \boldsymbol{\psi})'$ is asymptotically normal with mean $\mathbf{0}$ and covariance matrix $[\boldsymbol{D}_{\boldsymbol{\psi}}{}' \boldsymbol{\Sigma}_{\boldsymbol{\psi}}{}^{-1} \boldsymbol{D}_{\boldsymbol{\psi}}]^{-1}$. (Refer to Subsection 2.3.1.2).

Using the converged estimates from the GQL method, we derive the forecasting equations of the constrained BINAR(1) model (6.1). Following Sutradhar (2008), the conditional expectation and variance of the one-step ahead forecast $Y_{t+1}^{[k]}$ given $y_t^{[k]}$ are as follows:

$$
\begin{aligned}
E(Y_{t+1}^{[k]}|y_t^{[k]}) &= E(\alpha_{kk,t} \circ Y_t^{[k]}|y_t^{[k]}) + E(R_{t+1}^{[k]}) \\
&= \rho_{kk} y_t^{[k]} + \lambda_{t+1}^{[k]}
\end{aligned}
\tag{6.32}
$$

and

$$
\begin{aligned}
&\mathrm{Var}(Y_{t+1}^{[k]}|y_t^{[k]}) \\
&= \mathrm{Var}(\alpha_{kk,t} \circ Y_t^{[k]}|y_t^{[k]}) + \mathrm{Var}(R_{t+1}^{[k]}) \\
&= E_{\alpha_{kk,t}}[\mathrm{Var}(\alpha_{kk,t} \circ Y_t^{[k]}|y_t^{[k]}, \alpha_{kk,t})] \\
&\quad + \mathrm{Var}_{\alpha_{kk,t}}[E(\alpha_{kk,t} \circ Y_t^{[k]}|y_t^{[k]}, \alpha_{kk,t})] + \mathrm{Var}(R_{t+1}^{[k]}) \\
&= E_{\alpha_{kk,t}}\left[\alpha_{kk,t}(1-\alpha_{kk,t})y_t^{[k]}\right] + \mathrm{Var}_{\alpha_{kk,t}}[\alpha_{kk,t} y_t^{[k]}] + \mathrm{Var}(R_{t+1}^{[k]}) \\
&= \left[\rho_{kk} - (\frac{\rho_{kk}(1-\rho_{kk})c_k}{1+c_k} + \rho_{kk}^2)\right] y_t^{[k]} + \frac{\rho_{kk}(1-\rho_{kk})c_k}{1+c_k}(y_t^{[k]})^2 \\
&\quad + \left[\lambda_{t+1}^{[k]} + c_k(\lambda_{t+1}^{[k]})^2\right] \\
&= \frac{\rho_{kk}(1-\rho_{kk})}{1+c_k} y_t^{[k]} + \frac{\rho_{kk}(1-\rho_{kk})c_k}{1+c_k}(y_t^{[k]})^2 + [\lambda_{t+1}^{[k]} + c_k(\lambda_{t+1}^{[k]})^2] \\
&= \frac{\rho_{kk}(1-\rho_{kk})}{1+c_k} y_t^{[k]}(1 + c_k y_t^{[k]}) + \left[\lambda_{t+1}^{[k]} + c_k(\lambda_{t+1}^{[k]})^2\right].
\end{aligned}
\tag{6.33}
$$

## 6.4 Numerical Experiments

This section focuses on generating non-stationary BINAR(1) series of counts using the OD Equation (6.1) for $T = 60, 100, 600$. The innovations link predictor $\lambda_t^{[k]} = \exp(\boldsymbol{x}_t^{'}\boldsymbol{\beta}^{[k]})$, for $k \in \{1, 2\}$ is calculated assuming a $(2 \times 1)$ covariate structure where

$$x_{t1} = \begin{cases} -\cos(2\pi t) + 0.01 & (t = 1, \ldots, T/4) \\ \sin(2\pi t) + 0.05 & (t = (T/4) + 1, \ldots, 3T/4) \\ \cos(2\pi t) + 0.10 & (t = (3T/4) + 1, \ldots, T) \end{cases}$$

$$x_{t2} = \begin{cases} t + 0.02 & (t = 1, \ldots, T/2) \\ t - 0.02 & (t = (T/2) + 1, \ldots, T), \end{cases}$$

where $\beta_1^{[1]} = 0.5, \beta_2^{[1]} = 1$ and $\beta_1^{[2]} = 0.3, \beta_2^{[2]} = 1.5$. As for $c_1$ and $c_2$, we consider the following combinations: $[c_1, c_2] = [0.3, 0.7], [0.9, 1.5], [0.2, 0.5]$ and $[1.2, 1.9]$. The serial parameters $[\rho_{11}, \rho_{22}]$ are assumed $[\rho_{11}, \rho_{22}] = [0.9, 0.9]$, $[0.3, 0.9]$ and $[0.3, 0.3]$. The generation of $[R_t^{[1]}, R_t^{[2]}]'$ follows the same algorithm as in Ng et al. (2010) where:

1. $\theta_1 = \frac{c_1 \lambda_t^{[1]}}{1 + c_1 \lambda_t^{[1]} + c_2 \lambda_t^{[2]}}$, $\theta_2 = \frac{c_2 \lambda_t^{[2]}}{1 + c_1 \lambda_t^{[1]} + c_2 \lambda_t^{[2]}}$.

2. Simulate $U_1$ and $U_2$ where $U_k$ follows NB with size parameter $(c_k^{-1} - \nu)$ and probability $\theta_k$ for $k \in \{1, 2\}$.

3. Using Ng et al. (2010) algorithm, we generate the innovation terms $r_t^{[1]}$ and $r_t^{[2]}$ as follows:
   Let $S_k$ follows NB with size parameter $\nu$ and probability $\theta_k$ for $k \in \{1, 2\}$ and $\mathrm{Corr}(S_1, S_2) = \frac{\lambda_t^{[1]} \lambda_t^{[2]} \sqrt{c_1 c_2}}{\sqrt{\lambda_t^{[1]} + c_1 \lambda_t^{[1]2}} \sqrt{\lambda_t^{[2]} + c_2 \lambda_t^{[2]2}}}$, we obtain $r_t^{[1]} = u_1 + s_1$ and $r_t^{[2]} = u_2 + s_2$.

Using the values of $r_t^{[1]}$ and $r_t^{[2]}$, $\alpha_{11,t} \sim \mathrm{Beta}(\frac{\rho_{11}}{c_1}, \frac{1-\rho_{11}}{c_1})$, $\alpha_{22,t} \sim \mathrm{Beta}(\frac{\rho_{22}}{c_2}, \frac{1-\rho_{22}}{c_2})$ and $Y_{t-1}^{[k]} \sim \mathrm{NB}(\frac{1}{c_k^*}, c_k^* \mu_{t-1}^{[k]})$, the $y_t^{[1]}$ and $y_t^{[2]}$ values are generated from Equation (6.1) for $t = 1, 2, \ldots, T$. Note from Equation (6.5) and the combinations of $[c_1, c_2]$, the values of $\nu = 0.5, 0.6, 1.4, 1.9$ for the corresponding combinations $[1.2, 1.9], [0.9, 1.5], [0.3, 0.7]$ and $[0.2, 0.5]$.

Thereon, for the above combinations, 5000 simulated runs are made. Tables 6.1–6.4 display the simulated mean estimates of the different parameters under the GQL Equations (6.21) and (6.29):

**TABLE 6.1**
Estimates of the Parameters and Standard Errors under Non-Stationary BINAR(1) Process for $[c_1, c_2] = [0.3, 0.7]$

| $\rho_{11}$ | $\rho_{22}$ | T | $\hat{\beta}_1^{[1]}$ | $\hat{\beta}_2^{[1]}$ | $\hat{\beta}_1^{[2]}$ | $\hat{\beta}_2^{[2]}$ | $\hat{c}_1 = 0.3$ | $\hat{c}_2 = 0.7$ | $\hat{\rho}_{11}$ | $\hat{\rho}_{22}$ | $\hat{\nu} = 1.4$ |
|---|---|---|---|---|---|---|---|---|---|---|---|
| 0.9 | 0.9 | 60 | 0.5189 | 0.9857 | 0.3175 | 1.4821 | 0.3140 | 0.6864 | 0.8887 | 0.8817 | 1.3868 |
| | | | (0.1167) | (0.1238) | (0.1062) | (0.1220) | (0.1454) | (0.1314) | (0.0982) | (0.0911) | (0.1361) |
| | | 100 | 0.4931 | 1.0092 | 0.2916 | 1.5093 | 0.2928 | 0.6922 | 0.8949 | 0.8939 | 1.3915 |
| | | | (0.0939) | (0.0857) | (0.1020) | (0.0826) | (0.1214) | (0.1116) | (0.0736) | (0.0849) | (0.1120) |
| | | 600 | 0.4976 | 0.9964 | 0.2958 | 1.4990 | 0.2972 | 0.6966 | 0.8964 | 0.8958 | 1.3970 |
| | | | (0.0409) | (0.0560) | (0.0467) | (0.0250) | (0.0894) | (0.0511) | (0.0228) | (0.0212) | (0.06817) |
| 0.3 | 0.9 | 60 | 0.5192 | 0.9853 | 0.2831 | 1.4828 | 0.3174 | 0.6864 | 0.2845 | 0.9107 | 1.3854 |
| | | | (0.1313) | (0.1047) | (0.1236) | (0.1134) | (0.1290) | (0.1386) | (0.1088) | (0.0913) | (0.1363) |
| | | 100 | 0.4914 | 1.0092 | 0.3056 | 1.4935 | 0.2919 | 0.6935 | 0.2943 | 0.8922 | 1.3928 |
| | | | (0.0787) | (0.0822) | (0.0794) | (0.0869) | (0.1011) | (0.1234) | (0.0603) | (0.0759) | (0.1143) |
| | | 600 | 0.4952 | 1.0018 | 0.2955 | 1.4961 | 0.3014 | 0.7013 | 0.2964 | 0.8974 | 1.3942 |
| | | | (0.0383) | (0.0467) | (0.0325) | (0.0214) | (0.0675) | (0.0851) | (0.0223) | (0.0127) | (0.0729) |
| 0.3 | 0.3 | 60 | 0.5168 | 1.0156 | 0.2833 | 1.5109 | 0.2813 | 0.6830 | 0.3118 | 0.2870 | 1.3896 |
| | | | (0.1119) | (0.1036) | (0.1276) | (0.1051) | (0.1146) | (0.1332) | (0.1037) | (0.0978) | (0.1289) |
| | | 100 | 0.5043 | 0.9924 | 0.3062 | 1.4937 | 0.2940 | 0.6927 | 0.2929 | 0.2930 | 1.3948 |
| | | | (0.0752) | (0.0862) | (0.0713) | (0.0851) | (0.1002) | (0.1214) | (0.0771) | (0.0652) | (0.1085) |
| | | 600 | 0.4990 | 0.9993 | 0.2978 | 1.5015 | 0.2996 | 0.7011 | 0.2980 | 0.2964 | 1.3978 |
| | | | (0.0239) | (0.0425) | (0.0352) | (0.0416) | (0.0725) | (0.0808) | (0.0145) | (0.0193) | (0.0739) |

**TABLE 6.2**
Estimates of the Parameters and Standard Errors under Non-Stationary BINAR(1) Process for $[c_1, c_2] = [0.2, 0.5], [1.2, 1.9]$

| $\rho_{11}$ | $\rho_{22}$ | T | $\hat{\beta}_1^{[1]}$ | $\hat{\beta}_2^{[1]}$ | $\hat{\beta}_1^{[2]}$ | $\hat{\beta}_2^{[2]}$ | $\hat{c}_1$=0.2 | $\hat{c}_2$=0.5 | $\hat{\rho}_{11}$ | $\hat{\rho}_{22}$ | $\hat{\nu} = 1.9$ |
|---|---|---|---|---|---|---|---|---|---|---|---|
| 0.9 | 0.9 | 60 | 0.5120 | 0.9835 | 0.2899 | 1.4815 | 0.2157 | 0.5199 | 0.8810 | 0.9115 | 1.8855 |
| | | | (0.0920) | (0.1008) | (0.0915) | (0.0940) | (0.1014) | (0.1346) | (0.0909) | (0.0829) | (0.1441) |
| | | 100 | 0.4966 | 0.9934 | 0.2991 | 1.5095 | 0.2068 | 0.4920 | 0.8974 | 0.8973 | 1.9090 |
| | | | (0.0812) | (0.0908) | (0.0882) | (0.0869) | (0.0948) | (0.1225) | (0.0720) | (0.0697) | (0.1264) |
| | | 600 | 0.5019 | 0.9966 | 0.2995 | 1.5012 | 0.2015 | 0.4972 | 0.8985 | 0.8984 | 1.8980 |
| | | | (0.0444) | (0.0193) | (0.0411) | (0.0215) | (0.0653) | (0.0795) | (0.0140) | (0.0294) | (0.0725) |
| 0.3 | 0.9 | 60 | 0.5190 | 0.9890 | 0.3143 | 1.4862 | 0.1866 | 0.4826 | 0.2852 | 0.8857 | 1.8853 |
| | | | (0.0826) | (0.0831) | (0.0909) | (0.0906) | (0.0928) | (0.1117) | (0.0813) | (0.0869) | (0.1158) |
| | | 100 | 0.5064 | 0.9927 | 0.2934 | 1.4918 | 0.1902 | 0.5088 | 0.2939 | 0.8912 | 1.8955 |
| | | | (0.0722) | (0.0739) | (0.0812) | (0.0764) | (0.0892) | (0.1047) | (0.0640) | (0.0694) | (0.0914) |
| | | 600 | 0.4972 | 0.9978 | 0.2965 | 1.4979 | 0.1981 | 0.4959 | 0.3011 | 0.8994 | 1.8975 |
| | | | (0.0335) | (0.0317) | (0.0450) | (0.0297) | (0.0586) | (0.0643) | (0.0220) | (0.0112) | (0.0552) |
| 0.3 | 0.3 | 60 | 0.4866 | 0.9828 | 0.2818 | 1.4878 | 0.1886 | 0.5186 | 0.3148 | 0.2802 | 1.8831 |
| | | | (0.0850) | (0.0906) | (0.0824) | (0.0865) | (0.0922) | (0.1160) | (0.0873) | (0.0811) | (0.1112) |
| | | 100 | 0.5080 | 0.9911 | 0.2950 | 1.4938 | 0.2093 | 0.4929 | 0.2950 | 0.2945 | 1.8952 |
| | | | (0.0631) | (0.0862) | (0.0709) | (0.0766) | (0.0883) | (0.1040) | (0.0765) | (0.0602) | (0.0916) |
| | | 600 | 0.4984 | 0.9987 | 0.3005 | 1.4951 | 0.1979 | 0.5012 | 0.2971 | 0.2964 | 1.8981 |
| | | | (0.0247) | (0.0187) | (0.0298) | (0.0213) | (0.0564) | (0.0741) | (0.0110) | (0.0133) | (0.0563) |
| $\rho_{11}$ | $\rho_{22}$ | T | $\hat{\beta}_1^{[1]}$ | $\hat{\beta}_2^{[1]}$ | $\hat{\beta}_1^{[2]}$ | $\hat{\beta}_2^{[2]}$ | $\hat{c}_1$=1.2 | $\hat{c}_2$=1.9 | $\hat{\rho}_{11}$ | $\hat{\rho}_{22}$ | $\hat{\nu} = 0.5$ |
| 0.9 | 0.9 | 60 | 0.5113 | 0.9818 | 0.3106 | 1.4824 | 1.1810 | 1.8889 | 0.9184 | 0.8844 | 0.4891 |
| | | | (0.1491) | (0.1076) | (0.1218) | (0.1020) | (0.1551) | (0.1323) | (0.0918) | (0.0828) | (0.1465) |
| | | 100 | 0.5074 | 0.9947 | 0.2929 | 1.5078 | 1.1910 | 1.8951 | 0.8914 | 0.8904 | 0.4918 |
| | | | (0.0761) | (0.0780) | (0.0969) | (0.0871) | (0.1376) | (0.1288) | (0.0631) | (0.0611) | (0.1220) |
| | | 600 | 0.4958 | 0.9997 | 0.2966 | 1.4959 | 1.1957 | 1.9005 | 0.9017 | 0.8998 | 0.4951 |
| | | | (0.0235) | (0.0304) | (0.0421) | (0.0415) | (0.0610) | (0.0821) | (0.0276) | (0.0188) | (0.0734) |

(*Continued*)

**TABLE 6.2**
*(Continued)*

| $\rho_{11}$ | $\rho_{22}$ | T | $\hat{\beta}_1^{[1]}$ | $\hat{\beta}_2^{[1]}$ | $\hat{\beta}_1^{[2]}$ | $\hat{\beta}_2^{[2]}$ | $\hat{c}_1$=1.2 | $\hat{c}_2$=1.9 | $\hat{\rho}_{11}$ | $\hat{\rho}_{22}$ | $\hat{\nu}=0.5$ |
|---|---|---|---|---|---|---|---|---|---|---|---|
| 0.3 | 0.9 | 60 | 0.4895 | 0.9883 | 0.3130 | 1.4871 | 1.2120 | 1.8887 | 0.3168 | 0.8877 | 0.4860 |
| | | | (0.1472) | (0.1057) | (0.1335) | (0.1013) | (0.1660) | (0.1348) | (0.0931) | (0.0936) | (0.1468) |
| | | 100 | 0.4911 | 1.0087 | 0.2934 | 1.5082 | 1.1909 | 1.8931 | 0.2928 | 0.8909 | 0.5065 |
| | | | (0.0746) | (0.0829) | (0.0904) | (0.0807) | (0.1315) | (0.1261) | (0.0713) | (0.0719) | (0.1237) |
| | | 600 | 0.5013 | 0.9946 | 0.3011 | 1.4982 | 1.1970 | 1.8980 | 0.3018 | 0.8982 | 0.4970 |
| | | | (0.0217) | (0.0451) | (0.0163) | (0.0424) | (0.0614) | (0.0760) | (0.0164) | (0.0183) | (0.0616) |
| 0.3 | 0.3 | 60 | 0.5150 | 0.9830 | 0.3181 | 1.4803 | 1.2153 | 1.8875 | 0.2830 | 0.2889 | 0.4821 |
| | | | (0.1060) | (0.1233) | (0.1038) | (0.1403) | (0.1367) | (0.1670) | (0.1062) | (0.0916) | (0.1542) |
| | | 100 | 0.4929 | 0.9911 | 0.2940 | 1.5086 | 1.1936 | 1.9069 | 0.2922 | 0.2932 | 0.5018 |
| | | | (0.0837) | (0.0944) | (0.0801) | (0.0942) | (0.1365) | (0.1236) | (0.0833) | (0.0758) | (0.1317) |
| | | 600 | 0.4978 | 0.9985 | 0.3008 | 1.5012 | 1.1982 | 1.8971 | 0.3025 | 0.2973 | 0.5011 |
| | | | (0.0124) | (0.0478) | (0.0216) | (0.0465) | (0.0971) | (0.0822) | (0.0117) | (0.0165) | (0.0878) |

**TABLE 6.3**
Estimates of the Parameters and Standard Errors under Non-Stationary BINAR(1) Process for $[c_1, c_2] = [0.9, 1.5]$

| $\rho_{11}$ | $\rho_{22}$ | T | $\hat{\beta}_1^{[1]}$ | $\hat{\beta}_2^{[1]}$ | $\hat{\beta}_1^{[2]}$ | $\hat{\beta}_2^{[2]}$ | $\hat{c}_1$=0.9 | $\hat{c}_2$=1.5 | $\hat{\rho}_{11}$ | $\hat{\rho}_{22}$ | $\hat{\nu}=0.6$ |
|---|---|---|---|---|---|---|---|---|---|---|---|
| 0.9 | 0.9 | 60 | 0.5143 | 0.9838 | 0.3118 | 1.4854 | 0.9192 | 1.4815 | 0.8809 | 0.8895 | 0.6251 |
| | | | (0.1324) | (0.1040) | (0.1423) | (0.1190) | (0.1587) | (0.1371) | (0.0986) | (0.0950) | (0.1406) |
| | | 100 | 0.4926 | 0.9923 | 0.2920 | 1.4943 | 0.9029 | 1.5088 | 0.8928 | 0.8941 | 0.5930 |
| | | | (0.0939) | (0.0826) | (0.0961) | (0.0832) | (0.1449) | (0.1315) | (0.0616) | (0.0710) | (0.1347) |
| | | 600 | 0.4964 | 0.9951 | 0.3014 | 1.4972 | 0.9021 | 1.4997 | 0.8992 | 0.8970 | 0.5956 |
| | | | (0.0483) | (0.0158) | (0.0459) | (0.0276) | (0.0826) | (0.0775) | (0.0158) | (0.0160) | (0.0816) |
| 0.3 | 0.9 | 60 | 0.4857 | 0.9844 | 0.2816 | 1.4885 | 0.9131 | 1.5198 | 0.2851 | 0.8827 | 0.5890 |
| | | | (0.1227) | (0.1063) | (0.1311) | (0.1080) | (0.1472) | (0.1315) | (0.0975) | (0.0923) | (0.1391) |
| | | 100 | 0.4946 | 0.9953 | 0.2931 | 1.4942 | 0.9096 | 1.4914 | 0.2915 | 0.8948 | 0.6081 |
| | | | (0.0850) | (0.0721) | (0.0898) | (0.0915) | (0.1327) | (0.1261) | (0.0755) | (0.0649) | (0.1288) |
| | | 600 | 0.5018 | 0.9987 | 0.2981 | 1.4972 | 0.9019 | 1.4983 | 0.2998 | 0.8980 | 0.5936 |
| | | | (0.0436) | (0.0396) | (0.0416) | (0.0266) | (0.0654) | (0.0869) | (0.0153) | (0.0181) | (0.0879) |
| 0.3 | 0.3 | 60 | 0.5110 | 0.9851 | 0.3176 | 1.4885 | 0.8854 | 1.4866 | 0.2859 | 0.2870 | 0.5852 |
| | | | (0.1193) | (0.1002) | (0.1359) | (0.1067) | (0.1417) | (0.1393) | (0.0931) | (0.0925) | (0.1360) |
| | | 100 | 0.4930 | 0.9951 | 0.2939 | 1.4932 | 0.9011 | 1.4945 | 0.2958 | 0.2912 | 0.5927 |
| | | | (0.0731) | (0.0881) | (0.0785) | (0.0894) | (0.1151) | (0.1230) | (0.0679) | (0.0686) | (0.1232) |
| | | 600 | 0.4970 | 0.9974 | 0.2984 | 1.4981 | 0.8967 | 1.4964 | 0.2978 | 0.2980 | 0.5971 |
| | | | (0.0483) | (0.0111) | (0.0426) | (0.0231) | (0.0692) | (0.0716) | (0.0168) | (0.0120) | (0.0763) |

In the above set of results, the (.) stands for the standard errors for the respective model parameters. From Tables 6.1–6.3, we deduce that the estimates of the different model parameters are consistent and as expected, the standard errors decrease as the time size increases. Table 6.4 provides details on the number of non-convergence cases for the different combinations:

**TABLE 6.4**
Number of Non-Convergent Simulations under BINAR(1) Process under Different Combinations

| $\hat{c}_1$ | $\hat{c}_2$ | $\hat{\nu}$ | $\hat{\rho}_{11}$ | $\hat{\rho}_{22}$ | T | Failures | $\hat{c}_1$ | $\hat{c}_2$ | $\hat{\nu}$ | $\hat{\rho}_{11}$ | $\hat{\rho}_{22}$ | T | Failures |
|---|---|---|---|---|---|---|---|---|---|---|---|---|---|
| 0.3 | 0.7 | 1.4 | 0.9 | 0.9 | 60 | 240 | 0.2 | 0.5 | 1.9 | 0.9 | 0.9 | 60 | 230 |
| | | | | | 100 | 215 | | | | | | 100 | 205 |
| | | | | | 600 | 150 | | | | | | 600 | 140 |
| 0.3 | 0.7 | 1.4 | 0.3 | 0.9 | 60 | 225 | 0.2 | 0.5 | 1.9 | 0.3 | 0.9 | 60 | 215 |
| | | | | | 100 | 200 | | | | | | 100 | 190 |
| | | | | | 600 | 135 | | | | | | 600 | 125 |
| 0.3 | 0.7 | 1.4 | 0.3 | 0.3 | 60 | 215 | 0.2 | 0.5 | 1.9 | 0.3 | 0.3 | 60 | 205 |
| | | | | | 100 | 190 | | | | | | 100 | 180 |
| | | | | | 600 | 125 | | | | | | 600 | 120 |
| 0.9 | 1.5 | 0.6 | 0.9 | 0.9 | 60 | 260 | 1.2 | 1.9 | 0.5 | 0.9 | 0.9 | 60 | 285 |
| | | | | | 100 | 225 | | | | | | 100 | 250 |
| | | | | | 600 | 175 | | | | | | 600 | 195 |
| 0.9 | 1.5 | 0.6 | 0.3 | 0.9 | 60 | 250 | 1.2 | 1.9 | 0.5 | 0.3 | 0.9 | 60 | 275 |
| | | | | | 100 | 230 | | | | | | 100 | 240 |
| | | | | | 600 | 160 | | | | | | 600 | 180 |
| 0.9 | 1.5 | 0.6 | 0.3 | 0.3 | 60 | 245 | 1.2 | 1.9 | 0.5 | 0.3 | 0.3 | 60 | 260 |
| | | | | | 100 | 225 | | | | | | 100 | 230 |
| | | | | | 600 | 155 | | | | | | 600 | 175 |

The following remarks are made:

(a) As $c_1$ and $c_2$ increase, we observe an increase in the number of non-convergent simulations.

(b) For high auto-correlation pair of values $[0.9, 0.9]$, we notice a slight increase of approximately 15 percent in the number of non-convergent simulations compared to low or medium auto-correlation.

(c) In some simulations, the values of $\rho_{11}, \rho_{22}$ or $\nu$ deviated by large from the population values or were out of the relevant range and hence were considered as non-convergent.

(d) The Hessian component in Equation (6.31) turned out to be rather ill-conditioned in some cases.

## 6.5 Summary

In this chapter, we have considered the modeling of a non-stationary BINAR(1) process with NB innovations wherein the innovations are jointly distributed under a novel BVNB model with different over-dispersion indices. The serial and cross auto-covariance functions have been derived. However, the challenging part of this chapter has been to estimate the set of unknown parameters that include the regression, over-dispersion and dependence effects. To enable the estimation process, two GQL estimating equations have been considered. The first GQL equation estimates the regression and serial dependence parameters using a simple mean score function while the second GQL equation estimates the over-dispersion indices. The latter GQL has led to evaluating some higher-order moment expressions which have been solved exactly using the thinning properties. The Monte Carlo simulation results have illustrated that this method yields consistent estimates of the unknown model parameters. The next chapter provides a similar analysis for the constrained BINMA(1) model with NB innovations.

# 7

# *Constrained BINMA(1) Model with Correlated NB Innovations*

## 7.1 Introduction

In Chapter 3, a constrained BINMA(1) model has been developed by directly extending two INMA(1) models such that the cross-correlation between the series is induced only by the related Poisson innovations. As for the inference procedure, the GQL and FGLS approaches have been used to estimate the unknown parameters of the model, where it has been shown in the Monte Carlo experiments that GQL provides more efficient estimates than FGLS based on the standard errors. This constrained BINMA(1) model has been extended to an unconstrained BINMA(1) model in Chapter 5 under Poisson innovations, where it has been shown that the counting series are marginally Poisson. However, these BINMA(1) models can only be used to model equi-dispersed bivariate data and hence the need to construct over-dispersed BINMA(1) models that can analyze different levels of over-dispersed data.

In fact, Quoreshi (2006a, 2008) proposed a BINMA model that could cater for such needs but in their model construction, these authors did not clearly specify the distribution of the innovations and even in their application to some real-life series on Ericsson B and AstraZeneca, the distribution of the innovations was simply assumed Poisson marginally.

Hence, this chapter considers developing a constrained BINMA(1) model under the Ng et al. (2010) BVNB representation. In particular, in Section 7.2, a constrained BINMA(1) model with NB innovations is introduced and the marginal and joint moments are derived. In Section 7.3, the GQL approach is developed together with the forecasting equations and this is followed by a simulation study where the performance of the GQL is monitored. The conclusion is presented in the last section.

DOI: 10.1201/9781003600435-7

## 7.2 The Construction of the Constrained BINMA(1) Model

The random coefficient BINMA(1) model $\{[Y_t^{[1]}, Y_t^{[2]}]', \ t \in \mathbb{Z}\}$ is specified as:

$$\begin{pmatrix} Y_t^{[1]} \\ Y_t^{[2]} \end{pmatrix} = \begin{pmatrix} \alpha_{11,t} & 0 \\ 0 & \alpha_{22,t} \end{pmatrix} \circ \begin{pmatrix} R_{t-1}^{[1]} \\ R_{t-1}^{[2]} \end{pmatrix} + \begin{pmatrix} R_t^{[1]} \\ R_t^{[2]} \end{pmatrix}, \tag{7.1}$$

where $Y_t^{[k]}$ is the random counting variable for the $k^{th}$ series while $R_t^{[k]}$ is the corresponding randomly distributed innovation term. The sequence $R_t^{[k]}$ is identically and independently distributed such that $\text{Corr}(R_t^{[k]}, R_{t+h}^{[k]}) = 0$ for $h \neq 0$, $k \in \{1, 2\}$. The paired innovation terms $(R_t^{[1]}, R_t^{[2]})$ follows the BVNB model developed by Ng et al. (2010) where $R_t^{[1]}$ and $R_t^{[2]}$ exhibit different levels of over-dispersion. The pgf of the BVNB is specified as in Equation (6.5) and marginally $R_t^{[k]} \sim \text{NB}(1/c_k, c_k\lambda_t^{[k]})$, with pgf similar to Equation (6.6). The link predictor function of the innovation term, $\lambda_t^{[k]} = \exp(\boldsymbol{x}_t'\boldsymbol{\beta}^{[k]})$, for $\boldsymbol{x}_t = [x_{t1}, x_{t2}, \ldots, x_{tj}, \ldots x_{tp}]'$, while the vector of regression parameters is represented as $\boldsymbol{\beta}^{[k]} = [\beta_1^{[k]}, \beta_2^{[k]}, \ldots, \beta_j^{[k]}, \ldots, \beta_p^{[k]}]'$ for $k \in \{1, 2\}$.
In the above,

$$\text{Cov}(R_t^{[1]}, R_{t+h}^{[2]}) = \begin{cases} c_1 c_2 \nu \lambda_t^{[1]} \lambda_t^{[2]} & h = 0 \\ 0 & h \neq 0. \end{cases} \tag{7.2}$$

Some other assumptions of the BINMA(1) model (7.1) are:

(a) $\{\alpha_{11,t}\}$ and $\{\alpha_{22,t}\}$ are independent random variables where $\{\alpha_{kk,t}\}$ is identically Beta distributed such that $\alpha_{kk,t} \sim \text{Beta}\left(\frac{\rho_{kk}}{c_k}, \frac{1-\rho_{kk}}{c_k}\right)$, for $k \in \{1, 2\}$ and $\rho_{kk} \in [0, 1]$. Thus, $E(\alpha_{kk,t}) = \rho_{kk}$ and $\text{Var}(\alpha_{kk,t}) = c_k\rho_{kk}(1 - \rho_{kk})/(1 + c_k)$ for $k \in \{1, 2\}$.

(b) $\alpha_{kk,t}$ and $R_{t-1}^{[k]}$ are independent random variables for all $k \in \{1, 2\}$ and $t \in \mathbb{Z}$.

(c) '$\circ$' is the binomial thinning operator such that $\{\alpha_{kk,t} \circ R_{t-1}^{[k]}\}$ for given $R_{t-1}^{[k]}$ and $\alpha_{kk,t}$ has the binomial distribution with parameters $R_{t-1}^{[k]}$ and $\alpha_{kk,t}$, for $k \in \{1, 2\}$.
From (b) and (c), $E(\alpha_{kk,t} \circ R_{t-1}^{[k]}) = \rho_{kk} E(R_{t-1}^{[k]})$.

(d) The random variables $R_t^{[k]}$ and $R_u^{[j]}$ are independent for $t \neq u$ and for $k, j \in \{1, 2\}$.

**Lemma 7.1.** *Suppose that the conditions (a), (b), (c) and (d) hold. Then,*

(i) (a) $\text{Var}(\alpha_{kk,t} \circ R_{t-1}^{[k]} | R_{t-1}^{[k]}, \alpha_{kk,t}) = \alpha_{kk,t}(1-\alpha_{kk,t})R_{t-1}^{[k]}, k \in \{1,2\}$.

(b)
$$\text{Var}(\alpha_{kk,t} \circ R_{t-1}^{[k]}) = \frac{\rho_{kk}(c_k+\rho_{kk})}{1+c_k} E({R_{t-1}^{[k]}}^2) - \rho_{kk}^2 \left(E(R_{t-1}^{[k]})\right)^2 + \frac{\rho_{kk}(1-\rho_{kk})}{1+c_k} E(R_{t-1}^{[k]}), k \in \{1,2\}.$$

(ii) (a) $\text{Cov}(Y_t^{[k]}, \alpha_{jj,t+h+1} \circ R_{t+h}^{[j]}) = \rho_{jj}\text{Cov}(Y_t^{[k]}, R_{t+h}^{[j]})$.

(b) $\text{Cov}(\alpha_{kk,t+1} \circ R_t^{[k]}, \alpha_{jj,t+1} \circ R_{t+h}^{[j]}) = \rho_{jj}\rho_{kk}\text{Cov}(R_t^{[k]}, R_{t+h}^{[j]})$, *for* $k,j \in \{1,2\}$ *and* $h \geq 0$.

(iii) (a) $\text{Cov}(Y_t^{[k]}, R_t^{[k]}) = \text{Var}(R_t^{[k]}), k \in \{1,2\}$.

(b) $\text{Cov}(Y_t^{[k]}, R_{t+h}^{[j]})$
$$= \begin{cases} \text{Cov}(R_t^{[k]}, R_t^{[j]}) & h = 0 \\ 0 & h \neq 0, \quad \textit{for any} \quad k,j \in \{1,2\}, k \neq j. \end{cases}$$

*Proof:*

(i) (a) *The proof follows from the fact that the random variable* $\{\alpha_{kk,t} \circ R_{t-1}^{[k]} | R_{t-1}^{[k]}, \alpha_{kk,t}\} \sim \text{Binomial}(R_{t-1}^{[k]}, \alpha_{kk,t})$.

(b)

$$\begin{aligned}
&\text{Var}(\alpha_{kk,t} \circ R_{t-1}^{[k]}) \\
&= \text{Var}[E(\alpha_{kk,t} \circ R_{t-1}^{[k]} | R_{t-1}^{[k]}, \alpha_{kk,t})] + E[\text{Var}(\alpha_{kk,t} \circ R_{t-1}^{[k]} | R_{t-1}^{[k]}, \alpha_{kk,t})] \\
&= \text{Var}[\alpha_{kk,t} R_{t-1}^{[k]}] + E[\alpha_{kk,t}(1-\alpha_{kk,t})R_{t-1}^{[k]}] \\
&= E(\alpha_{kk,t}^2 {R_{t-1}^{[k]}}^2) - E(\alpha_{kk,t})^2 E(R_{t-1}^{[k]})^2 + [E(\alpha_{kk,t}) - E(\alpha_{kk,t}^2)] E(R_{t-1}^{[k]}) \\
&= E(\alpha_{kk,t}^2) E({R_{t-1}^{[k]}}^2) - E(\alpha_{kk,t})^2 E(R_{t-1}^{[k]})^2 + E(\alpha_{kk,t}) E(R_{t-1}^{[k]}) \\
&- E(\alpha_{kk,t}^2) E(R_{t-1}^{[k]}) \\
&= \left[\frac{\rho_{kk}(1-\rho_{kk})c_k}{1+c_k} + \rho_{kk}^2\right] E({R_{t-1}^{[k]}}^2) - \rho_{kk}^2 E(R_{t-1}^{[k]})^2 \\
&+ \left[\rho_{kk} - \frac{\rho_{kk}(1-\rho_{kk})c_k}{1+c_k} - \rho_{kk}^2\right] E(R_{t-1}^{[k]}) \\
&= \frac{\rho_{kk}(c_k+\rho_{kk})}{1+c_k} E({R_{t-1}^{[k]}}^2) - \rho_{kk}^2 \left(E(R_{t-1}^{[k]})\right)^2 + \frac{\rho_{kk}(1-\rho_{kk})}{1+c_k} E(R_{t-1}^{[k]}).
\end{aligned}$$

(ii) (a) $\text{Cov}(Y_t^{[k]}, \alpha_{jj,t+h+1} \circ R_{t+h}^{[j]})$

$$
\begin{aligned}
&= E(Y_t^{[k]} \alpha_{jj,t+h+1} \circ R_{t+h}^{[j]}) - E(Y_t^{[k]})E(\alpha_{jj,t+h+1} \circ R_{t+h}^{[j]}) \\
&= E[E(Y_t^{[k]} \alpha_{jj,t+h+1} \circ R_{t+h}^{[j]} | R_{t+h}^{[j]}, \alpha_{jj,t+h+1})] \\
&- E(Y_t^{[k]})E(\alpha_{jj,t+h+1})E(R_{t+h}^{[j]}) \\
&= E(Y_t^{[k]} \alpha_{jj,t+h+1} R_{t+h}^{[j]}) - E(Y_t^{[k]})E(\alpha_{jj,t+h+1})E(R_{t+h}^{[j]}) \\
&= E(\alpha_{jj,t+h+1})E(Y_t^{[k]} R_{t+h}^{[j]}) - E(\alpha_{jj,t+h+1})E(Y_t^{[k]})E(R_{t+h}^{[j]}) \\
&= \rho_{jj} E(Y_t^{[k]} R_{t+h}^{[j]}) - \rho_{jj} E(Y_t^{[k]})E(R_{t+h}^{[j]}) \\
&= \rho_{jj} \text{Cov}(Y_t^{[k]}, R_{t+h}^{[j]}).
\end{aligned}
$$

(b) $\text{Cov}(\alpha_{kk,t+1} \circ R_t^{[k]}, \alpha_{jj,t+1} \circ R_t^{[j]})$

$$
\begin{aligned}
&= E(\alpha_{kk,t+1} \circ R_t^{[k]} \alpha_{jj,t+1} \circ R_t^{[j]}) - E(\alpha_{kk,t+1} \circ R_t^{[k]})E(\alpha_{jj,t+1} \circ R_t^{[j]}) \\
&= E[E(\alpha_{kk,t+1} \circ R_t^{[k]} \alpha_{jj,t+1} \circ R_t^{[j]} | R_t^{[k]}, R_t^{[j]}, \alpha_{kk,t+1}, \alpha_{jj,t+1})] \\
&- E(\alpha_{kk,t+1})E(R_t^{[k]})E(\alpha_{jj,t+1})E(R_t^{[j]}) \\
&= E(\alpha_{kk,t+1} R_t^{[k]} \alpha_{jj,t+1} R_t^{[j]}) - E(\alpha_{kk,t+1})E(R_t^{[k]})E(\alpha_{jj,t+1})E(R_t^{[j]}) \\
&= E(\alpha_{kk,t+1})E(\alpha_{jj,t+1})E(R_t^{[k]} R_t^{[j]}) \\
&- E(\alpha_{kk,t+1})E(\alpha_{jj,t+1})E(R_t^{[k]})E(R_t^{[j]}) \\
&= \rho_{kk}\rho_{jj} E(R_t^{[k]} R_t^{[j]}) - \rho_{kk}\rho_{jj} E(R_t^{[k]})E(R_t^{[j]}) \\
&= \rho_{jj}\rho_{kk} \text{Cov}(R_t^{[k]}, R_t^{[j]}).
\end{aligned}
$$

(iii) (a)

$$
\begin{aligned}
\text{Cov}(Y_t^{[k]}, R_t^{[k]}) &= \text{Cov}(\alpha_{kk,t} \circ R_{t-1}^{[k]} + R_t^{[k]}, R_t^{[k]}) \\
&= \text{Cov}(\alpha_{kk,t} \circ R_{t-1}^{[k]}, R_t^{[k]}) + \text{Cov}(R_t^{[k]}, R_t^{[k]}) \\
&= \text{Var}(R_t^{[k]}).
\end{aligned}
$$

(b)

$$
\begin{aligned}
\text{Cov}(Y_t^{[k]}, R_{t+h}^{[j]}) &= \text{Cov}(\alpha_{kk,t} \circ R_{t-1}^{[k]} + R_t^{[k]}, R_{t+h}^{[j]}) \\
&= \text{Cov}(\alpha_{kk,t} \circ R_{t-1}^{[k]}, R_{t+h}^{[j]}) + \text{Cov}(R_t^{[k]}, R_{t+h}^{[j]}) \\
&= \text{Cov}(R_t^{[k]}, R_t^{[j]}) \quad for \quad h = 0.
\end{aligned}
$$

Using the above conditions and Lemma 7.1, it is proved that

$$
\begin{aligned}
E(Y_t^{[k]}) \equiv \mu_t^{[k]} &= E(\alpha_{kk,t} \circ R_{t-1}^{[k]}) + E(R_t^{[k]}) \\
&= E_{\alpha_{kk,t}} E_{R_{t-1}^{[k]}} E(\alpha_{kk,t} \circ R_{t-1}^{[k]} | R_{t-1}^{[k]}, \alpha_{kk,t}) + E(R_t^{[k]}) \\
&= \rho_{kk}\lambda_{t-1}^{[k]} + \lambda_t^{[k]} \qquad (7.3)
\end{aligned}
$$

while

$$\begin{aligned}
\mathrm{Var}(Y_t^{[k]}) &= \mathrm{Var}(\alpha_{kk,t} \circ R_{t-1}^{[k]}) + \mathrm{Var}(R_t^{[k]}) \\
&= \frac{\rho_{kk}(c_k + \rho_{kk})}{1+c_k} E({R_{t-1}^{[k]}}^2) - \rho_{kk}^2 \left(E(R_{t-1}^{[k]})\right)^2 \\
&+ \frac{\rho_{kk}(1-\rho_{kk})}{1+c_k} E(R_{t-1}^{[k]}) + \mathrm{Var}(R_t^{[k]}) \\
&= \frac{\rho_{kk}(c_k + \rho_{kk})}{1+c_k} \left[\lambda_{t-1}^{[k]} + c_k {\lambda_{t-1}^{[k]}}^2 + {\lambda_{t-1}^{[k]}}^2\right] \\
&- \rho_{kk}^2 {\lambda_{t-1}^{[k]}}^2 + \frac{\rho_{kk}(1-\rho_{kk})}{1+c_k} \lambda_{t-1}^{[k]} \\
&+ \lambda_t^{[k]} + c_k {\lambda_t^{[k]}}^2 \\
&= \rho_{kk}\lambda_{t-1}^{[k]}(1 + c_k\lambda_{t-1}^{[k]}) + \lambda_t^{[k]}(1 + c_k\lambda_t^{[k]}) \\
&= \lambda_t^{[k]} + \rho_{kk}\lambda_{t-1}^{[k]} + c_k^*(\lambda_t^{[k]} + \rho_{kk}\lambda_{t-1}^{[k]})^2
\end{aligned} \tag{7.4}$$

such that

$$c_k^* = \frac{c_k\left({\lambda_t^{[k]}}^2 + \rho_{kk}{\lambda_{t-1}^{[k]}}^2\right)}{\left(\lambda_t^{[k]} + \rho_{kk}\lambda_{t-1}^{[k]}\right)^2}. \tag{7.5}$$

Such a procedure to derive $c_k^*$ has also been considered by Mallick & Sutradhar (2008). Hence, $Y_t^{[k]} \sim \mathrm{NB}(1/c_k^*, c_k^*(\lambda_t^{[k]} + \rho_{kk}\lambda_{t-1}^{[k]}))$, where $c_k^*$ is given by Equation (7.5).

The lag-$h$ serial correlation is simply derived as follows:

$$\begin{aligned}
\mathrm{Cov}(Y_t^{[k]}, Y_{t+h}^{[k]}) &= \mathrm{Cov}(Y_t^{[k]}, \alpha_{kk,t+h} \circ R_{t+h-1}^{[k]} + R_{t+h}^{[k]}) \\
&= \rho_{kk}\mathrm{Cov}(Y_t^{[k]}, R_{t+h-1}^{[k]}) + \mathrm{Cov}(Y_t^{[k]}, R_{t+h}^{[k]}) \\
&= \rho_{kk}\mathrm{Cov}(Y_t^{[k]}, R_{t+h-1}^{[k]}) \\
&= \begin{cases} \rho_{kk}(\lambda_t^{[k]} + c_k{\lambda_t^{[k]}}^2) & h = 1 \\ 0 & h > 1 \end{cases}
\end{aligned} \tag{7.6}$$

and hence

$$\mathrm{Corr}(Y_t^{[k]}, Y_{t+h}^{[k]}) = \begin{cases} \frac{\rho_{kk}[\lambda_t^{[k]} + c_k{\lambda_t^{[k]}}^2]}{\sqrt{\mu_t^{[k]} + c_k^*(\mu_t^{[k]})^2}\sqrt{\mu_{t+h}^{[k]} + c_k^*(\mu_{t+h}^{[k]})^2}} & h = 1, \\ 0 & h > 1. \end{cases} \tag{7.7}$$

As for the joint covariance $\text{Cov}(Y_t^{[1]}, Y_t^{[2]})$,

$$\begin{aligned}
\text{Cov}(Y_t^{[1]}, Y_t^{[2]}) &= \text{Cov}(\alpha_{11,t} \circ R_{t-1}^{[1]} + R_t^{[1]}, \alpha_{22,t} \circ R_{t-1}^{[2]} + R_t^{[2]}) \\
&= \text{Cov}(\alpha_{11,t} \circ R_{t-1}^{[1]}, \alpha_{22,t} \circ R_{t-1}^{[2]}) + \text{Cov}(\alpha_{11,t} \circ R_{t-1}^{[1]}, R_t^{[2]}) \\
&+ \text{Cov}(R_t^{[1]}, \alpha_{22,t} \circ R_{t-1}^{[2]}) + \text{Cov}(R_t^{[1]}, R_t^{[2]}) \\
&= \rho_{11}\rho_{22}\text{Cov}(R_{t-1}^{[1]}, R_{t-1}^{[2]}) + \text{Cov}(R_t^{[1]}, R_t^{[2]}) \\
&= c_1 c_2 \nu(\rho_{11}\rho_{22}\lambda_{t-1}^{[1]}\lambda_{t-1}^{[2]} + \lambda_t^{[1]}\lambda_t^{[2]})
\end{aligned} \tag{7.8}$$

and hence

$$\text{Corr}(Y_t^{[1]}, Y_t^{[2]}) = \frac{c_1 c_2 \nu(\rho_{11}\rho_{22}\lambda_{t-1}^{[1]}\lambda_{t-1}^{[2]} + \lambda_t^{[1]}\lambda_t^{[2]})}{\sqrt{\mu_t^{[1]} + c_1^*(\mu_t^{[1]})^2}\sqrt{\mu_t^{[2]} + c_2^*(\mu_t^{[2]})^2}}. \tag{7.9}$$

As for the other off-diagonal cross-covariances,

$$\begin{aligned}
\text{Cov}(Y_t^{[1]}, Y_{t+h}^{[2]}) &= \text{Cov}(Y_t^{[1]}, \alpha_{22,t+h} \circ R_{t+h-1}^{[2]} + R_{t+h}^{[2]}) \\
&= \rho_{22}\text{Cov}(Y_t^{[1]}, R_{t+h-1}^{[2]}) + \text{Cov}(Y_t^{[1]}, R_{t+h}^{[2]}) \\
&= \rho_{22}\text{Cov}(R_t^{[1]}, R_{t+h-1}^{[2]}) \\
&= \begin{cases} \rho_{22} c_1 c_2 \nu \lambda_t^{[1]}\lambda_t^{[2]} & h = 1 \\ 0 & h > 1 \end{cases}
\end{aligned} \tag{7.10}$$

and similarly

$$\text{Cov}(Y_{t+h}^{[1]}, Y_t^{[2]}) = \begin{cases} \rho_{11} c_1 c_2 \nu \lambda_t^{[1]}\lambda_t^{[2]} & h = 1, \\ 0 & h > 1. \end{cases} \tag{7.11}$$

**Remark 7.1.** Since $c_k = 0 \Rightarrow c_k^* = 0$, Equations (7.3) and (7.4) reduce to

$$E(Y_t^{[k]}) = \text{Var}(Y_t^{[k]}) = \rho_{kk}\lambda_{t-1}^{[k]} + \lambda_t^{[k]}. \tag{7.12}$$

Hence, $Y_t^{[k]} \sim \text{Poisson}(\rho_{kk}\lambda_{t-1}^{[k]} + \lambda_t^{[k]})$.

**Remark 7.2.** Under stationary moment condition,

$$E(Y_t^{[k]}) \equiv \mu_1^{[k]} = (1 + \rho_{kk})\lambda_1^{[k]}, \tag{7.13}$$

$$\text{Var}(Y_t^{[k]}) = (1 + \rho_{kk})\lambda_1^{[k]} + c_k^* \left[(1 + \rho_{kk})\lambda_1^{[k]}\right]^2 \tag{7.14}$$

and

$$\text{Cov}(Y_t^{[1]}, Y_t^{[2]}) = c_1 c_2 \nu \lambda_1^{[1]}\lambda_1^{[2]}(\rho_{11}\rho_{22} + 1). \tag{7.15}$$

## 7.3 Estimation of Parameters

Two GQL equations are formulated in the following subsections to estimate the model parameters.

### 7.3.1 GQL-I

The first GQL equation estimates the vector of unknown parameters $\boldsymbol{\psi} = [\beta_1^{[1]}, \beta_2^{[1]}, \beta_1^{[2]}, \beta_2^{[2]}, \rho_{11}, \rho_{22}]$ and is specified as

$$\boldsymbol{D}_{\boldsymbol{\psi}}^{'} \boldsymbol{\Sigma}_{\boldsymbol{\psi}}{}^{-1}(\boldsymbol{y} - \boldsymbol{\mu}) = 0, \tag{7.16}$$

$$\boldsymbol{y} = \begin{pmatrix} \boldsymbol{y}^{[1]} \\ \boldsymbol{y}^{[2]} \end{pmatrix}_{2T\times 1}, \boldsymbol{y}^{[1]} = [y_1^{[1]}, y_2^{[1]}, \ldots, y_T^{[1]}]', \boldsymbol{y}^{[2]} = [y_1^{[2]}, y_2^{[2]}, \ldots, y_T^{[2]}]' \tag{7.17}$$

and

$$\boldsymbol{\mu} = \begin{pmatrix} \boldsymbol{\mu}^{[1]} \\ \boldsymbol{\mu}^{[2]} \end{pmatrix}_{2T\times 1}, \boldsymbol{\mu}^{[1]} = [\mu_1^{[1]}, \mu_2^{[1]}, \ldots, \mu_T^{[1]}]', \boldsymbol{\mu}^{[2]} = [\mu_1^{[2]}, \mu_2^{[2]}, \ldots, \mu_T^{[2]}]'. \tag{7.18}$$

The covariance matrix $\boldsymbol{\Sigma}_{\boldsymbol{\psi}}$ is tri-diagonal given the covariance structure representation and is given by

$$\begin{pmatrix} \boldsymbol{\Sigma}_{Y_t^{[1]}} & \vdots & \boldsymbol{\Sigma}_{Y_t^{[1]}, Y_t^{[2]}} \\ \cdots & \cdots & \cdots \\ \boldsymbol{\Sigma}_{Y_t^{[2]}, Y_t^{[1]}} & \vdots & \boldsymbol{\Sigma}_{Y_t^{[2]}} \end{pmatrix}, \tag{7.19}$$

where

$$\boldsymbol{\Sigma}_{Y_t^{[k]}} = \begin{pmatrix} \mathrm{Var}(Y_1^{[k]}) & \mathrm{Cov}(Y_1^{[k]}, Y_2^{[k]}) & & \\ \mathrm{Cov}(Y_2^{[k]}, Y_1^{[k]}) & \mathrm{Var}(Y_2^{[k]}) & \mathrm{Cov}(Y_2^{[k]}, Y_3^{[k]}) & \\ & & \ddots & \\ & & \mathrm{Cov}(Y_T^{[k]}, Y_{T-1}^{[k]}) & \mathrm{Var}(Y_T^{[k]}) \end{pmatrix}, \tag{7.20}$$

$$\boldsymbol{\Sigma}_{Y_t^{[2]}, Y_t^{[1]}} = \begin{pmatrix} \mathrm{Cov}(Y_1^{[2]}, Y_1^{[1]}) & \mathrm{Cov}(Y_1^{[2]}, Y_2^{[1]}) & & \\ \mathrm{Cov}(Y_2^{[2]}, Y_1^{[1]}) & \mathrm{Cov}(Y_2^{[2]}, Y_2^{[1]}) & \mathrm{Cov}(Y_2^{[2]}, Y_3^{[1]}) & \\ & & \ddots & \\ & & \mathrm{Cov}(Y_T^{[2]}, Y_{T-1}^{[1]}) & \mathrm{Cov}(Y_T^{[2]}, Y_T^{[1]}) \end{pmatrix} \tag{7.21}$$

and $\boldsymbol{\Sigma}_{Y_t^{[1]},Y_t^{[2]}} = \boldsymbol{\Sigma}'_{Y_t^{[2]},Y_t^{[1]}}$, where the entries are computed using Equations (7.4)–(7.11), assuming $\hat{\lambda}_0^{[k]} = \hat{\lambda}_1^{[k]}$.

As for the derivative matrix, $\boldsymbol{D}_{\boldsymbol{\psi}} = [\boldsymbol{D}_1, \boldsymbol{D}_2, \ldots, \boldsymbol{D}_t, \ldots, \boldsymbol{D}_T]'$ with

$$\boldsymbol{D}_t = \begin{pmatrix} \frac{\partial \mu_t^{[1]}}{\partial \boldsymbol{\beta}^{[1]}} & 0 \\ 0 & \frac{\partial \mu_t^{[2]}}{\partial \boldsymbol{\beta}^{[2]}} \\ \frac{\partial \mu_t^{[1]}}{\partial \rho_{11}} & 0 \\ 0 & \frac{\partial \mu_t^{[2]}}{\partial \rho_{22}} \end{pmatrix}_{(2p+2)\times 2} \tag{7.22}$$

where

$$\frac{\partial \mu_t^{[k]}}{\partial \boldsymbol{\beta}^{[l]}} = \left[\frac{\partial \mu_t^{[k]}}{\partial \beta_1^{[l]}}, \frac{\partial \mu_t^{[k]}}{\partial \beta_2^{[l]}}, \ldots, \frac{\partial \mu_t^{[k]}}{\partial \beta_j^{[l]}}, \ldots, \frac{\partial \mu_t^{[k]}}{\partial \beta_p^{[l]}}\right]', \quad k, k', l \in \{1,2\}, k \neq k',$$

$$\frac{\partial \mu_t^{[k]}}{\partial \beta_j^{[k]}} = \rho_{kk}\lambda_{t-1}^{[k]} x_{t-1,j} + \lambda_t^{[k]} x_{tj} \quad \text{and} \quad \frac{\partial \mu_t^{[k]}}{\partial \rho_{kk}} = \lambda_{t-1}^{[k]}.$$

Equation (7.16) is then solved iteratively until convergence using the Newton-Raphson as follows:

$$\left(\ \hat{\boldsymbol{\psi}}_{r+1}\ \right) = \left(\ \hat{\boldsymbol{\psi}}_r\ \right) + [\boldsymbol{D}_{\boldsymbol{\psi}}{}'\boldsymbol{\Sigma}_{\boldsymbol{\psi}}{}^{-1}\boldsymbol{D}_{\boldsymbol{\psi}}]_r^{-1}[\boldsymbol{D}_{\boldsymbol{\psi}}{}'\boldsymbol{\Sigma}_{\boldsymbol{\psi}}{}^{-1}(\boldsymbol{y} - \boldsymbol{\mu})]_r \tag{7.23}$$

where $\hat{\boldsymbol{\psi}}_r$ is the vector of estimates at the $r^{th}$ iteration and $[.]_r$ are the values of the expression at the $r^{th}$ iteration. These estimates are consistent and under mild regularity conditions, it is shown that $(\hat{\boldsymbol{\psi}} - \boldsymbol{\psi})'$ is asymptotic normal with mean $\mathbf{0}$ and covariance matrix $[\boldsymbol{D}_{\boldsymbol{\psi}}{}'\boldsymbol{\Sigma}_{\boldsymbol{\psi}}{}^{-1}\boldsymbol{D}_{\boldsymbol{\psi}}]^{-1}$ (Refer to Subsection 2.3.1.2).

### 7.3.2 GQL-II

The second GQL equation is specified as:

$$\boldsymbol{D}'_{\boldsymbol{c},\boldsymbol{\nu}}\boldsymbol{\Sigma}_{\boldsymbol{c},\boldsymbol{\nu}}{}^{-1}(\boldsymbol{Z} - \boldsymbol{m}) = 0, \tag{7.24}$$

where $\boldsymbol{Z} = [Z_1, Z_2, \ldots, Z_t, \ldots, Z_T]'_{T\times 1}$ with $Z_t = \{Y_t^{[1]} Y_t^{[2]}\}$ and $\boldsymbol{m} = [m_1, m_2, \ldots, m_t, \ldots, m_T]'_{T\times 1}$ with $m_t = E(Z_t) = E(Y_t^{[1]} Y_t^{[2]}) = c_1 c_2 \nu(\rho_{11}\rho_{22}\lambda_{t-1}^{[1]}\lambda_{t-1}^{[2]} + \lambda_t^{[1]}\lambda_t^{[2]}) + (\rho_{11}\lambda_{t-1}^{[1]} + \lambda_t^{[1]})(\rho_{22}\lambda_{t-1}^{[2]} + \lambda_t^{[2]})$.

Based on the above score, the entries of $\boldsymbol{\Sigma}_{\boldsymbol{c},\boldsymbol{\nu}}$ consist of $\text{Var}(Z_t)$ and $\text{Cov}(Z_t, Z_{t+1})$, that is

$$\boldsymbol{\Sigma}_{\boldsymbol{c},\boldsymbol{\nu}} = \begin{pmatrix} \mathrm{Var}(Z_1) & \mathrm{Cov}(Z_1,Z_2) & 0 & \dots & 0 & 0 \\ \mathrm{Cov}(Z_2,Z_1) & \mathrm{Var}(Z_2) & \mathrm{Cov}(Z_2,Z_3) & \dots & 0 & 0 \\ \vdots & \vdots & \vdots & \ddots & \vdots & \vdots \\ 0 & 0 & 0 & \dots & \mathrm{Var}(Z_{T-1}) & \mathrm{Cov}(Z_{T-1},Z_T) \\ 0 & 0 & 0 & \dots & \mathrm{Cov}(Z_T,Z_{T-1}) & \mathrm{Var}(Z_T) \end{pmatrix}_{T\times T} \tag{7.25}$$

To simplify the derivation of these entries, we make use of the following lemmas, by denoting $W_{kk,t} = \alpha_{kk,t+1} \circ R_t^{[k]}$ for $k \in \{1,2\}$:

**Lemma 7.2.**

(a) $\mathrm{Cov}(W_{kk,t-1}W_{jj,t-1}, R_t^{[k]}R_t^{[j]})=0$, $j,k \in \{1,2\}$, $j \neq k$.

(b) $\mathrm{Cov}(W_{kk,t-1}R_t^{[j]}, R_t^{[k]}R_{t+1}^{[j]})=0$, $j,k \in \{1,2\}$, $j \neq k$.

(c) $\mathrm{Cov}(R_t^{[1]}R_t^{[2]}, R_{t+1}^{[1]}R_{t+1}^{[2]}) = 0$ *for* $t \in \mathbb{Z}$.

*Proof: Since the product* $W_{kk,t-1}W_{jj,t-1}$ *is independent with* $R_t^{[k]}R_t^{[j]}$ *for all* $j,k \in \{1,2\}$ *and* $j \neq k$*, which implies that the corresponding covariance is 0 for the first statement. The remaining statements are proved in a similar way.*

**Lemma 7.3.**

(a) $\mathrm{Cov}(W_{kk,t-1}W_{jj,t-1}, W_{kk,t}R_{t+1}^{[i]})$
$= [\frac{\rho_{kk}(1-\rho_{kk})c_k}{1+c_k}]\rho_{jj}[E(R_{t-1}^{[k]}R_{t-1}^{[j]})\lambda_t^{[k]}\lambda_{t+1}^{[i]}]$, $\quad i,j,k \in \{1,2\}$, $j \neq k$.

(b) $\mathrm{Cov}(W_{kk,t-1}R_t^{[j]}, W_{kk,t}R_{t+1}^{[j]})$
$= [\frac{\rho_{kk}(1-\rho_{kk})c_k}{1+c_k}][\lambda_{t-1}^{[k]}\lambda_t^{[j]}\lambda_t^{[k]}\lambda_{t+1}^{[j]}]$, $\quad j,k \in \{1,2\}$, $j \neq k$.

*Proof:*

(a)
$$\begin{aligned}
&\mathrm{Cov}(W_{kk,t-1}W_{jj,t-1}, W_{kk,t}R_{t+1}^{[i]})\\
&=E(W_{kk,t-1}W_{jj,t-1}W_{kk,t}R_{t+1}^{[i]}) - E(W_{kk,t-1}W_{jj,t-1})E(W_{kk,t}R_{t+1}^{[i]})\\
&=E(\alpha_{kk,t+1}\alpha_{kk,t})E(\alpha_{jj,t})[E(R_{t-1}^{[k]}R_{t-1}^{[j]})\lambda_t^{[k]}\lambda_{t+1}^{[i]}]\\
&-E(\alpha_{kk,t+1})E(\alpha_{kk,t})E(\alpha_{jj})[E(R_{t-1}^{[k]}R_{t-1}^{[j]})\lambda_t^{[k]}\lambda_{t+1}^{[i]}]\\
&=[\tfrac{\rho_{kk}(1-\rho_{kk})c_k}{1+c_k}]\rho_{jj}[E(R_{t-1}^{[k]}R_{t-1}^{[j]})\lambda_t^{[k]}\lambda_{t+1}^{[i]}].
\end{aligned}$$

(b)
$$\begin{aligned}
&\mathrm{Cov}(W_{kk,t-1}R_t^{[j]}, W_{kk,t}R_{t+1}^{[j]})\\
&= E(W_{kk,t-1}R_t^{[j]}W_{kk,t}R_{t+1}^{[j]}) - E(W_{kk,t-1}R_t^{[j]})E(W_{kk,t}R_{t+1}^{[j]})\\
&=E(\alpha_{kk,t}\alpha_{kk,t+1})[\lambda_{t-1}^{[k]}\lambda_t^{[j]}\lambda_t^{[k]}\lambda_{t+1}^{[j]}]\\
&-E(\alpha_{kk,t})E(\alpha_{kk,t+1})[\lambda_{t-1}^{[k]}\lambda_t^{[j]}\lambda_t^{[k]}\lambda_{t+1}^{[j]}]\\
&=[\tfrac{\rho_{kk}(1-\rho_{kk})c_k}{1+c_k}][\lambda_{t-1}^{[k]}\lambda_t^{[j]}\lambda_t^{[k]}\lambda_{t+1}^{[j]}].
\end{aligned}$$

Hence,

$$\begin{aligned}
&\mathrm{Cov}(Z_t, Z_{t+1})\\
&= \mathrm{Cov}(Y_t^{[1]}Y_t^{[2]}, Y_{t+1}^{[1]}Y_{t+1}^{[2]})\\
&= \mathrm{Cov}[(W_{11,t-1}W_{22,t-1} + W_{11,t-1}R_t^{[2]} + W_{22,t-1}R_t^{[1]} + R_t^{[1]}R_t^{[2]}),\\
&\quad (W_{11,t}W_{22,t} + W_{11,t}R_{t+1}^{[2]} + W_{22,t}R_{t+1}^{[1]} + R_{t+1}^{[1]}R_{t+1}^{[2]})]. \qquad (7.26)
\end{aligned}$$

Expanding Equation (7.26) and using Lemmas 7.2 and 7.3, we obtain

$$\begin{aligned}
&\mathrm{Cov}(Z_t, Z_{t+1})\\
&= (\frac{\rho_{11}(1-\rho_{11})c_1}{1+c_1})(\frac{\rho_{22}(1-\rho_{22})c_2}{1+c_2})[E(R_{t-1}^{[1]}R_{t-1}^{[2]})E(R_t^{[1]}R_t^{[2]})]\\
&+ \frac{\rho_{11}(1-\rho_{11})c_1}{1+c_1}[\rho_{22}E(R_{t-1}^{[1]}R_{t-1}^{[2]})(\rho_{22}E(R_t^{[1]}R_t^{[2]}) + \lambda_t^{[1]}\lambda_{t+1}^{[2]})\\
&+ \lambda_{t-1}^{[1]}\lambda_t^{[1]}(\lambda_{t+1}^{[2]}\lambda_t^{[2]} + \rho_{22}(\lambda_t^{[2]} + c_2\lambda_t^{[2]^2} + \lambda_t^{[2]^2}))]\\
&+ \rho_{11}\rho_{22}\lambda_{t-1}^{[1]}(\lambda_t^{[2]} + c_2\lambda_t^{[2]^2})(\rho_{11}\lambda_t^{[1]} + \lambda_{t+1}^{[1]}) + \frac{\rho_{22}(1-\rho_{22})c_2}{1+c_2}\\
&\times [\rho_{11}E(R_{t-1}^{[1]}R_{t-1}^{[2]})(\rho_{11}E(R_t^{[1]}R_t^{[2]}) + \lambda_t^{[2]}\lambda_{t+1}^{[1]}) + \lambda_{t-1}^{[2]}\lambda_t^{[2]}(\lambda_{t+1}^{[1]}\lambda_t^{[1]}\\
&+ \rho_{11}(\lambda_t^{[1]} + c_1\lambda_t^{[1]^2} + \lambda_t^{[1]^2}))] + \rho_{11}\rho_{22}\lambda_{t-1}^{[2]}(\lambda_t^{[1]} + c_1\lambda_t^{[1]^2})\\
&\times (\rho_{22}\lambda_t^{[2]} + \lambda_{t+1}^{[2]}). \qquad (7.27)
\end{aligned}$$

**Lemma 7.4.** $\mathrm{Cov}(W_{kk,t-1}W_{jj,t-1}, R_t^{[k]}R_t^{[j]})=0,\ j,k \in \{1,2\},\ j \neq k.$

*Proof: Since the product* $W_{kk,t-1}W_{jj,t-1}$ *is independent with* $R_t^{[k]}R_t^{[j]}$ *for all* $j,k \in \{1,2\}$ *and* $j \neq k$*, which implies that the corresponding covariance is 0.*

**Lemma 7.5.**

(a)
$$\begin{aligned}
&\mathrm{Cov}(W_{kk,t-1}W_{jj,t-1}, W_{kk,t-1}R_t^{[i]})\\
&= [\frac{\rho_{kk}(1-\rho_{kk})c_k}{1+c_k} + \rho_{kk}^2]\rho_{jj}[E(R_{t-1}^{[k]^2}R_{t-1}^{[j]})\lambda_t^{[i]}]\\
&\quad - \rho_{11}^2\rho_{22}E(R_{t-1}^{[k]}R_{t-1}^{[j]})\lambda_{t-1}^{[k]}\lambda_t^{[i]}, \quad i,j,k \in \{1,2\}, j \neq k.
\end{aligned}$$

(b)
$$\begin{aligned}
&\mathrm{Cov}(W_{kk,t-1}R_t^{[j]}, W_{kk,t-1}R_t^{[j]})\\
&= [\frac{\rho_{kk}(1-\rho_{kk})c_k}{1+c_k} + \rho_{kk}^2](\lambda_{t-1}^{[k]} + c_k\lambda_{t-1}^{[k]^2} + \lambda_{t-1}^{[k]^2})\\
&\quad \times (\lambda_t^{[j]} + c_j\lambda_t^{[j]^2} + \lambda_t^{[j]^2}) - \rho_{11}^2\lambda_{t-1}^{[k]^2}\lambda_t^{[j]^2}, \quad j,k \in \{1,2\}, j \neq k.
\end{aligned}$$

(c)
$$\begin{aligned}
&\mathrm{Cov}(W_{kk,t-1}R_t^{[j]}, R_t^{[k]}R_t^{[j]})\\
&= \rho_{kk}\lambda_{t-1}^{[k]}E(R_t^{[k]}R_t^{[j]^2}) - \rho_{kk}\lambda_t^{[j]}\lambda_{t-1}^{[k]}E(R_t^{[k]}R_t^{[j]}), j,k \in \{1,2\}, j \neq k.
\end{aligned}$$

(d)
$$\begin{aligned}&\text{Cov}(W_{kk,t-1}R_t^{[j]}, W_{jj,t-1}R_t^{[k]})\\&= \rho_{kk}\rho_{jj}E(R_{t-1}^{[k]}R_{t-1}^{[j]})E(R_t^{[k]}R_t^{[j]})\\&\quad - \rho_{kk}\rho_{jj}\lambda_t^{[k]}\lambda_t^{[j]}\lambda_{t-1}^{[k]}\lambda_{t-1}^{[j]}, j,k \in \{1,2\}, j \neq k.\end{aligned}$$

*Proof:*

(a)
$$\begin{aligned}&\text{Cov}(W_{kk,t-1}W_{jj,t-1}, W_{kk,t-1}R_t^{[i]})\\&= E(W_{kk,t-1}W_{jj,t-1}W_{kk,t+h-1}R_t^{[i]})\\&\quad - E(W_{kk,t-1}W_{jj,t-1})E(W_{kk,t+h-1}R_t^{[i]})\\&= E(\alpha_{kk,t}\alpha_{kk,t+h})E(\alpha_{jj,t})[E({R_{t-1}^{[k]}}^2 R_{t-1}^{[j]})\lambda_t^{[i]}]\\&\quad - E(\alpha_{kk,t})E(\alpha_{kk,t+h})E(\alpha_{jj,t})[E(R_{t-1}^{[k]}R_{t-1}^{[j]})\lambda_{t-1}^{[j]}\lambda_t^{[i]}]\\&= [\frac{\rho_{kk}(1-\rho_{kk})c_k}{1+c_k} + \rho_{kk}^2]\rho_{jj}[E({R_{t-1}^{[k]}}^2 R_{t-1}^{[j]})\lambda_t^{[i]}]\\&\quad - \rho_{11}^2\rho_{22}E(R_{t-1}^{[k]}R_{t-1}^{[j]})\lambda_{t-1}^{[k]}\lambda_t^{[i]}.\end{aligned}$$

(b)
$$\begin{aligned}&\text{Cov}(W_{kk,t-1}R_t^{[j]}, W_{kk,t-1}R_t^{[j]})\\&= E(W_{kk,t-1}R_t^{[j]}W_{kk,t-1}R_t^{[j]}) - E(W_{kk,t-1}R_t^{[j]})E(W_{kk,t-1}R_t^{[j]})\\&= E(\alpha_{kk,t}^2)[E({R_{t-1}^{[k]}}^2)E({R_{t-1}^{[j]}}^2)] - E(\alpha_{kk,t})^2{\lambda_{t-1}^{[k]}}^2{\lambda_t^{[j]}}^2\\&= [\frac{\rho_{kk}(1-\rho_{kk})c_k}{1+c_k} + \rho_{kk}^2](\lambda_{t-1}^{[k]} + c_k{\lambda_{t-1}^{[k]}}^2 + {\lambda_{t-1}^{[k]}}^2)\\&\quad \times (\lambda_t^{[j]} + c_j{\lambda_t^{[j]}}^2 + {\lambda_t^{[j]}}^2) - \rho_{11}^2{\lambda_{t-1}^{[k]}}^2{\lambda_t^{[j]}}^2.\end{aligned}$$

(c)
$$\begin{aligned}&\text{Cov}(W_{kk,t-1}R_t^{[j]}, R_t^{[k]}R_t^{[j]})\\&= E(W_{kk,t-1}R_t^{[j]}R_t^{[k]}R_t^{[j]}) - E(W_{kk,t-1}R_t^{[j]})E(R_t^{[k]}R_t^{[j]})\\&= E(\alpha_{kk,t})\lambda_{t-1}^{[k]}E(R_t^{[k]}{R_t^{[j]}}^2) - E(\alpha_{kk,})\lambda_{t-1}^{[k]}\lambda_t^{[j]}E(R_t^{[k]}R_t^{[j]})\\&= \rho_{kk}\lambda_{t-1}^{[k]}E(R_t^{[k]}{R_t^{[j]}}^2) - \rho_{kk}\lambda_t^{[j]}\lambda_{t-1}^{[k]}E(R_t^{[k]}R_t^{[j]}).\end{aligned}$$

(d)
$$\begin{aligned}&\text{Cov}(W_{kk,t-1}R_t^{[j]}, W_{jj,t-1}R_t^{[k]})\\&= E(W_{kk,t-1}R_t^{[j]}, W_{jj,t-1}R_t^{[k]}) - E(W_{kk,t-1}R_t^{[j]})E(W_{jj,t-1}R_t^{[k]})\\&= E(\alpha_{kk,t})E(\alpha_{jj,t})[E(R_{t-1}^{[j]}, R_{t-1}^{[k]})E(R_t^{[j]}, R_t^{[k]})]\\&\quad - E(\alpha_{kk,t})E(\alpha_{jj,t})E(R_{t-1}^{[j]})E(R_{t-1}^{[k]})E(R_t^{[j]})E(R_t^{[k]})\\&= \rho_{kk}\rho_{jj}E(R_{t-1}^{[k]}R_{t-1}^{[j]})E(R_t^{[k]}R_t^{[j]}) - \rho_{kk}\rho_{jj}\lambda_t^{[k]}\lambda_t^{[j]}\lambda_{t-1}^{[k]}\lambda_{t-1}^{[j]}.\end{aligned}$$

Hence,

$$\begin{aligned}\mathrm{Var}(Z_t) &= \mathrm{Cov}(Z_t, Z_t)\\ &= \mathrm{Cov}(Y_t^{[1]}Y_t^{[2]}, Y_t^{[1]}Y_t^{[2]})\\ &= \mathrm{Cov}[(W_{11,t-1}W_{22,t-1} + W_{11,t-1}R_t^{[2]} + W_{22,t-1}R_t^{[1]} + R_t^{[1]}R_t^{[2]}),\\ &(W_{11,t-1}W_{22,t-1} + W_{11,t-1}R_t^{[2]} + W_{22,t-1}R_t^{[1]} + R_t^{[1]}R_t^{[2]})]. \qquad (7.28)\end{aligned}$$

By expanding Equation (7.28) and using Lemmas 7.4 and 7.5, we obtain

$$\begin{aligned}&\mathrm{Var}(Z_t)\\ &= [(\frac{\rho_{11}(1-\rho_{11})c_1}{1+c_1})(\frac{\rho_{22}(1-\rho_{22})c_2}{1+c_2}) + \rho_{11}^2\rho_{22}^2]E(R_{t-1}^{[1]\,2}, R_{t-1}^{[2]\,2})\\ &+ \frac{\rho_{11}(1-\rho_{11})c_1}{1+c_1}[\rho_{22}^2E(R_{t-1}^{[1]\,2}, R_{t-1}^{[2]\,2}) + 2\rho_{22}\lambda_t^{[2]}E(R_{t-1}^{[1]\,2}, R_{t-1}^{[2]})\\ &+ (\lambda_t^{[2]} + c_2\lambda_t^{[2]^2} + \lambda_t^{[2]^2})(\lambda_{t-1}^{[1]} + c_1\lambda_{t-1}^{[1]\,2} + \lambda_{t-1}^{[1]\,2})]\\ &+ \frac{\rho_{22}(1-\rho_{22})c_2}{1+c_2}[\rho_{11}^2E(R_{t-1}^{[1]\,2}, R_{t-1}^{[2]\,2}) + 2\rho_{11}\lambda_t^{[1]}E(R_{t-1}^{[1]}, R_{t-1}^{[2]\,2})\\ &+ (\lambda_t^{[1]} + c_1\lambda_t^{[1]^2} + \lambda_t^{[1]^2})(\lambda_{t-1}^{[2]} + c_2\lambda_{t-1}^{[2]\,2} + \lambda_{t-1}^{[2]\,2})]\\ &+ 2\rho_{22}E(R_{t-1}^{[1]\,2}, R_{t-1}^{[2]})[\rho_{11}^2\lambda_t^{[2]} + \lambda_{t-1}^{[2]}] + 2\rho_{11}E(R_{t-1}^{[1]}, R_{t-1}^{[2]\,2})[\rho_{22}^2\lambda_t^{[1]} + \lambda_{t-1}^{[1]}]\\ &+ 2\rho_{11}\rho_{22}E(R_{t-1}^{[1]}, R_{t-1}^{[2]})[E(R_t^{[1]}, R_t^{[2]}) - \rho_{22}\lambda_t^{[1]}\lambda_{t-1}^{[2]} - \rho_{11}\lambda_t^{[2]}\lambda_{t-1}^{[1]}]\\ &- 2E(R_t^{[1]}, R_t^{[2]})[\rho_{11}\lambda_t^{[2]}\lambda_{t-1}^{[1]} + \rho_{22}\lambda_t^{[1]}\lambda_{t-1}^{[2]}]\\ &+ \rho_{11}^2(\lambda_t^{[2]} + c_2\lambda_t^{[2]^2} + \lambda_t^{[2]^2})(\lambda_{t-1}^{[1]} + c_1\lambda_{t-1}^{[1]\,2} + \lambda_{t-1}^{[1]\,2}) - \rho_{11}^2\lambda_t^{[2]^2}\lambda_{t-1}^{[1]\,2}\\ &+ \rho_{22}^2(\lambda_t^{[1]} + c_1\lambda_t^{[1]^2} + \lambda_t^{[1]^2})(\lambda_{t-1}^{[2]} + c_2\lambda_{t-1}^{[2]\,2} + \lambda_{t-1}^{[2]\,2}) - \rho_{22}^2\lambda_t^{[1]^2}\lambda_{t-1}^{[2]\,2}\\ &- 2\rho_{11}\rho_{22}\lambda_t^{[1]}\lambda_t^{[1]}\lambda_{t-1}^{[1]}\lambda_{t-1}^{[1]} + E(R_t^{[1]^2}, R_t^{[2]^2}) - E(R_t^{[1]}, R_t^{[2]})^2 \qquad (7.29)\end{aligned}$$

where by differentiating the pgf (6.5) repeatedly, we obtain

$$E(R_t^{[1]}R_t^{[2]}) = \lambda_t^{[1]}\lambda_t^{[2]}(c_1c_2\nu + 1), \qquad (7.30)$$

$$E(R_t^{[1]^2}R_t^{[2]}) = \lambda_t^{[1]^2}\lambda_t^{[2]}(2c_1^2c_2\nu + 2c_1c_2\nu + c_1 + 1) + \lambda_t^{[1]}\lambda_t^{[2]}(c_1c_2\nu + 1), \qquad (7.31)$$

$$E(R_t^{[1]}R_t^{[2]^2}) = \lambda_t^{[1]}\lambda_t^{[2]^2}(2c_1c_2^2\nu + 2c_1c_2\nu + c_2 + 1) + \lambda_t^{[1]}\lambda_t^{[2]}(c_1c_2\nu + 1) \qquad (7.32)$$

and

$$E(R_t^{[1]^2} R_t^{[2]^2}) = \lambda_t^{[1]^2} \lambda_t^{[2]^2}(2c_1^2c_2^2\nu^2 + 6c_1^2c_2^2\nu + 4c_1^2c_2\nu + 4c_1c_2^2\nu + 4c_1c_2\nu + c_1c_2 + c_1 + c_2 + 1) + \lambda_t^{[1]^2}\lambda_t^{[2]}(2c_1^2c_2\nu + 2c_1c_2\nu + c_1 + 1) + \lambda_t^{[1]}\lambda_t^{[2]}(c_1c_2\nu + 1) + \lambda_t^{[1]}\lambda_t^{[2]^2}(2c_1c_2^2\nu + 2c_1c_2\nu + c_2 + 1). \tag{7.33}$$

The derivative matrix $\boldsymbol{D}'_{\boldsymbol{c},\boldsymbol{\nu}}$ has the following components:

1. $\frac{\partial E(Y_t^{[1]}Y_t^{[2]})}{\partial \nu} = c_1c_2(\rho_{11}\rho_{22}\lambda_{t-1}^{[1]}\lambda_{t-1}^{[2]} + \lambda_t^{[1]}\lambda_t^{[2]}).$
2. $\frac{\partial E(Y_t^{[1]}Y_t^{[2]})}{\partial c_1} = c_2\nu(\rho_{11}\rho_{22}\lambda_{t-1}^{[1]}\lambda_{t-1}^{[2]} + \lambda_t^{[1]}\lambda_t^{[2]}).$
3. $\frac{\partial E(Y_t^{[1]}Y_t^{[2]})}{\partial c_2} = c_1\nu(\rho_{11}\rho_{22}\lambda_{t-1}^{[1]}\lambda_{t-1}^{[2]} + \lambda_t^{[1]}\lambda_t^{[2]}).$

The Newton-Raphson iteration yields

$$(\hat{\boldsymbol{\Theta}}_{r+1}) = (\hat{\boldsymbol{\Theta}}_r) + [\boldsymbol{D}_{\boldsymbol{\Theta}}{}'\boldsymbol{\Sigma}_{\boldsymbol{\Theta}}{}^{-1}\boldsymbol{D}_{\boldsymbol{\Theta}}]_r^{-1}[\boldsymbol{D}_{\boldsymbol{\Theta}}{}'\boldsymbol{\Sigma}_{\boldsymbol{\Theta}}{}^{-1}(\boldsymbol{Z} - \boldsymbol{m})]_r, \tag{7.34}$$

where $\hat{\boldsymbol{\Theta}} = [c_1, c_2, \nu]$ and $\hat{\boldsymbol{\Theta}}_r$ are the estimates at the $r^{th}$ iteration and $[.]_r$ are the values of the expression at the $r^{th}$ iteration.

The estimation algorithm works as follows: For an initial estimate of the regression, dependence and the over-dispersion parameters, the first GQL Equation (7.16) is solved iteratively until convergence to obtain an updated regression and dependence vectors. Using the converged estimates of the regression and dependence parameters, the second GQL Equation (7.24) is then solved iteratively to obtain an updated value of the over-dispersion parameters using the regression and dependence values obtained from the first GQL equation. After obtaining the updated over-dispersion values, these are in turn substituted in the first GQL Equation (7.16) to obtain another updated regression and dependence vectors. Thereafter the cycle of iteration continues until convergence of all sets of parameters with their corresponding standard errors obtained from the respective Hessian structures. Under mild regularity conditions, it is shown that $(\hat{\boldsymbol{\Theta}} - \boldsymbol{\Theta})'$ is asymptotically normal with mean $\mathbf{0}$ and covariance matrix $[\boldsymbol{D}_{\boldsymbol{\Theta}}{}'\boldsymbol{\Sigma}_{\boldsymbol{\Theta}}{}^{-1}\boldsymbol{D}_{\boldsymbol{\Theta}}]^{-1}$ (Refer to Subsection 2.3.1.2).

Using the converged estimates from the GQL method, we derive the forecasting equations of the constrained BINMA(1) model. The conditional expectation and variance of the one-step ahead forecast $Y_{t+1}^{[k]}$ given $r_t^{[k]}$ are as follows:

$$E(Y_{t+1}^{[k]}|r_t^{[k]}) = E(\alpha_{kk,t} \circ r_t^{[k]}|r_t^{[k]}) + E(R_{t+1}^{[k]}) = \rho_{kk}r_t^{[k]} + \lambda_{t+1}^{[k]} \tag{7.35}$$

and

$$\begin{aligned}
&\text{Var}(Y_{t+1}^{[k]}|r_t^{[k]}) \\
&= \text{Var}(\alpha_{kk,t} \circ R_t^{[k]}|r_t^{[k]}) + \text{Var}(R_{t+1}^{[k]}) \\
&= E_{\alpha_{kk,t}}[\text{Var}(\alpha_{kk,t} \circ R_t^{[k]}|r_t^{[k]}, \alpha_{kk,t})] \\
&\quad + \text{Var}_{\alpha_{kk,t}}[E(\alpha_{kk,t} \circ R_t^{[k]}|r_t^{[k]}, \alpha_{kk,t})] + \text{Var}(R_{t+1}^{[k]}) \\
&= E_{\alpha_{kk,t}}[\alpha_{kk,t}(1-\alpha_{kk,t})r_t^{[k]}] + \text{Var}_{\alpha_{kk,t}}[\alpha_{kk,t}r_t^{[k]}] + \text{Var}(R_{t+1}^{[k]}) \\
&= \left[\rho_{kk} - (\frac{\rho_{kk}(1-\rho_{kk})c_k}{1+c_k} + \rho_{kk}^2)\right] r_t^{[k]} + \frac{\rho_{kk}(1-\rho_{kk})c_k}{1+c_k}(r_t^{[k]})^2 \\
&\quad + \left[\lambda_{t+1}^{[k]} + c_k(\lambda_{t+1}^{[k]})^2\right] \\
&= \frac{\rho_{kk}(1-\rho_{kk})}{1+c_k} r_t^{[k]} + \frac{\rho_{kk}(1-\rho_{kk})c_k}{1+c_k}(r_t^{[k]})^2 + [\lambda_{t+1}^{[k]} + c_k(\lambda_{t+1}^{[k]})^2] \\
&= \frac{\rho_{kk}(1-\rho_{kk})}{1+c_k} r_t^{[k]}(1 + c_k r_t^{[k]}) + \left[\lambda_{t+1}^{[k]} + c_k(\lambda_{t+1}^{[k]})^2\right], \qquad (7.36)
\end{aligned}$$

with $r_t^{[k]}$ approximated by their corresponding means.

## 7.4 Simulation Study

Some Monte Carlo simulations are performed in this section with two covariate structures:

$$x_{t1} = \begin{cases} -cos(2\pi t) + 0.01 & (t = 1, \ldots, T/4) \\ sin(2\pi t) + 0.05 & (t = (T/4) + 1, \ldots, 3T/4) \\ cos(2\pi t) + 0.10 & (t = (3T/4) + 1, \ldots, T) \end{cases}$$

$$x_{t2} = \begin{cases} -t + 1 & (t = 1, \ldots, T/4) \\ 0 & (t = (T/4) + 1, \ldots, 3T/4) \\ t + 1 & (t = (3T/4) + 1, \ldots, T) \end{cases}$$

for population parameter $\beta_1^{[1]} = \beta_2^{[1]} = 0.3$ and $\beta_1^{[2]} = \beta_2^{[2]} = 0.5$. The over-dispersion coefficients $c_1$ and $c_2$ are assumed to be $[c_1, c_2]$=[0.3,0.7], [1.5,0.2] and [1.8,0.9], to illustrate the performance of the GQL under low and high over-dispersion. Furthermore, the different pair of serial-correlation $[\rho_{11}, \rho_{22}]$ are assumed at [0.9, 0.9], [0.3,0.9],[0.3,0.3] to highlight the performance of GQL under low and high serial-correlation combination, for all time size. The pair of innovation terms are simulated using a similar algorithm as illustrated in Section 6.4. Using the values of $r_t^{[1]}$ and $r_t^{[2]}$, $\alpha_{11,t} \sim \text{Beta}(\frac{\rho_{11}}{c_1}, \frac{1-\rho_{11}}{c_1})$,

$\alpha_{22,t} \sim \text{Beta}(\frac{\rho_{22}}{c_2}, \frac{1-\rho_{22}}{c_2})$, the $y_t^{[1]}$ and $y_t^{[2]}$ values are generated from Equation (7.1) for $T = 60, 100, 600$. Note from Equation (6.5) and the combinations of $[c_1, c_2]$, the values of $\nu = 0.5, 0.6, 1.4$ for the corresponding combinations [1.8,0.9], [1.5,0.2] and [0.3,0.7]. Tables 7.1–7.2 shows the simulated mean estimates of the different parameters based on 5000 Monte-Carlo replications for each combination:

**TABLE 7.1**
Estimates of the Parameters and Standard Errors under Non-Stationary BINMA(1) Process for $[c_1, c_2] = [0.3, 0.7], [1.5, 0.2]$

| $\rho_{11}$ | $\rho_{22}$ | T | $\hat{\beta}_1^{[1]}$ | $\hat{\beta}_2^{[1]}$ | $\hat{\beta}_1^{[2]}$ | $\hat{\beta}_2^{[2]}$ | $\hat{c}_1 = 0.3$ | $\hat{c}_2 = 0.7$ | $\hat{\rho}_{11}$ | $\hat{\rho}_{22}$ | $\hat{\nu} = 1.4$ |
|---|---|---|---|---|---|---|---|---|---|---|---|
| 0.9 | 0.9 | 60 | 0.2882 | 0.3195 | 0.4879 | 0.5121 | 0.2846 | 0.6832 | 0.8889 | 0.8836 | 1.3823 |
| | | | (0.1064) | (0.1187) | (0.1080) | (0.1135) | (0.1293) | (0.1384) | (0.0847) | (0.0960) | (0.1330) |
| | | 100 | 0.2908 | 0.2924 | 0.5069 | 0.4911 | 0.2922 | 0.6917 | 0.8909 | 0.8949 | 1.3941 |
| | | | (0.0909) | (0.0974) | (0.0860) | (0.0892) | (0.1061) | (0.1153) | (0.0636) | (0.0767) | (0.1147) |
| | | 600 | 0.2955 | 0.3033 | 0.4993 | 0.5021 | 0.2972 | 0.6962 | 0.8988 | 0.8985 | 1.3953 |
| | | | (0.0116) | (0.0206) | (0.0232) | (0.0373) | (0.0569) | (0.0770) | (0.0181) | (0.0221) | (0.0751) |
| 0.3 | 0.9 | 60 | 0.3163 | 0.2826 | 0.4831 | 0.5110 | 0.2832 | 0.7171 | 0.2829 | 0.8841 | 1.4183 |
| | | | (0.1071) | (0.1111) | (0.1228) | (0.1086) | (0.1362) | (0.1327) | (0.0966) | (0.0995) | (0.1409) |
| | | 100 | 0.2904 | 0.3094 | 0.4922 | 0.4902 | 0.2910 | 0.6911 | 0.2938 | 0.9078 | 1.3901 |
| | | | (0.0710) | (0.0958) | (0.0879) | (0.0776) | (0.1195) | (0.1012) | (0.0746) | (0.0682) | (0.1251) |
| | | 600 | 0.2967 | 0.2976 | 0.4985 | 0.5020 | 0.2962 | 0.6962 | 0.2955 | 0.9044 | 1.3993 |
| | | | (0.0237) | (0.0340) | (0.0103) | (0.0422) | (0.0739) | (0.0645) | (0.0111) | (0.0223) | (0.0842) |
| 0.3 | 0.3 | 60 | 0.2849 | 0.2860 | 0.5190 | 0.4824 | 0.2842 | 0.6887 | 0.2869 | 0.2893 | 1.3864 |
| | | | (0.1219) | (0.1180) | (0.1114) | (0.1003) | (0.1332) | (0.1402) | (0.1031) | (0.0948) | (0.1382) |
| | | 100 | 0.2906 | 0.2919 | 0.4921 | 0.4918 | 0.2912 | 0.7057 | 0.2907 | 0.2915 | 1.3903 |
| | | | (0.0718) | (0.0805) | (0.0814) | (0.0907) | (0.1051) | (0.1152) | (0.0723) | (0.0611) | (0.1161) |
| | | 600 | 0.2993 | 0.2981 | 0.4964 | 0.5002 | 0.2971 | 0.6991 | 0.2986 | 0.3014 | 1.3996 |
| | | | (0.0346) | (0.0260) | (0.0112) | (0.0386) | (0.0610) | (0.0791) | (0.0127) | (0.0282) | (0.0774) |
| $\rho_{11}$ | $\rho_{22}$ | T | $\hat{\beta}_1^{[1]}$ | $\hat{\beta}_2^{[1]}$ | $\hat{\beta}_1^{[2]}$ | $\hat{\beta}_2^{[2]}$ | $\hat{c}_1 = 1.5$ | $\hat{c}_2 = 0.2$ | $\hat{\rho}_{11}$ | $\hat{\rho}_{22}$ | $\hat{\nu} = 0.6$ |
| 0.9 | 0.9 | 60 | 0.2811 | 0.3116 | 0.4860 | 0.4885 | 1.5134 | 0.1806 | 0.8820 | 0.8846 | 0.5884 |
| | | | (0.0731) | (0.0708) | (0.0656) | (0.0748) | (0.1209) | (0.0844) | (0.0778) | (0.0607) | (0.1210) |
| | | 100 | 0.2935 | 0.2921 | 0.4935 | 0.4913 | 1.5066 | 0.1941 | 0.8902 | 0.8927 | 0.5910 |
| | | | (0.0676) | (0.0668) | (0.0651) | (0.0511) | (0.1061) | (0.0732) | (0.0634) | (0.0585) | (0.1082) |
| | | 600 | 0.2993 | 0.3015 | 0.4958 | 0.4960 | 1.5002 | 0.2964 | 0.8966 | 0.9019 | 0.5982 |
| | | | (0.0315) | (0.0323) | (0.0256) | (0.0242) | (0.0710) | (0.0591) | (0.0207) | (0.0115) | (0.0792) |
| 0.3 | 0.9 | 60 | 0.2823 | 0.2837 | 0.4953 | 0.5175 | 1.5114 | 0.2149 | 0.2855 | 0.8872 | 0.5809 |
| | | | (0.0784) | (0.0619) | (0.0605) | (0.0796) | (0.1015) | (0.0824) | (0.0776) | (0.0697) | (0.1211) |
| | | 100 | 0.2908 | 0.2926 | 0.5056 | 0.4936 | 1.4912 | 0.1922 | 0.2947 | 0.8933 | 0.5939 |
| | | | (0.0612) | (0.0558) | (0.0550) | (0.0637) | (0.0884) | (0.0713) | (0.0542) | (0.0557) | (0.0904) |
| | | 600 | 0.2970 | 0.2964 | 0.4977 | 0.5014 | 1.4991 | 0.2022 | 0.2990 | 0.8967 | 0.5958 |
| | | | (0.0123) | (0.0394) | (0.0114) | (0.0178) | (0.0403) | (0.0369) | (0.0242) | (0.0102) | (0.0612) |
| 0.3 | 0.3 | 60 | 0.2876 | 0.2864 | 0.5138 | 0.4860 | 1.4857 | 0.2112 | 0.2845 | 0.2829 | 0.5814 |
| | | | (0.0573) | (0.0669) | (0.0604) | (0.0780) | (0.1083) | (0.0861) | (0.0542) | (0.0592) | (0.1016) |
| | | 100 | 0.2924 | 0.3081 | 0.4919 | 0.5094 | 1.4931 | 0.1936 | 0.2914 | 0.2933 | 0.5911 |
| | | | (0.0412) | (0.0509) | (0.0569) | (0.0557) | (0.0801) | (0.0673) | (0.0419) | (0.0467) | (0.0849) |
| | | 600 | 0.2956 | 0.2985 | 0.4960 | 0.4966 | 1.4976 | 0.1979 | 0.2972 | 0.2956 | 0.5990 |
| | | | (0.0116) | (0.0219) | (0.0122) | (0.0274) | (0.0407) | (0.0349) | (0.0261) | (0.0158) | (0.0577) |

**TABLE 7.2**
Estimates of the Parameters and Standard Errors under Non-Stationary BINMA(1) Process for $[c_1, c_2] = [1.8, 0.9]$

| $\rho_{11}$ | $\rho_{22}$ | T | $\hat{\beta}_1^{[1]}$ | $\hat{\beta}_2^{[1]}$ | $\hat{\beta}_1^{[2]}$ | $\hat{\beta}_2^{[2]}$ | $\hat{c}_1 = 1.8$ | $\hat{c}_2 = 0.9$ | $\hat{\rho}_{11}$ | $\hat{\rho}_{22}$ | $\hat{\nu} = 0.5$ |
|---|---|---|---|---|---|---|---|---|---|---|---|
| 0.9 | 0.9 | 60 | 0.2873 | 0.2894 | 0.4824 | 0.5125 | 1.8151 | 0.8887 | 0.8848 | 0.8874 | 0.4843 |
| | | | (0.1153) | (0.1034) | (0.1391) | (0.1206) | (0.1449) | (0.1441) | (0.1005) | (0.1146) | (0.1387) |
| | | 100 | 0.2935 | 0.2902 | 0.5082 | 0.4932 | 1.7920 | 0.8943 | 0.9068 | 0.8907 | 0.4935 |
| | | | (0.0876) | (0.0741) | (0.0970) | (0.0969) | (0.1045) | (0.1114) | (0.0722) | (0.0710) | (0.1113) |
| | | 600 | 0.2955 | 0.2977 | 0.4951 | 0.5013 | 1.7980 | 0.8954 | 0.9010 | 0.8990 | 0.4971 |
| | | | (0.0327) | (0.0447) | (0.0334) | (0.0212) | (0.0704) | (0.0647) | (0.0303) | (0.0362) | (0.0738) |
| 0.3 | 0.9 | 60 | 0.2856 | 0.2823 | 0.5013 | 0.5108 | 1.7875 | 0.8861 | 0.2820 | 0.8825 | 0.5018 |
| | | | (0.1142) | (0.1164) | (0.1248) | (0.1116) | (0.1268) | (0.1376) | (0.1050) | (0.1091) | (0.1339) |
| | | 100 | 0.2916 | 0.3058 | 0.4904 | 0.4913 | 1.8075 | 0.8911 | 0.2912 | 0.8917 | 0.4924 |
| | | | (0.0802) | (0.0831) | (0.0925) | (0.0868) | (0.1103) | (0.1108) | (0.0771) | (0.0723) | (0.1042) |
| | | 600 | 0.2988 | 0.3011 | 0.4997 | 0.4995 | 1.8027 | 0.8973 | 0.2978 | 0.8980 | 0.4969 |
| | | | (0.0235) | (0.0205) | (0.0357) | (0.0150) | (0.0516) | (0.0666) | (0.0216) | (0.0137) | (0.0673) |
| 0.3 | 0.3 | 60 | 0.2849 | 0.2853 | 0.5147 | 0.5129 | 1.7836 | 0.8866 | 0.2890 | 0.3049 | 0.5117 |
| | | | (0.1217) | (0.1074) | (0.1289) | (0.1009) | (0.1370) | (0.1366) | (0.1030) | (0.0998) | (0.1231) |
| | | 100 | 0.2938 | 0.3086 | 0.4930 | 0.4928 | 1.8061 | 0.8925 | 0.2948 | 0.2916 | 0.4930 |
| | | | (0.0765) | (0.0786) | (0.0889) | (0.0811) | (0.1085) | (0.1121) | (0.0783) | (0.0551) | (0.1086) |
| | | 600 | 0.3017 | 0.2951 | 0.5028 | 0.4985 | 1.7952 | 0.8998 | 0.2972 | 0.2976 | 0.4970 |
| | | | (0.0257) | (0.0334) | (0.0229) | (0.0296) | (0.0571) | (0.0558) | (0.0379) | (0.0152) | (0.0678) |

Note that the initial values of the model parameters were as follows: $[\beta_1^{[1]} = \beta_2^{[1]} = 0.01, \beta_1^{[2]} = \beta_2^{[2]} = 0.02, \rho_{11} = 0.1, \rho_{22} = 0.15, c_1 = 0.12, c_2 = 0.17, \nu = 5]$. The results in Tables 7.1–7.2 show that the GQL method provides consistent estimates and as time size increases, the standard errors decrease irrespective of the high or low serial- and cross-dependence parameters. We also note that the over-dispersion parameters exhibit slightly higher standard errors compared to other parameters. However, some non-convergent simulations are noted for almost all cluster sizes. For instance, when $c_1 = 0.3, c_2 = 0.7, \rho_{11} = 0.3, \rho_{22} = 0.3$ and $\nu = 1.4$, the non-convergent simulations turn around 245 for $T = 60$, 220 for $T = 100$ and 165 for $T = 600$. As the values of $\rho_{11}$ and $\rho_{22}$ increase, that is, $\rho_{11} = 0.9, \rho_{22} = 0.9$, the number of non-convergent simulations is approximately 280 for $T = 60$, 245 for $T = 100$ and 180 for $T = 600$. We also observe that as the values of the over-dispersion parameters increase, the corresponding standard errors are higher compared to the other parameters and there is a slight increase in the number of non-convergent simulations. For instance, when $c_1 = 1.8, c_2 = 0.9, \rho_{11} = 0.9, \rho_{22} = 0.9$ and $\nu = 0.5$, the non-convergent simulations turn around 310 for $T = 60$, 270 for $T = 100$ and 195 for $T = 600$. These non-convergence were due mainly to the auto-covariance matrix (7.34) being close to singularity and where the values of $\nu$ and the serial parameters were out of range. Overall the performance of GQL approach is satisfactory.

## 7.5 Summary

In this chapter, the constrained BINMA(1) model with NB innovations has been developed by deriving the marginal and joint moments. The estimation methodology has been based on a two-phased GQL approach where the first GQL equation estimates the regression effects and the serial-correlation while the second GQL equation computes the vector of over-dispersion parameters based on some higher-order moments. The Monte Carlo results have illustrated that the GQL method yields consistent estimates of the different parameters. The next chapter extends the constrained bivariate models with NB innovations to an unconstrained bivariate models with NB innovations, by allowing for a full correlation structure.

# 8

# *Unconstrained BINAR(1) Model with NB Innovations*

## 8.1 Introduction

Following Chapter 5 and motivated by the findings of Nastic et al. (2016b), we develop the unconstrained BINAR(1) model with NB innovations where the cross-correlation is considered from the relationship of the counting series at the present time point with the previous-lagged survival elements of the series. Moreover, it is important to investigate the quality of fit under the constrained BINAR(1) model introduced in Chapter 6 with the proposed model in this chapter. Thus, this chapter extends the model developed in Chapter 6, but by relaxing the assumption of dependence between the innovation terms. We also develop a GQL estimation method to estimate the unknown model parameters.

An overview of this chapter is given as follows: In Section 8.2, the BINAR(1) model with NB innovations is developed under stationary and non-stationary moment conditions. This section is followed by the development of two GQL equations and the method to solve them. In the same section, we present the one-step ahead forecasting equations under the new BINAR(1) model. The conclusion is presented in the last section.

## 8.2 Model Construction

Let the random coefficient BINAR(1) model be expressed as:

$$Y_t^{[1]} = \alpha_{11,t} \circ Y_{t-1}^{[1]} + \alpha_{12,t} \circ Y_{t-1}^{[2]} + R_t^{[1]} \quad \text{and} \tag{8.1}$$

$$Y_t^{[2]} = \alpha_{21,t} \circ Y_{t-1}^{[1]} + \alpha_{22,t} \circ Y_{t-1}^{[2]} + R_t^{[2]}, \tag{8.2}$$

 DOI: 10.1201/9781003600435-8

where $Y_t^{[k]}$ is the counting series and $R_t^{[k]}$ is its corresponding innovation term for $k \in \{1,2\}$ and $t \in \mathbb{Z}$. $R_t^{[k]}$ follows marginally the NB model specified as:

$$P_{R_t^{[k]}}(r^{[k]}) = \frac{\Gamma({c_k}^{-1} + r^{[k]})}{\Gamma({c_k}^{-1}) r^{[k]}!} \left(\frac{1}{1 + c_k \lambda_t^{[k]}}\right)^{{c_k}^{-1}} \left(\frac{c_k \lambda_t^{[k]}}{1 + c_k \lambda_t^{[k]}}\right)^{r^{[k]}},$$
$$c_k, \lambda_t^{[k]} > 0, \quad r^{[k]} \geq 0 \tag{8.3}$$

with $\mathrm{E}(R_t^{[k]}) = \lambda_t^{[k]}$ and $\mathrm{Var}(R_t^{[k]}) = \lambda_t^{[k]}(1 + c_k \lambda_t^{[k]})$. For convenience, we write $R_t^{[k]} \sim \mathrm{NB}(\frac{1}{c_k}, c_k \lambda_t^{[k]})$. As for the pair of innovation terms $(R_t^{[1]}, R_t^{[2]})$, we assume $\mathrm{Corr}(R_t^{[1]}, R_t^{[2]}) = 0$. In the event there is a common set of explanatory variables, of let's say $p$ dimension, influencing $Y_t^{[1]}$ and $Y_t^{[2]}$, $\boldsymbol{x}_t = [x_{t1}, x_{t2}, \ldots, x_{tj}, \ldots, x_{tp}]'$ with regression effects $\boldsymbol{\beta}^{[k]} = [\beta_1^{[k]}, \beta_2^{[k]}, \ldots, \beta_j^{[k]}, \ldots, \beta_p^{[k]}]'$, then we assume $\lambda_t^{[k]} = \exp(\boldsymbol{x}_t' \boldsymbol{\beta}^{[k]})$ for $k \in \{1,2\}$.

'$\circ$' denotes the binomial thinning operator (Steutel & Van Harn (1979)), which indicates

$$\alpha_{jk,t} \circ Y_{t-1}^{[k]} = \begin{cases} \sum_{l=1}^{Y_{t-1}^{[k]}} b_l(\alpha_{jk,t}), & Y_{t-1}^{[k]} > 0, \\ 0 & Y_{t-1}^{[k]} = 0 \end{cases} \tag{8.4}$$

that is, $\{\alpha_{jk,t} \circ Y_{t-1}^{[k]} \mid Y_{t-1}^{[k]}\} \sim \mathrm{Binomial}(Y_{t-1}^{[k]}, \alpha_{jk,t})$. In the above, $\alpha_{jk,t}$ is assumed a random Beta-distributed variable with $\alpha_{jk,t} \sim \mathrm{Beta}(\frac{\rho_{jk}}{c_k}, \frac{1-\rho_{jk}}{c_k})$, such that $\mathrm{E}(\alpha_{jk,t}) = \rho_{jk}$, $\mathrm{Var}(\alpha_{jk,t}) = \frac{\rho_{jk}(1-\rho_{jk})c_k}{1+c_k}$ for $\rho_{jk} \in [0,1]$, $j,k \in \{1,2\}$ (See McKenzie (1986) and Jowaheer & Sutradhar (2002)).

Below, we present the assumptions related to Equations (8.1) and (8.2):

(a) The random variables $\{\alpha_{jk,t}\}$, $j,k \in \{1,2\}$, are mutually independent sequences and identically distributed.

(b) The random variables $\alpha_{jk,t}$ and $Y_{t-1}^{[k]}$ are independent for all $j,k \in \{1,2\}$.

Equations (8.1) and (8.2) can be re-written in matrix form as

$$\boldsymbol{Y}_t = \boldsymbol{A}_t \circ \boldsymbol{Y}_{t-1} + \boldsymbol{R}_t, \tag{8.5}$$

where $\boldsymbol{A}_t = \begin{pmatrix} \alpha_{11,t} & \alpha_{12,t} \\ \alpha_{21,t} & \alpha_{22,t} \end{pmatrix}$, $\boldsymbol{R}_t = [R_t^{[1]}, R_t^{[2]}]'$ and $\boldsymbol{Y}_t = [Y_t^{[1]}, Y_t^{[2]}]'$.

Based on the above conditions and using Lemma 6.1, we derive the marginal and joint moments of the BINAR(1) model:

$$\begin{aligned} E(\boldsymbol{Y}_t) &= E(\boldsymbol{A}_t \circ \boldsymbol{Y}_{t-1}) + E(\boldsymbol{R}_t) \\ &= E(\boldsymbol{A}_t) E(\boldsymbol{Y}_{t-1}) + E(\boldsymbol{R}_t) \\ &= \boldsymbol{A} \boldsymbol{\mu}_{t-1} + \boldsymbol{\lambda}_t, \end{aligned} \tag{8.6}$$

where $\boldsymbol{A} = \begin{pmatrix} \rho_{11} & \rho_{12} \\ \rho_{21} & \rho_{22} \end{pmatrix}$, $\boldsymbol{\mu}_{t-1} = [\mu_{t-1}^{[1]}, \mu_{t-1}^{[2]}]'$ and $\boldsymbol{\lambda}_t = [\lambda_t^{[1]}, \lambda_t^{[2]}]'$.
Expanding Equation (8.6), we obtain

$$\mu_t^{[1]} \equiv E(Y_t^{[1]}) = \rho_{11}\mu_{t-1}^{[1]} + \rho_{12}\mu_{t-1}^{[2]} + \lambda_t^{[1]}, \quad \mu_t^{[2]} \equiv E(Y_t^{[2]}) \\ = \rho_{21}\mu_{t-1}^{[1]} + \rho_{22}\mu_{t-1}^{[2]} + \lambda_t^{[2]}. \tag{8.7}$$

As for the variance-covariance matrix, it is denoted by

$$\text{Var}(\boldsymbol{Y_t}) = \boldsymbol{\Sigma}_{0,t} = \begin{bmatrix} \text{Var}(Y_t^{[1]}) & \text{Cov}(Y_t^{[1]}, Y_t^{[2]}) \\ \text{Cov}(Y_t^{[2]}, Y_t^{[1]}) & \text{Var}(Y_t^{[2]}) \end{bmatrix}, \tag{8.8}$$

where

$$\text{Var}(\boldsymbol{Y}_t) = \text{Var}(\boldsymbol{A}_t \circ \boldsymbol{Y}_{t-1}) + \text{Var}(\boldsymbol{R}_t). \tag{8.9}$$

Using the law of total variance, we obtain

$$\text{Var}(\boldsymbol{A}_t \circ \boldsymbol{Y}_{t-1}) = \text{Var}[E(\boldsymbol{A}_t \circ \boldsymbol{Y}_{t-1} | \boldsymbol{Y}_{t-1}, \boldsymbol{A}_t)] + E[\text{Var}(\boldsymbol{A}_t \circ \boldsymbol{Y}_{t-1} | \boldsymbol{Y}_{t-1}, \boldsymbol{A}_t)], \tag{8.10}$$

where $\text{Var}(\alpha_{ij,t} \circ Y_{t-1}^{[j]} | \boldsymbol{Y}_t, \boldsymbol{A}_t) = \alpha_{ij,t}(1 - \alpha_{ij,t})Y_{t-1}^{[j]}$ for $i, j \in \{1, 2\}$.

Hence, the entries of the variance-covariance matrix (8.8) are given by

$$\begin{aligned}
&\text{Var}(Y_t^{[1]}) \\
&\quad = \text{Var}(\alpha_{11,t}Y_{t-1}^{[1]}) + \text{Var}(\alpha_{12,t}Y_{t-1}^{[2]}) + E(\alpha_{11,t}(1 - \alpha_{11,t}))E(Y_{t-1}^{[1]}) \\
&\quad\quad + E(\alpha_{12,t}(1 - \alpha_{12,t}))E(Y_{t-1}^{[2]}) + \text{Var}(R_t^{[1]}) + 2\rho_{11}\rho_{12}\text{Cov}(Y_{t-1}^{[1]}, Y_{t-1}^{[2]}) \\
&\quad = \left[\frac{\rho_{11}(1 - \rho_{11})c_1}{1 + c_1} + \rho_{11}^2\right] [\text{Var}(Y_{t-1}^{[1]}) + {\mu_{t-1}^{[1]}}^2] - \rho_{11}^2 {\mu_{t-1}^{[1]}}^2 \\
&\quad\quad + \left[\frac{\rho_{12}(1 - \rho_{12})c_2}{1 + c_2} + \rho_{12}^2\right] [\text{Var}(Y_{t-1}^{[2]}) + {\mu_{t-1}^{[2]}}^2] - \rho_{12}^2 {\mu_{t-1}^{[2]}}^2 \\
&\quad\quad + \left[\rho_{11} - \frac{\rho_{11}(1 - \rho_{11})c_1}{1 + c_1} - \rho_{11}^2\right] \mu_{t-1}^{[1]} + \left[\rho_{12} - \frac{\rho_{12}(1 - \rho_{12})c_2}{1 + c_2} - \rho_{12}^2\right] \mu_{t-1}^{[2]} \\
&\quad\quad + \lambda_t^{[1]} + c_1 {\lambda_t^{[1]}}^2 + 2\rho_{11}\rho_{12}\text{Cov}(Y_{t-1}^{[1]}, Y_{t-1}^{[2]}) \\
&\quad = \left[\frac{\rho_{11}(\rho_{11} + c_1)}{1 + c_1}\right] \text{Var}(Y_{t-1}^{[1]}) + \left[\frac{\rho_{11}(1 - \rho_{11})c_1}{1 + c_1}\right] {\mu_{t-1}^{[1]}}^2 \\
&\quad\quad + \left[\frac{\rho_{12}(\rho_{12} + c_2)}{1 + c_2}\right] \text{Var}(Y_{t-1}^{[2]}) \\
&\left[\frac{\rho_{12}(1 - \rho_{12})c_2}{1 + c_2}\right] {\mu_{t-1}^{[2]}}^2 - \left[\frac{\rho_{11}(\rho_{11} + c_1)}{1 + c_1}\right] \mu_{t-1}^{[1]} - \left[\frac{\rho_{12}(\rho_{12} + c_2)}{1 + c_2}\right] \mu_{t-1}^{[2]}
\end{aligned}$$

$$
\begin{aligned}
&+ (\rho_{11}\mu_{t-1}^{[1]} + \rho_{11}\mu_{t-1}^{[2]} + \lambda_t^{[1]}) + c_1\lambda_t^{[1]^2} + 2\rho_{11}\rho_{12}\text{Cov}(Y_{t-1}^{[1]}, Y_{t-1}^{[2]}) \\
&= \mu_t^{[1]} + 2\rho_{11}\rho_{12}\text{Cov}(Y_{t-1}^{[1]}, Y_{t-1}^{[2]}) + c_1\lambda_t^{[1]^2} \\
&+ \left[\frac{\rho_{11}(1-\rho_{11})}{1+c_1}\right] c_1\mu_{t-1}^{[1]^{\ 2}} + \frac{\rho_{11}(\rho_{11}+c_1)}{1+c_1}[\text{Var}(Y_{t-1}^{[1]}) - \mu_{t-1}^{[1]}] \\
&+ \left[\frac{\rho_{12}(1-\rho_{12})}{1+c_2}\right] c_2\mu_{t-1}^{[2]^{\ 2}} + \frac{\rho_{12}(\rho_{12}+c_2)}{1+c_2}[\text{Var}(Y_{t-1}^{[2]}) - \mu_{t-1}^{[2]}].
\end{aligned} \tag{8.11}
$$

Similarly,

$$
\begin{aligned}
\text{Var}(Y_t^{[2]}) = \mu_t^{[2]} &+ 2\rho_{21}\rho_{22}\text{Cov}(Y_{t-1}^{[1]}, Y_{t-1}^{[2]}) + c_2\lambda_t^{[2]^2} \\
&+ \left[\frac{\rho_{21}(1-\rho_{21})}{1+c_1}\right] c_1\mu_{t-1}^{[1]^{\ 2}} + \frac{\rho_{21}(\rho_{21}+c_1)}{1+c_1}[\text{Var}(Y_{t-1}^{[1]}) - \mu_{t-1}^{[1]}] \\
&+ \left[\frac{\rho_{22}(1-\rho_{22})}{1+c_2}\right] c_2\mu_{t-1}^{[2]^{\ 2}} + \frac{\rho_{22}(\rho_{22}+c_2)}{1+c_2}[\text{Var}(Y_{t-1}^{[2]}) - \mu_{t-1}^{[2]}]
\end{aligned} \tag{8.12}
$$

and

$$
\begin{aligned}
&\text{Cov}(Y_t^{[1]}, Y_t^{[2]}) \\
&= (\rho_{11}\rho_{22} + \rho_{12}\rho_{21})\text{Cov}(Y_{t-1}^{[1]}, Y_{t-1}^{[2]}) + \rho_{11}\rho_{21}\text{Var}(Y_{t-1}^{[1]}) + \rho_{12}\rho_{22}\text{Var}(Y_{t-1}^{[2]}).
\end{aligned} \tag{8.13}
$$

**Remark 8.1.** Under (8.5), we can use the self-decomposability property by Steutel & Van Harn (1986) and McKenzie (1988, 1991), $G_{\boldsymbol{Y_t}}(s) = G_{\boldsymbol{Y_{t-1}}}(\boldsymbol{A_t}s)G_{\boldsymbol{R_t}}(s)$. From Pedeli & Karlis (2013a), $\boldsymbol{Y_t}$ in Equation (8.5) can be expressed as $\boldsymbol{Y}_t = \sum_{i=0}^{\infty} \boldsymbol{A}_t^i \circ \boldsymbol{R}_{t-1} + \boldsymbol{R}_{t-i} = \sum_{i=0}^{\infty} \boldsymbol{Z}_i$, then in deriving $G_{\boldsymbol{Y_t}}(s)$, we require $G_{\boldsymbol{Z_i}|\boldsymbol{R_t}}$. The derivation of $G_{\boldsymbol{Z_i}|\boldsymbol{R_t}}$ is extremely difficult here because $\boldsymbol{A}_t$ is a random beta-distributed matrix and $\boldsymbol{R}_t$ is non-stationary and hence an equation of the form Equation (10) in Pedeli & Karlis (2013a) is almost impossible to derive. Thus, the marginal distribution of $Y_t^{[k]}$ is difficult to obtain under the non-stationary assumption.

**Remark 8.2.** Under stationary moment condition,

$$
E(Y_t^{[1]}) = \mu_1^{[1]} = \frac{\rho_{12}\mu_1^{[2]} + \lambda_1^{[1]}}{1-\rho_{11}}, \quad E(Y_t^{[2]}) = \mu_1^{[2]} = \frac{\rho_{12}\mu_1^{[1]} + \lambda_1^{[2]}}{1-\rho_{22}}, \tag{8.14}
$$

$$
\begin{aligned}
\text{Var}(Y_t^{[1]}) = &\left[\frac{1+c_1}{1+c_1-\rho_{11}(\rho_{11}+c_1)}\right]\{\mu_1^{[1]} + 2\rho_{11}\rho_{12}\text{Cov}(Y_t^{[1]}, Y_t^{[2]}) + c_1\lambda_1^{[1]^2} \\
&+ \left[\frac{\rho_{11}(1-\rho_{11})}{1+c_1}\right] c_1\mu_1^{[1]^2} - \frac{\rho_{11}(\rho_{11}+c_1)}{1+c_1}\mu_1^{[1]} + \left[\frac{\rho_{12}(1-\rho_{12})}{1+c_2}\right] c_2\mu_1^{[2]^2} \\
&+ \frac{\rho_{12}(\rho_{12}+c_2)}{1+c_2}[\text{Var}(Y_t^{[2]}) - \mu_1^{[2]}]\},
\end{aligned} \tag{8.15}
$$

$$\begin{aligned}\mathrm{Var}(Y_t^{[2]}) &= \left[\frac{1+c_2}{1+c_2-\rho_{22}(\rho_{22}+c_2)}\right]\{\mu_1^{[2]} + 2\rho_{21}\rho_{22}\mathrm{Cov}(Y_t^{[1]}, Y_t^{[2]}) + c_2{\lambda_1^{[2]}}^2 \\ &+ \left[\frac{\rho_{21}(1-\rho_{21})}{1+c_1}\right] c_1{\mu_1^{[1]}}^2 + \frac{\rho_{21}(\rho_{21}+c_1)}{1+c_1}[\mathrm{Var}(Y_t^{[1]}) - \mu_1^{[1]}] \\ &+ \left[\frac{\rho_{22}(1-\rho_{22})}{1+c_2}\right] c_2{\mu_1^{[2]}}^2 - \frac{\rho_{22}(\rho_{22}+c_2)}{1+c_2}\mu_1^{[2]}\},\end{aligned} \tag{8.16}$$

$$\mathrm{Cov}(Y_t^{[1]}, Y_t^{[2]}) = \frac{\rho_{11}\rho_{21}\mathrm{Var}(Y_t^{[1]}) + \rho_{12}\rho_{22}\mathrm{Var}(Y_t^{[2]})}{1-\rho_{11}\rho_{22}-\rho_{12}\rho_{21}} \tag{8.17}$$

and when solving simultaneously, we obtain

$$\mu_1^{[1]} = \frac{(1-\rho_{22})\lambda_1^{[1]} + \rho_{12}\lambda_1^{[2]}}{(1-\rho_{11})(1-\rho_{22})-\rho_{12}\rho_{21}}, \quad \mu_1^{[2]} = \frac{(1-\rho_{11})\lambda_1^{[2]} + \rho_{21}\lambda_1^{[1]}}{(1-\rho_{11})(1-\rho_{22})-\rho_{12}\rho_{21}}, \tag{8.18}$$

$$\begin{aligned}\mathrm{Var}(Y_t^{[1]}) &= \{(\delta_1)[2\rho_{12}\rho_{21}\rho_{22}^2 + (\frac{\rho_{22}(\rho_{22}+c_2)}{1+c_2})(1-\rho_{11}\rho_{22}+\rho_{12}\rho_{21}) \\ &+ \rho_{11}\rho_{22}+\rho_{12}\rho_{21}-1] + (\delta_2)[(\frac{\rho_{22}(\rho_{22}+c_2)}{1+c_2})(\rho_{11}\rho_{22}+\rho_{12}\rho_{21}-1) \\ &- 2\rho_{11}\rho_{12}^2\rho_{22}]\}/[\delta_3],\end{aligned} \tag{8.19}$$

$$\begin{aligned}\mathrm{Var}(Y_t^{[2]}) &= \{(\delta_1)[(\frac{\rho_{21}(\rho_{21}+c_1)}{1+c_1})(\rho_{11}\rho_{22}+\rho_{12}\rho_{21}-1) - 2\rho_{11}\rho_{21}^2\rho_{22}] \\ &+ (\delta_2)[2\rho_{11}^2\rho_{12}\rho_{21} + (\frac{\rho_{11}(\rho_{11}+c_1)}{1+c_1})(1-\rho_{11}\rho_{22}+\rho_{12}\rho_{21}) \\ &+ \rho_{11}\rho_{22}+\rho_{12}\rho_{21}-1]\}/[\delta_3],\end{aligned} \tag{8.20}$$

$$\begin{aligned}&\mathrm{Cov}(Y_t^{[1]}, Y_t^{[2]}) \\ &= \{(\delta_1)[(\frac{\rho_{22}(\rho_{22}+c_2)}{1+c_2})\rho_{11}\rho_{21} - \rho_{11}\rho_{21} - (\frac{\rho_{21}(\rho_{21}+c_1)}{1+c_1})\rho_{12}\rho_{22}] \\ &+ (\delta_2)[\rho_{12}\rho_{22}(\frac{\rho_{11}(\rho_{11}+c_1)}{1+c_1}) - \rho_{12}\rho_{22} - (\frac{\rho_{12}(\rho_{12}+c_2)}{1+c_2})\rho_{11}\rho_{21}]\}/[\delta_3],\end{aligned} \tag{8.21}$$

with

$$\delta_1 = [(\frac{\rho_{11}(1-\rho_{11})}{1+c_1})\mu_1^{[1]}(1+c_1\mu_1^{[1]}) + (\frac{\rho_{12}(1-\rho_{12})}{1+c_2})\mu_1^{[2]}(1+c_2\mu_1^{[2]}) + \lambda_1^{[1]} + c_1{\lambda_1^{[1]}}^2], \tag{8.22}$$

$$\delta_2=[(\frac{\rho_{21}(1-\rho_{21})}{1+c_1})\mu_1^{[1]}(1+c_1\mu_1^{[1]})+(\frac{\rho_{22}(1-\rho_{22})}{1+c_2})\mu_1^{[2]}(1+c_2\mu_1^{[2]})+\lambda_1^{[2]}+c_2{\lambda_1^{[2]}}^2], \tag{8.23}$$

$$\begin{aligned}\delta_3 = \{&(\frac{\rho_{11}(\rho_{11}+c_1)}{1+c_1})[(\frac{\rho_{22}(\rho_{22}+c_2)}{1+c_2})(\rho_{11}\rho_{22}+\rho_{12}\rho_{21}-1)\\ &-2\rho_{12}\rho_{21}\rho_{22}^2-\rho_{11}\rho_{22}-\rho_{12}\rho_{21}+1]\\ &+(\frac{\rho_{12}(\rho_{12}+c_2)}{1+c_2})[2\rho_{11}\rho_{21}^2\rho_{22}+(\frac{\rho_{21}(\rho_{21}+c_1)}{1+c_1})(1-\rho_{11}\rho_{22}+\rho_{12}\rho_{21})]\\ &+2\rho_{11}\rho_{12}[(\frac{\rho_{21}(\rho_{21}+c_1)}{1+c_1})\rho_{12}\rho_{22}+\rho_{11}\rho_{21}(1-\frac{\rho_{22}(\rho_{22}+c_2)}{1+c_2})]\\ &+(\frac{\rho_{22}(\rho_{22}+c_2)}{1+c_2})(1-\rho_{11}\rho_{22}-\rho_{12}\rho_{21})+2\rho_{12}\rho_{21}\rho_{22}^2+\rho_{11}\rho_{22}+\rho_{12}\rho_{21}+1\}.\end{aligned} \tag{8.24}$$

In general, we express the covariance matrix at lag-$h$ as follows:

$$\boldsymbol{\Sigma}_{h,t} = \begin{bmatrix} \text{Cov}(Y_t^{[1]}, Y_{t+h}^{[1]}) & \text{Cov}(Y_{t+h}^{[1]}, Y_t^{[2]}) \\ \text{Cov}(Y_t^{[1]}, Y_{t+h}^{[2]}) & \text{Cov}(Y_t^{[2]}, Y_{t+h}^{[2]}) \end{bmatrix} = \begin{bmatrix} \Sigma_{11,h,t} & \Sigma_{12,h,t} \\ \Sigma_{21,h,t} & \Sigma_{22,h,t} \end{bmatrix}. \tag{8.25}$$

When $h = 1$,

$$\begin{aligned}&\text{Cov}(Y_t^{[1]}, Y_{t+1}^{[1]})\\ &\quad= \text{Cov}[Y_t^{[1]}, (\alpha_{11,t}\circ Y_t^{[1]} + \alpha_{12,t}\circ Y_t^{[2]} + R_{t+1}^{[1]})]\\ &\quad= \text{Cov}(Y_t^{[1]}, \alpha_{11,t}\circ Y_t^{[1]}) + \text{Cov}(Y_t^{[1]}, \alpha_{12,t}\circ Y_t^{[2]}) + \text{Cov}(Y_t^{[1]}, R_{t+1}^{[1]})\\ &\quad= \rho_{11}\text{Cov}(Y_t^{[1]}, Y_t^{[1]}) + \rho_{12}\text{Cov}(Y_t^{[1]}Y_t^{[2]})\\ &= \rho_{11}\text{Var}(Y_t^{[1]}) + \rho_{12}\text{Cov}(Y_t^{[1]}, Y_t^{[2]}).\end{aligned} \tag{8.26}$$

Similarly,

$$\text{Cov}(Y_t^{[2]}, Y_{t+1}^{[2]}) = \rho_{22}\text{Var}(Y_t^{[2]}) + \rho_{21}\text{Cov}(Y_t^{[2]}, Y_t^{[1]}). \tag{8.27}$$

Thus,

$$\Sigma_{11,1,t} = \rho_{11}\Sigma_{11,0,t} + \rho_{12}\Sigma_{21,0,t}$$
$$\Sigma_{22,1,t} = \rho_{22}\Sigma_{22,0,t} + \rho_{21}\Sigma_{12,0,t}$$

As for the cross-covariances,

$$\begin{aligned}\text{Cov}(Y_t^{[1]}, Y_{t+1}^{[2]}) &= \text{Cov}[Y_t^{[1]}, (\alpha_{21,t}\circ Y_t^{[1]} + \alpha_{22,t}\circ Y_t^{[2]} + R_{t+1}^{[2]})]\\ &= \rho_{21}\text{Var}(Y_t^{[1]}) + \rho_{22}\text{Cov}(Y_t^{[1]}, Y_t^{[2]}).\end{aligned} \tag{8.28}$$

Similarly,

$$\text{Cov}(Y_t^{[2]}, Y_{t+1}^{[1]}) = \rho_{12}\text{Var}(Y_t^{[2]}) + \rho_{11}\text{Cov}(Y_t^{[2]}, Y_t^{[1]}). \tag{8.29}$$

Thus,

$$\Sigma_{21,1,t} = \rho_{21}\Sigma_{11,0,t} + \rho_{22}\Sigma_{21,0,t}$$
$$\Sigma_{12,1,t} = \rho_{12}\Sigma_{22,0,t} + \rho_{11}\Sigma_{12,0,t}$$

$$\boldsymbol{\Sigma}_{1,t} = \begin{bmatrix} \Sigma_{11,1,t} & \Sigma_{12,1,t} \\ \Sigma_{21,1,t} & \Sigma_{22,1,t} \end{bmatrix} = \begin{bmatrix} E(\alpha_{11,t}) & E(\alpha_{12,t}) \\ E(\alpha_{21,t}) & E(\alpha_{22,t}) \end{bmatrix} \begin{bmatrix} \Sigma_{11,0,t} & \Sigma_{12,0,t} \\ \Sigma_{21,0,t} & \Sigma_{22,0,t} \end{bmatrix}. \tag{8.30}$$

Hence, $\boldsymbol{\Sigma}_{1,t} = \begin{bmatrix} \rho_{11} & \rho_{12} \\ \rho_{21} & \rho_{22} \end{bmatrix} \boldsymbol{\Sigma}_{0,t}$.

When $h = 2$,
$\Sigma_{11,2,t} = \rho_{11}\Sigma_{11,1,t} + \rho_{12}\Sigma_{21,1,t}$
$\Sigma_{22,2,t} = \rho_{22}\Sigma_{22,1,t} + \rho_{21}\Sigma_{12,1,t}$
and
$\Sigma_{21,2,t} = \rho_{21}\Sigma_{11,1,t} + \rho_{22}\Sigma_{21,1,t}$
$\Sigma_{12,2,t} = \rho_{12}\Sigma_{22,1,t} + \rho_{11}\Sigma_{12,1,t}$

Hence, $\boldsymbol{\Sigma}_{2,t} = \begin{bmatrix} \rho_{11} & \rho_{12} \\ \rho_{21} & \rho_{22} \end{bmatrix} \boldsymbol{\Sigma}_{1,t} = \begin{bmatrix} \rho_{11} & \rho_{12} \\ \rho_{21} & \rho_{22} \end{bmatrix}^2 \boldsymbol{\Sigma}_{0,t}$.

In general,

$$\Sigma_{11,h,t} = \rho_{11}\Sigma_{11,h-1,t} + \rho_{12}\Sigma_{21,h-1,t}$$
$$\Sigma_{22,h,t} = \rho_{22}\Sigma_{22,h-1,t} + \rho_{21}\Sigma_{12,h-1,t}$$

and

$$\Sigma_{21,h,t} = \rho_{21}\Sigma_{11,h-1,t} + \rho_{22}\Sigma_{21,h-1,t}$$
$$\Sigma_{12,h,t} = \rho_{12}\Sigma_{22,h-1,t} + \rho_{11}\Sigma_{12,h-1,t}$$

Hence,

$$\boldsymbol{\Sigma}_{h,t} = \begin{bmatrix} E(\alpha_{11,t}) & E(\alpha_{12,t}) \\ E(\alpha_{21,t}) & E(\alpha_{22,t}) \end{bmatrix}^h \boldsymbol{\Sigma}_{0,t} \quad \text{or} \quad \boldsymbol{\Sigma}_{h,t} = \begin{bmatrix} \rho_{11} & \rho_{12} \\ \rho_{21} & \rho_{22} \end{bmatrix}^h \boldsymbol{\Sigma}_{0,t}. \tag{8.31}$$

## 8.3 Estimation of Parameters

As discussed in Section 8.1, the estimation of parameters is done using the GQL estimating equation. The regression effects are introduced through the link predictor of the innovation term as follows: $\lambda_t^{[k]} = \exp(\boldsymbol{x}_t^{'}\boldsymbol{\beta}^{[k]})$. In the first stage, GQL-I is implemented using the scores $\boldsymbol{f}_t = [Y_t^{[1]}, Y_t^{[2]}]'$ to estimate the regression and dependence parameters while in the second phase, the updated estimates from GQL-I are used to estimate the over-dispersion parameters via GQL-II. For GQL-II, the scores are conditioned on the previous set of observations as $\boldsymbol{f}^*_{t\,|t-1} = [{Y_t^{[1]}}^2 | y_{t-1}^{[1]}, y_{t-1}^{[2]}, {Y_t^{[2]}}^2 | y_{t-1}^{[1]}, y_{t-1}^{[2]}]'$. However, the computation of the auto-covariance function of $\boldsymbol{f}^*_{t\,|t-1}$ is not easily derivable. In this context, we propose some new lemmas related to the thinning mechanism to derive these entries. More details are provided in the following subsections.

### 8.3.1 GQL-I

GQL-I is specified as:

$$\left(\frac{\partial \boldsymbol{\mu}}{\partial \boldsymbol{\beta}, \partial \boldsymbol{\rho}}\right)' \boldsymbol{\Sigma}_{\boldsymbol{\beta},\boldsymbol{\rho}}{}^{-1}(\boldsymbol{f} - \boldsymbol{\mu}) = 0 \tag{8.32}$$

with $\boldsymbol{f} = [\boldsymbol{f}_1, \boldsymbol{f}_2, \ldots, \boldsymbol{f}_t, \ldots, \boldsymbol{f}_T]'$ where $\boldsymbol{f}_t = [Y_t^{[1]}, Y_t^{[2]}]'$ and $E(\boldsymbol{f}) = \boldsymbol{\mu} = [\boldsymbol{\mu}_1, \boldsymbol{\mu}_2, \ldots, \boldsymbol{\mu}_t, \ldots, \boldsymbol{\mu}_T]'$ $\boldsymbol{\mu}_t = [\mu_t^{[1]}, \mu_t^{[2]}]'$.
Details on the above components are expressed as follows:

- $\left(\frac{\partial \boldsymbol{\mu}}{\partial \boldsymbol{\beta}, \partial \boldsymbol{\rho}}\right) = \left[\left(\frac{\partial \boldsymbol{\mu_1}}{\partial \boldsymbol{\beta}, \partial \boldsymbol{\rho}}\right), \left(\frac{\partial \boldsymbol{\mu_2}}{\partial \boldsymbol{\beta}, \partial \boldsymbol{\rho}}\right), \ldots, \left(\frac{\partial \boldsymbol{\mu_t}}{\partial \boldsymbol{\beta}, \partial \boldsymbol{\rho}}\right), \ldots, \left(\frac{\partial \boldsymbol{\mu_T}}{\partial \boldsymbol{\beta}, \partial \boldsymbol{\rho}}\right)\right]'$ is the derivative matrix with

$$\left(\frac{\partial \boldsymbol{\mu_t}}{\partial \boldsymbol{\beta}, \partial \boldsymbol{\rho}}\right) = \begin{pmatrix} \frac{\partial \mu_t^{[1]}}{\partial \boldsymbol{\beta}^{[1]}} & \frac{\partial \mu_t^{[2]}}{\partial \boldsymbol{\beta}^{[1]}} \\ \frac{\partial \mu_t^{[1]}}{\partial \boldsymbol{\beta}^{[2]}} & \frac{\partial \mu_t^{[2]}}{\partial \boldsymbol{\beta}^{[2]}} \\ \frac{\partial \mu_t^{[1]}}{\partial \rho_{11}} & \frac{\partial \mu_t^{[2]}}{\partial \rho_{11}} \\ \frac{\partial \mu_t^{[1]}}{\partial \rho_{12}} & \frac{\partial \mu_t^{[2]}}{\partial \rho_{12}} \\ \frac{\partial \mu_t^{[1]}}{\partial \rho_{21}} & \frac{\partial \mu_t^{[2]}}{\partial \rho_{21}} \\ \frac{\partial \mu_t^{[1]}}{\partial \rho_{22}} & \frac{\partial \mu_t^{[2]}}{\partial \rho_{22}} \end{pmatrix}_{(2p+4)\times 2} \tag{8.33}$$

where

$$\frac{\partial \mu_t^{[k]}}{\partial \boldsymbol{\beta}^{[l]}} = \left[\frac{\partial \mu_t^{[k]}}{\partial \beta_1^{[l]}}, \frac{\partial \mu_t^{[k]}}{\partial \beta_2^{[l]}}, \ldots, \frac{\partial \mu_t^{[k]}}{\partial \beta_j^{[l]}}, \ldots, \frac{\partial \mu_t^{[k]}}{\partial \beta_p^{[l]}}\right]', \quad k, l \in \{1, 2\}.$$

The derivative expressions with respect to $\beta_j^{[k]}$ for $k \in \{1, 2\}$ are specified as follows:

$$\frac{\partial \mu_t^{[1]}}{\partial \beta_j^{[1]}} = \rho_{11}\frac{\partial \mu_{t-1}^{[1]}}{\partial \beta_j^{[1]}} + \rho_{12}\frac{\partial \mu_{t-1}^{[2]}}{\partial \beta_j^{[1]}} + \lambda_t^{[1]} x_{tj}, \quad \frac{\partial \mu_t^{[2]}}{\partial \beta_j^{[1]}} = \rho_{21}\frac{\partial \mu_{t-1}^{[1]}}{\partial \beta_j^{[1]}} + \rho_{22}\frac{\partial \mu_{t-1}^{[2]}}{\partial \beta_j^{[1]}},$$

$$\frac{\partial \mu_t^{[1]}}{\partial \beta_j^{[2]}} = \rho_{11}\frac{\partial \mu_{t-1}^{[1]}}{\partial \beta_j^{[2]}} + \rho_{12}\frac{\partial \mu_{t-1}^{[2]}}{\partial \beta_j^{[2]}}, \quad \frac{\partial \mu_t^{[2]}}{\partial \beta_j^{[2]}} = \rho_{21}\frac{\partial \mu_{t-1}^{[1]}}{\partial \beta_j^{[2]}} + \rho_{22}\frac{\partial \mu_{t-1}^{[2]}}{\partial \beta_j^{[2]}} + \lambda_t^{[2]} x_{tj}. \tag{8.34}$$

Equation (8.34) is evaluated iteratively for $t = 2, \ldots, T$, using the following initial derivatives from Equation (8.18):

$$
\begin{aligned}
\frac{\partial \mu_1^{[1]}}{\partial \beta_j^{[1]}} &= \frac{(1-\rho_{22})\lambda_1^{[1]} x_{1j}}{(1-\rho_{11})(1-\rho_{22})-\rho_{12}\rho_{21}}, & \frac{\partial \mu_1^{[2]}}{\partial \beta_j^{[1]}} &= \frac{\rho_{21}\lambda_1^{[1]} x_{1j}}{(1-\rho_{11})(1-\rho_{22})-\rho_{12}\rho_{21}}, \\
\frac{\partial \mu_1^{[1]}}{\partial \beta_j^{[2]}} &= \frac{\rho_{12}\lambda_1^{[2]} x_{1j}}{(1-\rho_{11})(1-\rho_{22})-\rho_{12}\rho_{21}}, & \frac{\partial \mu_1^{[2]}}{\partial \beta_j^{[2]}} &= \frac{(1-\rho_{11})\lambda_1^{[2]} x_{1j}}{(1-\rho_{11})(1-\rho_{22})-\rho_{12}\rho_{21}}.
\end{aligned}
\tag{8.35}
$$

Similarly, the derivatives with respect to the dependence parameters $\boldsymbol{\rho}$ are computed iteratively for $t = 2, \ldots, T$ using

$$
\begin{aligned}
\frac{\partial \mu_t^{[1]}}{\partial \rho_{11}} &= \rho_{11}\frac{\partial \mu_{t-1}^{[1]}}{\partial \rho_{11}} + \mu_{t-1}^{[1]} + \rho_{12}\frac{\partial \mu_{t-1}^{[2]}}{\partial \rho_{11}}, & \frac{\partial \mu_t^{[2]}}{\partial \rho_{11}} &= \rho_{21}\frac{\partial \mu_{t-1}^{[1]}}{\partial \rho_{11}} + \rho_{22}\frac{\partial \mu_{t-1}^{[2]}}{\partial \rho_{11}}, \\
\frac{\partial \mu_t^{[1]}}{\partial \rho_{12}} &= \rho_{11}\frac{\partial \mu_{t-1}^{[1]}}{\partial \rho_{12}} + \rho_{12}\frac{\partial \mu_{t-1}^{[2]}}{\partial \rho_{12}} + \mu_{t-1}^{[2]}, & \frac{\partial \mu_t^{[2]}}{\partial \rho_{12}} &= \rho_{21}\frac{\partial \mu_{t-1}^{[1]}}{\partial \rho_{12}} + \rho_{22}\frac{\partial \mu_{t-1}^{[2]}}{\partial \rho_{12}}, \\
\frac{\partial \mu_t^{[1]}}{\partial \rho_{21}} &= \rho_{11}\frac{\partial \mu_{t-1}^{[1]}}{\partial \rho_{21}} + \rho_{12}\frac{\partial \mu_{t-1}^{[2]}}{\partial \rho_{21}}, & \frac{\partial \mu_t^{[2]}}{\partial \rho_{21}} &= \rho_{21}\frac{\partial \mu_{t-1}^{[1]}}{\partial \rho_{21}} + \mu_{t-1}^{[1]} + \rho_{22}\frac{\partial \mu_{t-1}^{[2]}}{\partial \rho_{21}}, \\
\frac{\partial \mu_t^{[1]}}{\partial \rho_{22}} &= \rho_{11}\frac{\partial \mu_{t-1}^{[1]}}{\partial \rho_{22}} + \rho_{12}\frac{\partial \mu_{t-1}^{[2]}}{\partial \rho_{22}}, & \frac{\partial \mu_t^{[2]}}{\partial \rho_{22}} &= \rho_{21}\frac{\partial \mu_{t-1}^{[1]}}{\partial \rho_{22}} + \rho_{22}\frac{\partial \mu_{t-1}^{[2]}}{\partial \rho_{22}} + \mu_{t-1}^{[2]},
\end{aligned}
\tag{8.36}
$$

where the initial derivatives are given by

$$
\begin{aligned}
\frac{\partial \mu_1^{[1]}}{\partial \rho_{11}} &= \frac{[(1-\rho_{22})\lambda_1^{[1]}+\rho_{12}\lambda_1^{[2]}][1-\rho_{22}]}{\theta_1^2}, & \frac{\partial \mu_1^{[2]}}{\partial \rho_{11}} &= \frac{[(1-\rho_{11})\lambda_1^{[2]}+\rho_{21}\lambda_1^{[1]}][1-\rho_{22}]-\lambda_1^{[2]}\theta_1}{\theta_1^2}, \\
\frac{\partial \mu_1^{[1]}}{\partial \rho_{12}} &= \frac{\lambda_1^{[2]}\theta_1+[(1-\rho_{22})\lambda_1^{[1]}+\rho_{12}\lambda_1^{[2]}][\rho_{21}]}{\theta_1^2}, & \frac{\partial \mu_1^{[2]}}{\partial \rho_{12}} &= \frac{[(1-\rho_{11})\lambda_1^{[2]}+\rho_{21}\lambda_1^{[1]}][\rho_{21}]}{\theta_1^2}, \\
\frac{\partial \mu_1^{[1]}}{\partial \rho_{21}} &= \frac{[(1-\rho_{22})\lambda_1^{[1]}+\rho_{12}\lambda_1^{[2]}][\rho_{12}]}{\theta_1^2}, & \frac{\partial \mu_1^{[2]}}{\partial \rho_{21}} &= \frac{\lambda_1^{[1]}\theta_1+[(1-\rho_{11})\lambda_1^{[2]}+\rho_{21}\lambda_1^{[1]}][\rho_{12}]}{\theta_1^2}, \\
\frac{\partial \mu_1^{[1]}}{\partial \rho_{22}} &= \frac{[(1-\rho_{22})\lambda_1^{[1]}+\rho_{12}\lambda_1^{[2]}][1-\rho_{11}]-\lambda_1^{[1]}\theta_1}{\theta_1^2}, & \frac{\partial \mu_1^{[2]}}{\partial \rho_{22}} &= \frac{[(1-\rho_{11})\lambda_1^{[2]}+\rho_{21}\lambda_1^{[1]}][1-\rho_{11}]}{\theta_1^2},
\end{aligned}
\tag{8.37}
$$

where $\theta_1 = (1-\rho_{11})(1-\rho_{22}) - \rho_{12}\rho_{21}$.

- $\boldsymbol{\Sigma}_{\boldsymbol{\beta},\boldsymbol{\rho}}$ represents the $(2T \times 2T)$ auto-covariance matrix of the form

$$
\boldsymbol{\Sigma}_{\boldsymbol{\beta},\boldsymbol{\rho}} = \begin{pmatrix}
\boldsymbol{\Sigma}_{0,1} & \boldsymbol{\Sigma}'_{1,1} & \boldsymbol{\Sigma}'_{2,1} & \cdots & \boldsymbol{\Sigma}'_{t-1,1} & \cdots & \boldsymbol{\Sigma}'_{t+h-1,1} & \cdots & \boldsymbol{\Sigma}'_{T-1,1} \\
\boldsymbol{\Sigma}_{1,1} & \boldsymbol{\Sigma}_{0,2} & \boldsymbol{\Sigma}'_{1,2} & \cdots & \boldsymbol{\Sigma}'_{t-2,2} & \cdots & \boldsymbol{\Sigma}'_{t+h-2,2} & \cdots & \boldsymbol{\Sigma}'_{T-2,2} \\
\boldsymbol{\Sigma}_{2,1} & \boldsymbol{\Sigma}_{1,2} & \boldsymbol{\Sigma}_{0,3} & \cdots & \boldsymbol{\Sigma}'_{t-3,3} & \cdots & \boldsymbol{\Sigma}'_{t+h-3,3} & \cdots & \boldsymbol{\Sigma}'_{T-3,3} \\
\vdots & \vdots & \vdots & \ddots & \vdots & \vdots & \vdots & \vdots & \vdots \\
\boldsymbol{\Sigma}_{t-1,1} & \boldsymbol{\Sigma}_{t-2,2} & \boldsymbol{\Sigma}_{t-3,3} & \cdots & \boldsymbol{\Sigma}_{0,t} & \cdots & \boldsymbol{\Sigma}'_{t+h-t,t} & \cdots & \boldsymbol{\Sigma}'_{T-t,t} \\
\vdots & \vdots & \vdots & \vdots & \vdots & \ddots & \vdots & \vdots & \vdots \\
\boldsymbol{\Sigma}_{t+h-1,1} & \boldsymbol{\Sigma}_{t+h-2,2} & \boldsymbol{\Sigma}_{t+h-3,3} & \cdots & \boldsymbol{\Sigma}_{t+h-t,t} & \cdots & \boldsymbol{\Sigma}_{0,t+h} & \cdots & \boldsymbol{\Sigma}'_{T-t,t} \\
\vdots & \vdots & \vdots & \vdots & \vdots & \vdots & \vdots & \ddots & \vdots \\
\boldsymbol{\Sigma}_{T-1,1} & \boldsymbol{\Sigma}_{T-2,2} & \boldsymbol{\Sigma}_{T-3,3} & \cdots & \boldsymbol{\Sigma}_{T-t,t} & \cdots & \boldsymbol{\Sigma}_{T-t+h,t} & \cdots & \boldsymbol{\Sigma}_{0,T}
\end{pmatrix}_{(2T \times 2T)}
\tag{8.38}
$$

where the above entries are computed iteratively for $t = 2, \ldots, T$ using Equation (8.31), with initial variances and covariances given by Equations (8.19)–(8.21).

The Newton-Raphson technique is used to solve the first GQL (8.32)

$$(\hat{\boldsymbol{\psi}}_{r+1}) = (\hat{\boldsymbol{\psi}}_r) + \left[\left(\frac{\partial \boldsymbol{\mu_t}}{\partial \boldsymbol{\psi}}\right)' \boldsymbol{\Sigma}_{\boldsymbol{\psi}}{}^{-1} \left(\frac{\partial \boldsymbol{\mu_t}}{\partial \boldsymbol{\psi}}\right)\right]_r^{-1} \left[\left(\frac{\partial \boldsymbol{\mu_t}}{\partial \boldsymbol{\psi}}\right)' \boldsymbol{\Sigma}_{\boldsymbol{\psi}}{}^{-1}(\boldsymbol{f} - \boldsymbol{\mu})\right]_r \tag{8.39}$$

where $\hat{\boldsymbol{\psi}} = [\hat{\beta}_1^{[1]}, \hat{\beta}_2^{[1]}, \hat{\beta}_1^{[2]}, \hat{\beta}_2^{[2]}, \hat{\rho}_{11}, \hat{\rho}_{12}, \hat{\rho}_{21}, \hat{\rho}_{22}]$ and $\hat{\boldsymbol{\psi}}_r$ are the estimates at the $r^{th}$ iteration and $[.]_r$ are the values of the expression at the $r^{th}$ iteration. Thus, for a given $\hat{c}_1$ and $\hat{c}_2$ and initial $\hat{\boldsymbol{\psi}}$, the iterative Equation (8.39) is solved until convergence. Under mild regulatory conditions, it is shown that $(\hat{\boldsymbol{\psi}} - \boldsymbol{\psi})'$ is asymptotically normal with mean $\mathbf{0}$ and covariance matrix $\left[\left(\frac{\partial \boldsymbol{\mu}}{\partial \boldsymbol{\psi}}\right)' [\boldsymbol{\Sigma}(\boldsymbol{\psi})]^{-1} \left(\frac{\partial \boldsymbol{\mu}}{\partial \boldsymbol{\psi}}\right)\right]^{-1}$ (Refer to Subsection 2.3.1.2).

### 8.3.2 GQL-II

GQL-II is specified as:

$$\left(\frac{\partial \boldsymbol{\mu}}{\partial \boldsymbol{c}}\right)' \boldsymbol{\Sigma_c}{}^{-1}(\boldsymbol{f}^* - \boldsymbol{m}) = 0, \tag{8.40}$$

with $\boldsymbol{f}^* = [\boldsymbol{f}^*_{1|0}, \boldsymbol{f}^*_{2|1} \ldots, \boldsymbol{f}^*_{t|t-1}, \ldots, \boldsymbol{f}^*_{T|T-1}]'$ where $\boldsymbol{f}^*_{t|t-1} = [{Y_t^{[1]}}^2 | y_{t-1}^{[1]}, y_{t-1}^{[2]}, {Y_t^{[2]}}^2 | y_{t-1}^{[1]}, y_{t-1}^{[2]}]'$ and $\mathrm{E}(\boldsymbol{f}^*) = \boldsymbol{m} = [\boldsymbol{m}_1, \boldsymbol{m}_2, \ldots, \boldsymbol{m}_t, \ldots, \boldsymbol{m}_T]'$ where $\boldsymbol{m}_t = [E({Y_t^{[1]}}^2 | y_{t-1}^{[1]}, y_{t-1}^{[2]}), E({Y_t^{[2]}}^2 | y_{t-1}^{[1]}, y_{t-1}^{[2]})]'$.
The above components are expressed as follows:

- The derivative matrix $\left(\frac{\partial \boldsymbol{\mu}}{\partial \boldsymbol{c}}\right) = \left[\left(\frac{\partial \boldsymbol{\mu_1}}{\partial \boldsymbol{c}}\right), \left(\frac{\partial \boldsymbol{\mu_2}}{\partial \boldsymbol{c}}\right), \ldots, \left(\frac{\partial \boldsymbol{\mu_t}}{\partial \boldsymbol{c}}\right), \ldots, \left(\frac{\partial \boldsymbol{\mu_T}}{\partial \boldsymbol{c}}\right)\right]'$ where

$$\left(\frac{\partial \boldsymbol{\mu_t}}{\partial \boldsymbol{c}}\right) = \begin{bmatrix} \frac{\partial m_{t|t-1}^{[1]}}{\partial c_1} & \frac{\partial m_{t|t-1}^{[2]}}{\partial c_1} \\ \frac{\partial m_{t|t-1}^{[1]}}{\partial c_2} & \frac{\partial m_{t|t-1}^{[2]}}{\partial c_2} \end{bmatrix}_{2\times 2} \tag{8.41}$$

where

$$\begin{aligned} m_{t|t-1}^{[1]} &= \frac{\rho_{11}(1-\rho_{11})}{1+c_1} Y_{t-1}^{[1]}[1 + c_1 Y_{t-1}^{[1]}] + \frac{\rho_{12}(1-\rho_{12})}{1+c_2} Y_{t-1}^{[2]}[1 + c_2 Y_{t-1}^{[2]}] \\ &\quad + \lambda_t^{[1]} + c_1 {\lambda_t^{[1]}}^2 + (\rho_{11} Y_{t-1}^{[1]} + \rho_{12} Y_{t-1}^{[2]} + \lambda_t^{[1]})^2 \end{aligned} \tag{8.42}$$

and

$$m^{[2]}_{t|t-1} = \frac{\rho_{21}(1-\rho_{21})}{1+c_1} Y^{[1]}_{t-1}[1+c_1 Y^{[1]}_{t-1}] + \frac{\rho_{22}(1-\rho_{22})}{1+c_2} Y^{[2]}_{t-1}[1+c_2 Y^{[2]}_{t-1}] + \lambda^{[2]}_t + c_2 {\lambda^{[2]}_t}^2 + (\rho_{21} Y^{[1]}_{t-1} + \rho_{22} Y^{[2]}_{t-1} + \lambda^{[2]}_t)^2. \tag{8.43}$$

Thus, the entries of the derivative matrix are

1. $\frac{\partial m^{[1]}_{t|t-1}}{\partial c_1} = \rho_{11}(1-\rho_{11}) Y^{[1]}_{t-1}[(1+c_1 Y^{[1]}_{t-1}) + \frac{Y^{[1]}_{t-1}}{(1+c_1)}] + {\lambda^{[1]}_t}^2.$
2. $\frac{\partial m^{[1]}_{t|t-1}}{\partial c_2} = \rho_{12}(1-\rho_{12}) Y^{[2]}_{t-1}[(1+c_2 Y^{[2]}_{t-1}) + \frac{Y^{[2]}_{t-1}}{(1+c_2)}].$
3. $\frac{\partial m^{[2]}_{t|t-1}}{\partial c_1} = \rho_{21}(1-\rho_{21}) Y^{[1]}_{t-1}[(1+c_1 Y^{[1]}_{t-1}) + \frac{Y^{[1]}_{t-1}}{(1+c_1)}].$
4. $\frac{\partial m^{[2]}_{t|t-1}}{\partial c_2} = \rho_{22}(1-\rho_{22}) Y^{[2]}_{t-1}[(1+c_2 Y^{[2]}_{t-1}) + \frac{Y^{[2]}_{t-1}}{(1+c_2)}] + {\lambda^{[2]}_t}^2.$

- As for the $(2T \times 2T)$ covariance matrix, it is of the form

$$\begin{array}{c} \\ \boldsymbol{f}^*_1 \\ \boldsymbol{f}^*_2 \\ \boldsymbol{f}^*_3 \\ \vdots \\ \boldsymbol{f}^*_t \\ \vdots \\ \boldsymbol{f}^*_{t+h} \\ \vdots \\ \boldsymbol{f}^*_T \end{array} \begin{array}{c} \begin{array}{ccccccccc} \boldsymbol{f}^*_1 & \boldsymbol{f}^*_2 & \boldsymbol{f}^*_3 & \cdots & \boldsymbol{f}^*_t & \cdots & \boldsymbol{f}^*_{t+h} & \cdots & \boldsymbol{f}^*_T \end{array} \\ \begin{pmatrix} \boldsymbol{\Sigma}^*_{0,1} & {\boldsymbol{\Sigma}^*_{1,1}}' & {\boldsymbol{\Sigma}^*_{2,1}}' & \cdots & {\boldsymbol{\Sigma}^*_{t-1,1}}' & \cdots & {\boldsymbol{\Sigma}^*_{t+h-1,1}}' & \cdots & {\boldsymbol{\Sigma}^*_{T-1,1}}' \\ \boldsymbol{\Sigma}^*_{1,1} & \boldsymbol{\Sigma}^*_{0,2} & {\boldsymbol{\Sigma}^*_{1,2}}' & \cdots & {\boldsymbol{\Sigma}^*_{t-2,2}}' & \cdots & {\boldsymbol{\Sigma}^*_{t+h-2,2}}' & \cdots & {\boldsymbol{\Sigma}^*_{T-2,2}}' \\ \boldsymbol{\Sigma}^*_{2,1} & \boldsymbol{\Sigma}^*_{1,2} & \boldsymbol{\Sigma}^*_{0,3} & \cdots & {\boldsymbol{\Sigma}^*_{t-3,3}}' & \cdots & {\boldsymbol{\Sigma}^*_{t+h-3,3}}' & \cdots & {\boldsymbol{\Sigma}^*_{T-3,3}}' \\ \vdots & \vdots & \vdots & \ddots & \vdots & \vdots & \vdots & \vdots & \vdots \\ \boldsymbol{\Sigma}^*_{t-1,1} & \boldsymbol{\Sigma}^*_{t-2,2} & \boldsymbol{\Sigma}^*_{t-3,3} & \cdots & \boldsymbol{\Sigma}^*_{0,t} & \cdots & {\boldsymbol{\Sigma}^*_{t+h-t,t}}' & \cdots & {\boldsymbol{\Sigma}^*_{T-t,t}}' \\ \vdots & \vdots & \vdots & \vdots & \vdots & \ddots & \vdots & \vdots & \vdots \\ \boldsymbol{\Sigma}^*_{t+h-1,1} & \boldsymbol{\Sigma}^*_{t+h-2,2} & \boldsymbol{\Sigma}^*_{t+h-3,3} & \cdots & \boldsymbol{\Sigma}^*_{t+h-t,t} & \cdots & \boldsymbol{\Sigma}^*_{0,t+h} & \cdots & {\boldsymbol{\Sigma}^*_{T-t,t}}' \\ \vdots & \vdots & \vdots & \vdots & \vdots & \vdots & \vdots & \ddots & \vdots \\ \boldsymbol{\Sigma}^*_{T-1,1} & \boldsymbol{\Sigma}^*_{T-2,2} & \boldsymbol{\Sigma}^*_{T-3,3} & \cdots & \boldsymbol{\Sigma}^*_{T-t,t} & \cdots & \boldsymbol{\Sigma}^*_{T-t+h,t} & \cdots & \boldsymbol{\Sigma}^*_{0,T} \end{pmatrix} \end{array} \tag{8.44}$$

The components of covariance structure $\boldsymbol{\Sigma}_c$ is represented by the matrix below:

$$\boldsymbol{\Sigma}^*_{h,t} = \begin{bmatrix} \begin{array}{c} \mathrm{Cov}({Y^{[1]}_t}^2, {Y^{[1]}_{t+h}}^2 | y^{[1]}_{t-1}, y^{[2]}_{t-1}, \\ y^{[1]}_{t+h-1}, y^{[2]}_{t+h-1}) \end{array} & \begin{array}{c} \mathrm{Cov}({Y^{[1]}_{t+h}}^2, {Y^{[2]}_t}^2 | y^{[1]}_{t-1}, y^{[2]}_{t-1}, \\ y^{[1]}_{t+h-1}, y^{[2]}_{t+h-1}) \end{array} \\ \begin{array}{c} \mathrm{Cov}({Y^{[1]}_t}^2, {Y^{[2]}_{t+h}}^2 | y^{[1]}_{t-1}, y^{[2]}_{t-1}, \\ y^{[1]}_{t+h-1}, y^{[2]}_{t+h-1}) \end{array} & \begin{array}{c} \mathrm{Cov}({Y^{[2]}_t}^2, {Y^{[2]}_{t+h}}^2 | y^{[1]}_{t-1}, y^{[2]}_{t-1}, \\ y^{[1]}_{t+h-1}, y^{[2]}_{t+h-1}) \end{array} \end{bmatrix}. \tag{8.45}$$

The above entries are derived using the following lemmas. By denoting $W_{jk,t} = \alpha_{jk,t} \circ Y_{t-1}^{[k]}$ for $j, k \in \{1, 2\}$, we obtain the following results.

**Lemma 8.1.** *The conditional covariances* $\text{Cov}(Y_t^{[k]^2}, Y_{t+h}^{[j]^2} | y_{t-1}^{[k]}, y_{t-1}^{[j]}, y_{t+h-1}^{[k]}, y_{t+h-1}^{[j]})$ *are 0 for* $h > 0$ *and* $\{j, k\} \in \{1, 2\}$.

*Proof: For given* $y_{t-1}^{[k]}$, $y_{t-1}^{[j]}$, $y_{t+h-1}^{[k]}$ *and* $y_{t+h-1}^{[j]}$, *the random variables* $Y_t^{[k]^2}$ *and* $Y_{t+h}^{[j]^2}$ *are independent random variables which proves the lemma.*

**Lemma 8.2.** *The conditional covariance* $\text{Cov}(Y_t^{[1]^2}, Y_t^{[2]^2} | y_{t-1}^{[1]}, y_{t-1}^{[2]})$ *are 0.*

*Proof: For given* $y_{t-1}^{[1]}$ *and* $y_{t-1}^{[2]}$ *and the fact that* $\text{Cov}(R_t^{[1]}, R_t^{[2]}) = 0$, *the random variables* $Y_t^{[1]^2}$ *and* $Y_t^{[2]^2}$ *are independent random variables which proves the lemma.*

Thus, only the conditional covariance $\text{Cov}(Y_t^{[k]^2}, Y_t^{[k]^2} | y_{t-1}^{[1]}, y_{t-1}^{[2]}) = \text{Var}(Y_t^{[k]^2} | y_{t-1}^{[1]}, y_{t-1}^{[2]})$ needs to be derived and the derivation is facilitated by making use of the following lemmas:

**Lemma 8.3.**

(a) $\text{Cov}(W_{kk,t} W_{kk,t}, R_t^{[k]^2} | y_{t-1}^{[1]}, y_{t-1}^{[2]}) = 0$ *for* $k \in \{1, 2\}$.

(b) $\text{Cov}(W_{kk,t} R_t^{[k]}, W_{kj,t} R_t^{[k]} | y_{t-1}^{[1]}, y_{t-1}^{[2]}) = \rho_{kk}\rho_{kj} Y_{t-1}^{[k]} Y_{t-1}^{[j]} (\lambda_t^{[k]} + c_k \lambda_t^{[k]^2} + \lambda_t^{[k]^2})$, *for* $j, k \in \{1, 2\}$ *and* $j \neq k$.

(c) $\text{Cov}(W_{kk,t} R_t^{[k]}, R_t^{[k]^2} | y_{t-1}^{[1]}, y_{t-1}^{[2]}) = \rho_{kk} Y_{t-1}^{[k]} E(R_t^{[k]^3}) - \rho_{kk} Y_{t-1}^{[k]} \lambda_t^{[k]} (\lambda_t^{[k]} + c_k \lambda_t^{[k]^2} + \lambda_t^{[k]^2})$ *for* $k \in \{1, 2\}$.

(d) $\text{Cov}(W_{kk,t} W_{kk,t}, W_{kk,t} R_t^{[k]} | y_{t-1}^{[1]}, y_{t-1}^{[2]}) = E(\alpha_{kk,t}{}^3) \lambda_t^{[k]} Y_{t-1}^{[k]^3} - \frac{\rho_{kk}(1+\rho_{kk})}{1+c_k} \rho_{kk} \lambda_t^{[k]} Y_{t-1}^{[k]^3}$ *for* $k \in \{1, 2\}$.

*Proof:*

(a) *For given* $y_{t-1}^{[1]}$ *and* $y_{t-1}^{[2]}$, *the product* $W_{kk,t} W_{kk,t}$ *is independent of the product* $R_t^{[k]^2}$, *which implies that the corresponding covariance is 0.*

(b)
$$
\begin{aligned}
&\text{Cov}(W_{kk,t} R_t^{[k]}, W_{kj,t} R_t^{[k]} | y_{t-1}^{[1]}, y_{t-1}^{[2]}) \\
&= E(W_{kk,t} W_{kj,t} R_t^{[k]^2} | y_{t-1}^{[1]}, y_{t-1}^{[2]}) - E(W_{kk,t} R_t^{[k]} | y_{t-1}^{[1]}, y_{t-1}^{[2]}) \\
&\quad \times E(W_{kj,t} R_t^{[k]} | y_{t-1}^{[1]}, y_{t-1}^{[2]}) \\
&= \rho_{kk}\rho_{kj} Y_{t-1}^{[k]} Y_{t-1}^{[j]} (\lambda_t^{[k]} + c_k \lambda_t^{[k]^2} + \lambda_t^{[k]^2}).
\end{aligned}
$$

(c)
$$\begin{aligned}
&\mathrm{Cov}(W_{kk,t}R_t^{[k]}, R_t^{[k]^2}|y_{t-1}^{[1]}, y_{t-1}^{[2]}) \\
&= E(W_{kk,t}R_t^{[k]^3}|y_{t-1}^{[1]}, y_{t-1}^{[2]}) - E(W_{kk,t}R_t^{[k]}|y_{t-1}^{[1]}, y_{t-1}^{[2]}) \\
&\quad \times E(R_t^{[k]^2}|y_{t-1}^{[1]}, y_{t-1}^{[2]}) \\
&= \rho_{kk}Y_{t-1}^{[k]}E(R_t^{[k]^3}) - \rho_{kk}Y_{t-1}^{[k]}\lambda_t^{[k]}(\lambda_t^{[k]} + c_k\lambda_t^{[k]^2} + \lambda_t^{[k]^2}).
\end{aligned}$$

(d)
$$\begin{aligned}
&\mathrm{Cov}(W_{kk,t}W_{kk,t}, W_{kk,t}R_t^{[k]}|y_{t-1}^{[1]}, y_{t-1}^{[2]}) \\
&= E(W_{kk,t}W_{kk,t}W_{kk,t}R_t^{[k]}|y_{t-1}^{[1]}, y_{t-1}^{[2]}) - E(W_{kk,t}W_{kk,t}|y_{t-1}^{[1]}, y_{t-1}^{[2]}) \\
&\quad \times E(W_{kk,t}R_t^{[k]}|y_{t-1}^{[1]}, y_{t-1}^{[2]}) \\
&= E(\alpha_{kk,t}{}^3)\lambda_t^{[k]}Y_{t-1}^{[k]^3} - \frac{\rho_{kk}(1+\rho_{kk})}{1+c_k}\rho_{kk}\lambda_t^{[k]}Y_{t-1}^{[k]^3}.
\end{aligned}$$

Hence, using the above lemmas, we obtain

$$\begin{aligned}
&\mathrm{Var}(Y_t^{[k]^2}) \\
&= E(\alpha_{kk,t}{}^4)Y_{t-1}^{[k]^4} - \left(\frac{\rho_{kk}(1+\rho_{kk})}{1+c_k}\right)^2 Y_{t-1}^{[k]^4} \\
&\quad + 4E(\alpha_{kk,t}{}^3)Y_{t-1}^{[k]^3}[\rho_{kj}Y_{t-1}^{[j]} + \lambda_t^{[k]}] - 4\left(\frac{\rho_{kk}(1+\rho_{kk})}{1+c_k}\right)Y_{t-1}^{[k]^2} \\
&\quad \times [\rho_{kk}\rho_{kj}Y_{t-1}^{[j]}Y_{t-1}^{[k]} + \rho_{kk}\lambda_t^{[k]}Y_{t-1}^{[k]} - 2\rho_{kj}Y_{t-1}^{[j]}\lambda_t^{[k]}] \\
&\quad + 4\left(\frac{\rho_{kk}(1+\rho_{kk})}{1+c_k}\right)\left(\frac{\rho_{kj}(1+\rho_{kj})}{1+c_j}\right)Y_{t-1}^{[k]^2}Y_{t-1}^{[j]^2} \\
&\quad - 4\rho_{kk}^2\rho_{kj}Y_{t-1}^{[k]^2}Y_{t-1}^{[j]}[\rho_{kj}Y_{t-1}^{[j]} + 2\lambda_t^{[k]}] \\
&\quad + 4\left(\frac{\rho_{kj}(1+\rho_{kj})}{1+c_j}\right)[\rho_{kk}Y_{t-1}^{[k]}(2Y_{t-1}^{[j]^2}\lambda_t^{[k]} - \rho_{kj}Y_{t-1}^{[j]^3}) \\
&\quad + \lambda_t^{[k]^2}(Y_{t-1}^{[j]^2} + Y_{t-1}^{[k]^2}) - \rho_{kj}\lambda_t^{[k]}Y_{t-1}^{[j]^3}] \\
&\quad + 4E(\alpha_{kj,t}{}^3)Y_{t-1}^{[j]^2}[\rho_{kk}Y_{t-1}^{[k]} + \lambda_t^{[1]}] + 4\rho_{kk}Y_{t-1}^{[k]}E(R_t^{[k]^3}) \\
&\quad - 4\rho_kY_{t-1}^{[k]}(\lambda_t^{[k]} + c_k\lambda_t^{[k]^2} + \lambda_t^{[k]^2}) + 4\rho_{kk}Y_{t-1}^{[k]} \\
&\quad \times [2\rho_{kj}Y_{t-1}^{[j]}(\lambda_t^{[k]} + c_k\lambda_t^{[k]^2}) - \rho_{kk}Y_{t-1}^{[k]}\lambda_t^{[k]^2} - 2\rho_{kj}Y_{t-1}^{[j]^2}\lambda_t^{[k]}] \\
&\quad - 4\rho_{kk}^2Y_{t-1}^{[j]^2}\lambda_t^{[k]^2} + E(\alpha_{kj,t}{}^4)Y_{t-1}^{[j]^4} - \left(\frac{\rho_{kj}(1+\rho_{kj})}{1+c_j}\right)^2 Y_{t-1}^{[j]^4} \\
&\quad + 4\rho_{kj}Y_{t-1}^{[j]}E(R_t^{[k]^3}) - 4\rho_{kj}Y_{t-1}^{[j]}\lambda_t^{[k]}(\lambda_t^{[k]} + c_k\lambda_t^{[k]^2} + \lambda_t^{[k]^2}) \\
&\quad + E(R_t^{[k]^4}) - (\lambda_t^{[k]} + c_k\lambda_t^{[k]^2} + \lambda_t^{[k]^2})^2, \qquad (8.46)
\end{aligned}$$

where by differentiating the moment generating function (mgf) $M_{R_t^{[k]}}(q) = \{1 + c_k\lambda_t^{[k]} - c_k\lambda_t^{[k]}\exp(q)\}^{-1/c_k}$ repeatedly, we obtain

$$E({R_t^{[k]}}^3) = \lambda_t^{[k]} + 3{\lambda_t^{[k]}}^2 + 3c_k{\lambda_t^{[k]}}^2 + 3c_k{\lambda_t^{[k]}}^3 + 2c_k^2{\lambda_t^{[k]}}^3 + {\lambda_t^{[k]}}^3 \quad (8.47)$$

$$\begin{aligned} E({R_t^{[k]}}^4) &= \lambda_t^{[k]} + 7{\lambda_t^{[k]}}^2 + 7c_k{\lambda_t^{[k]}}^2 + 6{\lambda_t^{[k]}}^3 + 18c_k{\lambda_t^{[k]}}^3 \\ &+ 12c_k^2{\lambda_t^{[k]}}^3 + 6c_k{\lambda_t^{[k]}}^4 + 11c_k^2{\lambda_t^{[k]}}^4 + 6c_k^3{\lambda_t^{[k]}}^4 + {\lambda_t^{[k]}}^4 \end{aligned} \quad (8.48)$$

and since $\alpha_{jk,t} \sim \text{Beta}(\frac{\rho_{jk}}{c_k}, \frac{1-\rho_{jk}}{c_k})$ implies $E(\alpha_{jk,t}^a) = \left[\frac{\frac{\rho_{jk}}{c_k}+a-1}{\frac{\rho_{jk}}{c_k}+\frac{1-\rho_{jk}}{c_k}+a-1}\right]$,

we have

$$E(\alpha_{jk,t}^3) = \frac{(\rho_{jk} + 2c_k)(1 + \rho_{jk})\rho_{jk}}{(1 + 2c_k)(1 + c_k)} \quad (8.49)$$

$$E(\alpha_{jk,t}^4) = \frac{(\rho_{jk} + 3c_k)(\rho_{jk} + 2c_k)(1 + \rho_{jk})\rho_{jk}}{(1 + 3c_k)(1 + 2c_k)(1 + c_k)}. \quad (8.50)$$

The Newton-Raphson iteration technique is used to solve Equation (8.40) as follows:

$$\begin{pmatrix} \hat{c_1}_{r+1} \\ \hat{c_2}_{r+1} \end{pmatrix} = \begin{pmatrix} \hat{c_1}_r \\ \hat{c_2}_r \end{pmatrix} + \left[\left(\frac{\partial \boldsymbol{\mu}}{\partial \boldsymbol{c}}\right)' \boldsymbol{\Sigma_c}^{-1} \left(\frac{\partial \boldsymbol{\mu}}{\partial \boldsymbol{c}}\right)\right]_r^{-1} \left[\left(\frac{\partial \boldsymbol{\mu}}{\partial \boldsymbol{c}}\right)' \boldsymbol{\Sigma_c}^{-1}(\boldsymbol{f}^* - \boldsymbol{m})\right]_r \quad (8.51)$$

where $\hat{c_1}_r$ and $\hat{c_2}_r$ are the estimates at the $r^{th}$ iteration and $[.]_r$ are the values of the expression at the $r^{th}$ iteration.

The estimation algorithm works as follows: the iterative Equation (8.51) is run assuming an initial over-dispersion parameters and using the updated regression vector and dependence parameters obtained using the iterative Equation (8.39). Using the updated values of the over-dispersion parameters, the iterative Equation (8.39) is run again to obtain a new set of regression and dependence parameters which is then used in Equation (8.51) to obtain another set of over-dispersion parameters. Another cycle of iteration begins with these updated values until the cycle converges. Under mild regularity and asymptotically conditions, these estimates are consistent and it can be shown that $((\hat{c}_1, \hat{c}_2) - (c_1, c_2))'$ has an asymptotic normal distribution with mean $\mathbf{0}$ and covariance $\left[\left(\frac{\partial \boldsymbol{\mu}}{\partial \boldsymbol{c}}\right)' \boldsymbol{\Sigma}(\boldsymbol{c})^{-1} \left(\frac{\partial \boldsymbol{\mu}}{\partial \boldsymbol{c}}\right)\right]^{-1}$ (Refer to Subsection 2.3.1.2).

### 8.3.3 Forecasting Equations

We now derive the one-step ahead forecasting equation of the proposed BINAR(1) model. The conditional expectation and variance of the one-step ahead forecast $Y_{t+1}^{[k]}$ given $y_t^{[k]}$ are as follows:

$$\begin{aligned} E(Y_{t+1}^{[1]}|y_t^{[1]}, y_t^{[2]}) &= E(\alpha_{11,t} \circ Y_t^{[1]}|y_t^{[1]}) + E(\alpha_{12,t} \circ Y_t^{[2]}|y_t^{[2]}) + E(R_{t+1}^{[1]}) \\ &= \rho_{11} y_t^{[1]} + \rho_{12} y_t^{[2]} + \lambda_{t+1}^{[1]}, \end{aligned} \tag{8.52}$$

$$\begin{aligned} E(Y_{t+1}^{[2]}|y_t^{[1]}, y_t^{[2]}) &= E(\alpha_{12,t} \circ Y_t^{[1]}|y_t^{[1]}) + E(\alpha_{22,t} \circ Y_t^{[2]}|y_t^{[2]}) + E(R_{t+1}^{[2]}) \\ &= \rho_{21} y_t^{[1]} + \rho_{22} y_t^{[2]} + \lambda_{t+1}^{[2]}, \end{aligned} \tag{8.53}$$

with the corresponding variances:

$$\begin{aligned} &\mathrm{Var}(Y_{t+1}^{[1]}|y_t^{[1]}, y_t^{[2]}) \\ &= \mathrm{Var}(\alpha_{11,t} \circ Y_t^{[1]}|y_t^{[1]}) + \mathrm{Var}(\alpha_{12,t} \circ Y_t^{[2]}|y_t^{[2]}) + \mathrm{Var}(R_{t+1}^{[1]}) \\ &= E_{\alpha_{11,t}}[\mathrm{Var}(\alpha_{11,t} \circ Y_t^{[1]}|y_t^{[1]}, \alpha_{11,t})] + \mathrm{Var}_{\alpha_{11,t}}[E(\alpha_{11,t} \circ Y_t^{[1]}|y_t^{[1]}, \alpha_{11,t})] \\ &\quad + E_{\alpha_{12,t}}[\mathrm{Var}(\alpha_{12,t} \circ Y_t^{[2]}|y_t^{[2]}, \alpha_{12,t})] + \mathrm{Var}_{\alpha_{12,t}}[E(\alpha_{12,t} \circ Y_t^{[2]}|y_t^{[2]}, \alpha_{12,t})] \\ &\quad + \mathrm{Var}(R_{t+1}^{[1]}) \\ &= E_{\alpha_{11,t}}[\alpha_{11,t}(1-\alpha_{11,t}) y_t^{[1]}] + \mathrm{Var}_{\alpha_{11,t}}[\alpha_{11,t} y_t^{[1]}] \\ &\quad + E_{\alpha_{12,t}}[\alpha_{12,t}(1-\alpha_{12,t}) y_t^{[2]}] + \mathrm{Var}_{\alpha_{12,t}}[\alpha_{12,t} y_t^{[2]}] + \mathrm{Var}(R_{t+1}^{[1]}) \\ &= \frac{\rho_{11}(1-\rho_{11})}{1+c_1} y_t^{[1]}[1 + c_1 y_t^{[1]}] + \frac{\rho_{12}(1-\rho_{12})}{1+c_2} y_t^{[2]}[1 + c_2 y_t^{[2]}] \\ &\quad + \lambda_{t+1}^{[1]} + c_1(\lambda_{t+1}^{[1]})^2. \end{aligned} \tag{8.54}$$

Similarly,

$$\begin{aligned} \mathrm{Var}(Y_{t+1}^{[2]}|y_t^{[1]}, y_t^{[2]}) &= \frac{\rho_{21}(1-\rho_{21})}{1+c_1} y_t^{[1]}[1 + c_1 y_t^{[1]}] + \frac{\rho_{22}(1-\rho_{22})}{1+c_2} y_t^{[2]}[1 + c_2 y_t^{[2]}] \\ &\quad + \lambda_{t+1}^{[2]} + c_2(\lambda_{t+1}^{[2]})^2. \end{aligned} \tag{8.55}$$

## 8.4 Summary

In this chapter, we have developed a new unconstrained BINAR(1) model with NB innovations where the non-stationarity has been considered through specifying time variant covariates in the link predictor function of the innovation terms. In this model the cross-correlation is considered by relating the current series observation with the previous-lagged observation of the other series.

Interestingly, the marginal counting series exhibit significant over-dispersion but due to the complicated moment structures, it is rather difficult to specify the exact probability distribution. GQL method has been developed to estimate the regression, over-dispersion and dependence effects of the series. The asymptotic properties of the GQL estimators have also been derived. The next chapter presents a similar analysis for the unconstrained BINMA(1) model with NB innovations.

# 9

# *Unconstrained BINMA(1) Model with NB Innovations*

## 9.1 Introduction

On the same motivating note as in Chapter 8, the unconstrained BINMA(1) model with cross-correlation structure induced as in Chapter 5 is developed, that is, the intent of this chapter is to present the BINMA(1) model under the assumption that the cross-correlation is induced by relating the current series observation with the previous-lagged of the other innovation series, assuming the pair of innovations at current time point is uncorrelated. Inferential analysis is based on the GQL estimation approach which is developed to estimate the model parameters. The chapter is organized as follows: In Section 9.2, the unconstrained BINMA(1) model with NB innovations is developed and the marginal and joint moments are derived. In Section 9.3, the GQL approach is developed wherein two GQL equations are formulated. In particular, for the higher-order covariance entries in the second GQL, some new results related to the multivariate thinning properties are proposed. Section 9.4 focusses on some Monte Carlo simulation experiments to assess the performance of the GQL approach. The conclusion is presented in the last section.

## 9.2 The Unconstrained BINMA(1) Model

In this section, a new random coefficient BINMA(1) model with the NB innovations is introduced. The BINMA(1) model $\{[Y_t^{[1]}, Y_t^{[2]}]', \ t \in \mathbb{Z}\}$ is expressed as

$$Y_t^{[1]} = \alpha_{11,t} \circ R_{t-1}^{[1]} + \alpha_{12,t} \circ R_{t-1}^{[2]} + R_t^{[1]} \tag{9.1}$$

$$Y_t^{[2]} = \alpha_{21,t} \circ R_{t-1}^{[1]} + \alpha_{22,t} \circ R_{t-1}^{[2]} + R_t^{[2]} \tag{9.2}$$

with the following conditions:

(a) The innovation marginal processes $\{R_t^{[1]}\}$ and $\{R_t^{[2]}\}$ are random variables with the NB distribution which are given by


DOI: 10.1201/9781003600435-9

$$P_{R_t^{[k]}}(r^{[k]}) = \frac{\Gamma(c_k^{-1} + r^{[k]})}{\Gamma(c_k^{-1}) r^{[k]}!} \left( \frac{1}{1 + c_k \lambda_t^{[k]}} \right)^{c_k^{-1}} \left( \frac{c_k \lambda_t^{[k]}}{1 + c_k \lambda_t^{[k]}} \right)^{r^{[k]}},$$
$$c_k, \lambda_t^{[k]} > 0, \quad r^{[k]} \geq 0,$$

where $k \in \{1, 2\}$. As for the pair of innovation terms, $\mathrm{Corr}(R_t^{[1]}, R_t^{[2]})$ is assumed to be zero.

(b) The sequences $\{\alpha_{jk,t}\}$ are mutually independent and identically distributed random variables with Beta $\left(\frac{\rho_{jk}}{c_k}, \frac{1-\rho_{jk}}{c_k}\right)$ distribution, where $\rho_{jk} \in [0, 1]$ and $j, k \in \{1, 2\}$.

(c) The random variables $\alpha_{jk,t}$ and $R_u^{[l]}$ are independent random variables for all $j, k, l \in \{1, 2\}$ and $t, u \in \mathbb{Z}$.

(d) '$\circ$' is the binomial thinning operator indicating that $\{\alpha_{jk,t} \circ R_{t-1}^{[k]}\}$ for given $R_{t-1}^{[k]}$ and $\alpha_{jk,t}$ has the binomial distribution with parameters $R_{t-1}^{[k]}$ and $\alpha_{jk,t}$ for $j, k \in \{1, 2\}$. The counting sequences in $\{\alpha_{jk,t} \circ R_{t-1}^{[k]}\}$ and $\{\alpha_{il,t} \circ R_{t-1}^{[l]}\}$ are independent if $(j, k) \neq (i, l)$. Also the counting sequences are independent of the random variables and $\alpha_{jk,t}$ and $R_{t-1}^{[l]}$ for all $j, k, l \in \{1, 2\}$ and $t \in \mathbb{Z}$.

Under the first condition (a) we have that the random variables $R_t^{[1]}$ and $R_t^{[2]}$ have the first two central moments $E(R_t^{[k]}) = \lambda_t^{[k]}$ and $\mathrm{Var}(R_t^{[k]}) = \lambda_t^{[k]}(1 + c_k \lambda_t^{[k]})$, where $\lambda_t^{[k]} = \exp(\boldsymbol{x}_t' \boldsymbol{\beta}^{[k]})$, with $\boldsymbol{x}_t = [x_{t1}, x_{t2}, \ldots, x_{tj}, \ldots, x_{tp}]'$ and $\boldsymbol{\beta}^{[k]} = [\beta_1^{[k]}, \beta_2^{[k]}, \ldots, \beta_j^{[k]}, \ldots, \beta_p^{[k]}]'$ for $k \in \{1, 2\}$. The second condition (b) implies that $E(\alpha_{jk,t}) = \rho_{jk}$ and $\mathrm{Var}(\alpha_{jk,t}) = c_k \rho_{jk}(1 - \rho_{jk})/(1 + c_k)$.

The BINMA(1) model given by (9.1) and (9.2) can be rewritten in the matrix form as

$$\boldsymbol{Y}_t = \boldsymbol{A}_t \circ \boldsymbol{R}_{t-1} + \boldsymbol{R}_t, \tag{9.3}$$

where $\boldsymbol{A}_t = \begin{pmatrix} \alpha_{11,t} & \alpha_{12,t} \\ \alpha_{21,t} & \alpha_{22,t} \end{pmatrix}$ is a random matrix with random variables as elements, $\boldsymbol{R}_t = [R_t^{[1]}, R_t^{[2]}]'$ is the innovation vector and $\boldsymbol{Y}_t = [Y_t^{[1]}, Y_t^{[2]}]'$ is a random vector.

The matrix form is useful for easy derivation of some properties of the BINMA(1) model. First, let us consider derivation of the expectation of the random vector $\boldsymbol{Y}_t$. According to Zheng et al. (2007) and the independency of the random variables $\alpha_{jk,t}$ and $R_{t-1}^{[k]}$ we have that

$$\begin{aligned} E(\alpha_{jk,t} \circ R_{t-1}^{[k]}) &= E(E(\alpha_{jk,t} \circ R_{t-1}^{[k]} | R_{t-1}^{[k]}, \alpha_{jk,t})) \\ &= E(\alpha_{jk,t}) E(R_{t-1}^{[k]}), \ j, k \in \{1, 2\}. \end{aligned}$$

This implies that

$$\begin{aligned} E(\boldsymbol{Y}_t) &= E(\boldsymbol{A}_t \circ \boldsymbol{R}_{t-1}) + E(\boldsymbol{R}_t) \\ &= E(\boldsymbol{A}_t)E(\boldsymbol{R}_{t-1}) + E(\boldsymbol{R}_t) \\ &= \boldsymbol{A}\boldsymbol{\lambda}_{t-1} + \boldsymbol{\lambda}_t, \end{aligned} \tag{9.4}$$

where $\boldsymbol{A} = \begin{pmatrix} \rho_{11} & \rho_{12} \\ \rho_{21} & \rho_{22} \end{pmatrix}$ and $\boldsymbol{\lambda}_t = [\lambda_t^{[1]}, \lambda_t^{[2]}]'$.
Now, using the first two conditions (a) and (b) and expanding Equation (9.4), we obtain

$$\begin{aligned} \mu_t^{[1]} &\equiv E(Y_t^{[1]}) = \rho_{11}\lambda_{t-1}^{[1]} + \rho_{12}\lambda_{t-1}^{[2]} + \lambda_t^{[1]} \quad \text{and} \\ \mu_t^{[2]} &\equiv E(Y_t^{[2]}) = \rho_{21}\lambda_{t-1}^{[1]} + \rho_{22}\lambda_{t-1}^{[2]} + \lambda_t^{[2]}. \end{aligned} \tag{9.5}$$

Similarly, we can derive the covariance matrix of the random vector $\boldsymbol{Y}_t$. From the independency of the random matrix $\boldsymbol{A}_t$ and the innovation terms $\boldsymbol{R}_t$ and $\boldsymbol{R}_{t-1}$, we have that $\text{Var}(\boldsymbol{Y}_t) = \text{Var}(\boldsymbol{A}_t \circ \boldsymbol{R}_{t-1}) + \text{Var}(\boldsymbol{R}_t)$. Using the law of total variance we obtain that

$$\text{Var}(\boldsymbol{A}_t \circ \boldsymbol{R}_{t-1}) = \text{Var}[E(\boldsymbol{A}_t \circ \boldsymbol{R}_{t-1}|\boldsymbol{R}_{t-1}, \boldsymbol{A}_t)] + E[\text{Var}(\boldsymbol{A}_t \circ \boldsymbol{R}_{t-1}|\boldsymbol{R}_{t-1}, \boldsymbol{A}_t)].$$

The elements of the matrix $\text{Var}(\boldsymbol{A}_t \circ \boldsymbol{R}_{t-1}|\boldsymbol{R}_{t-1}, \boldsymbol{A}_t)$ are the conditional covariances of the random variables $\{\alpha_{i1,t} \circ R_{t-1}^{[1]} + \alpha_{i2,t} \circ R_{t-1}^{[2]}\}$ and $\{\alpha_{j1,t} \circ R_{t-1}^{[1]} + \alpha_{j2,t} \circ R_{t-1}^{[2]}\}$, $i, j \in \{1, 2\}$, for given $\boldsymbol{R}_{t-1}$ and $\boldsymbol{A}_t$. We derive these elements by using the following lemma.

**Lemma 9.1.** *Let us suppose that the conditions (a)–(d) are valid. Then:*

*a) For all $i, j \in \{1, 2\}$, we have that*
$\text{Var}(\alpha_{ij,t} \circ R_{t-1}^{[j]}|\boldsymbol{R}_t, \boldsymbol{A}_t) = \alpha_{ij,t}(1 - \alpha_{ij,t})R_{t-1}^{[j]}$.

*b) For all $i, j, k, l \in \{1, 2\}$ with $(i, k) \neq (j, l)$, we have that*
$\text{Cov}(\alpha_{ik,t} \circ R_{t-1}^{[k]}, \alpha_{jl,t} \circ R_{t-1}^{[l]}|\boldsymbol{R}_t, \boldsymbol{A}_t) = 0$.

*Proof: The proof follows from the fact that the random variable $\{\alpha_{jk,t} \circ R_{t-1}^{[k]}\}$ for given $R_{t-1}^{[k]}$ and $\alpha_{jk,t}$ has the binomial distribution with parameters $R_{t-1}^{[k]}$ and $\alpha_{jk,t}$, while the product of the random variables $\{\alpha_{ik,t} \circ R_{t-1}^{[k]}\}$ and $\{\alpha_{jl,t} \circ R_{t-1}^{[l]}\}$ with $k \neq l$, for given $\boldsymbol{R}_{t-1}$ and $\boldsymbol{A}_t$ are two independent binomial random variables with parameters $R_{t-1}^{[k]}$ and $\alpha_{ik,t}$, and $R_{t-1}^{[l]}$ and $\alpha_{jl,t}$, respectively.*

Now, using the results of the above lemma we are able to derive the elements of the covariance matrix $\boldsymbol{\Sigma}_{0,t} \equiv \text{Var}(\boldsymbol{Y}_t)$. The $(1, 1)$ element of this matrix represents the variance of the random variable $Y_t^{[1]}$ and it is expressed as

$$\begin{aligned} \text{Var}(Y_t^{[1]}) = \text{Var}(\alpha_{11,t}R_{t-1}^{[1]} + \alpha_{12,t}R_{t-1}^{[2]}) + E(\alpha_{11,t}(1 - \alpha_{11,t}))E(R_{t-1}^{[1]}) \\ + E(\alpha_{12,t}(1 - \alpha_{12,t}))E(R_{t-1}^{[2]}) + \text{Var}(R_t^{[1]}). \end{aligned}$$

By using Lemma 9.1 and expressions for moments of the random variables $R_t^{[k]}$ and $\alpha_{ij,t}$, we obtain that the variance of the random variable $Y_t^{[1]}$ is given by

$$\begin{aligned}
\mathrm{Var}(Y_t^{[1]}) &= \mathrm{Var}(\alpha_{11,t}R_{t-1}^{[1]}) + \mathrm{Var}(\alpha_{12,t}R_{t-1}^{[2]}) + E(\alpha_{11,t}(1-\alpha_{11,t}))E(R_{t-1}^{[1]}) \\
&\quad + E(\alpha_{12,t}(1-\alpha_{12,t}))E(R_{t-1}^{[2]}) + \mathrm{Var}(R_t^{[1]}) \\
&= \left[\frac{\rho_{11}c_1(1-\rho_{11})}{1+c_1} + \rho_{11}^2\right]\left[\lambda_{t-1}^{[1]} + c_1{\lambda_{t-1}^{[1]}}^2 + {\lambda_{t-1}^{[1]}}^2\right] - \rho_{11}^2{\lambda_{t-1}^{[1]}}^2 \\
&\quad + \left[\frac{\rho_{12}c_2(1-\rho_{12})}{1+c_2} + \rho_{12}^2\right]\left[\lambda_{t-1}^{[2]} + c_2{\lambda_{t-1}^{[2]}}^2 + {\lambda_{t-1}^{[2]}}^2\right] - \rho_{12}^2{\lambda_{t-1}^{[2]}}^2 \\
&\quad + \rho_{11}\lambda_{t-1}^{[1]} - \left[\frac{\rho_{11}c_1(1-\rho_{11})}{1+c_1} + \rho_{11}^2\right]\lambda_{t-1}^{[1]} + \rho_{12}\lambda_{t-1}^{[2]} \\
&\quad - \left[\frac{\rho_{12}c_2(1-\rho_{12})}{1+c_2} + \rho_{12}^2\right]\lambda_{t-1}^{[2]} + \lambda_t^{[1]} + c_1{\lambda_t^{[1]}}^2 \\
&= \sum_{i=1}^{2}\rho_{1i}\lambda_{t-1}^{[i]}(1+c_i\lambda_{t-1}^{[i]}) + \lambda_t^{[1]}(1+c_1\lambda_t^{[1]}).
\end{aligned} \tag{9.6}$$

In a similar way, we can derive the variance $\mathrm{Var}(Y_t^{[2]})$. It is given by

$$\mathrm{Var}(Y_t^{[2]}) = \sum_{i=1}^{2}\rho_{2i}\lambda_{t-1}^{[i]}(1+c_i\lambda_{t-1}^{[i]}) + \lambda_t^{[2]}(1+c_2\lambda_t^{[2]}). \tag{9.7}$$

Finally we derive $(1,2)$ and $(2,1)$ elements of the covariance matrix $\mathrm{Var}(\boldsymbol{Y}_t)$. They represent the covariance between the random variables $Y_t^{[1]}$ and $Y_t^{[2]}$ and they are given as

$$\begin{aligned}
\mathrm{Cov}(Y_t^{[1]}, Y_t^{[2]}) &= \mathrm{Cov}(\alpha_{11,t}R_{t-1}^{[1]} + \alpha_{12,t}R_{t-1}^{[2]}, \alpha_{21,t}R_{t-1}^{[1]} + \alpha_{22,t}R_{t-1}^{[2]}) \\
&\quad + \mathrm{Cov}(R_t^{[1]}, R_t^{[2]}) \\
&= \sum_{i=1}^{2}\rho_{1i}\rho_{2i}\lambda_{t-1}^{[i]}(1+c_i\lambda_{t-1}^{[i]}).
\end{aligned} \tag{9.8}$$

**Remark 9.1.** Under stationary moment condition,

$$E(Y_t^{[1]}) = (1+\rho_{11})\lambda_1^{[1]} + \rho_{12}\lambda_1^{[2]}, \quad E(Y_t^{[2]}) = \rho_{21}\lambda_1^{[1]} + (1+\rho_{22})\lambda_1^{[2]}, \tag{9.9}$$

$$\mathrm{Var}(Y_t^{[1]}) = \sum_{i=1}^{2}\rho_{1i}\lambda_1^{[i]}(1+c_i\lambda_1^{[i]}) + \lambda_1^{[1]}(1+c_1\lambda_1^{[1]}), \tag{9.10}$$

$$\mathrm{Var}(Y_t^{[2]}) = \sum_{i=1}^{2}\rho_{2i}\lambda_1^{[i]}(1+c_i\lambda_1^{[i]}) + \lambda_1^{[2]}(1+c_2\lambda_1^{[2]}), \tag{9.11}$$

and

$$\text{Cov}(Y_t^{[1]}, Y_t^{[2]}) = \sum_{i=1}^{2} \rho_{1i}\rho_{2i}\lambda_1^{[i]}(1 + c_i\lambda_1^{[i]}). \tag{9.12}$$

By denoting the covariances between the random vectors $\boldsymbol{Y}_t$ and $\boldsymbol{Y}_{t+h}$ at lag-$h$, $h \geq 1$, as $\boldsymbol{\Sigma}_{h,t}$, we have that

$$\boldsymbol{\Sigma}_{h,t} = \begin{bmatrix} \text{Cov}(Y_t^{[1]}, Y_{t+h}^{[1]}) & \text{Cov}(Y_{t+h}^{[1]}, Y_t^{[2]}) \\ \text{Cov}(Y_t^{[1]}, Y_{t+h}^{[2]}) & \text{Cov}(Y_t^{[2]}, Y_{t+h}^{[2]}) \end{bmatrix} \equiv \begin{bmatrix} \Sigma_{11,h,t} & \Sigma_{12,h,t} \\ \Sigma_{21,h,t} & \Sigma_{22,h,t} \end{bmatrix}.$$

When $h = 1$, we have that

$$\begin{aligned} \Sigma_{11,1,t} &\equiv \text{Cov}(Y_t^{[1]}, Y_{t+1}^{[1]}) = \text{Cov}(Y_t^{[1]}, \alpha_{11,t} \circ R_t^{[1]} + \alpha_{12,t} \circ R_t^{[2]} + R_{t+1}^{[1]}) \\ &= \rho_{11}\text{Var}(R_t^{[1]}). \end{aligned}$$

Similarly,

$$\Sigma_{22,1,t} \equiv \text{Cov}(Y_t^{[2]}, Y_{t+1}^{[2]}) = \rho_{22}\text{Var}(R_t^{[2]}).$$

The other two covariances are

$$\begin{aligned} \Sigma_{21,1,t} &\equiv \text{Cov}(Y_t^{[1]}, Y_{t+1}^{[2]}) = \text{Cov}(Y_t^{[1]}, \alpha_{22,t} \circ R_t^{[2]} + \alpha_{21,t} \circ R_t^{[1]} + R_{t+1}^{[2]}) \\ &= \rho_{21}\text{Var}(R_t^{[1]}) \end{aligned}$$

and

$$\Sigma_{12,1,t} \equiv \text{Cov}(Y_{t+1}^{[1]}, Y_t^{[2]}) = \rho_{12}\text{Var}(R_t^{[2]}).$$

In the case $h \geq 2$, we have that the random variables $Y_t^{[k]}$ and $Y_{t+h}^{[l]}$ are independent random variables which implies that the covariances $\Sigma_{11,h,t}$, $\Sigma_{12,h,t}$, $\Sigma_{21,h,t}$ and $\Sigma_{22,h,t}$ are equal 0.

Hence,

$$\boldsymbol{\Sigma}_{h,t} = \begin{cases} \begin{bmatrix} \rho_{11} & \rho_{12} \\ \rho_{21} & \rho_{22} \end{bmatrix} \begin{bmatrix} \lambda_t^{[1]}(1 + c_1\lambda_t^{[1]}) & 0 \\ 0 & \lambda_t^{[2]}(1 + c_2\lambda_t^{[2]}) \end{bmatrix}, & \text{for} \quad h = 1 \\ 0, & h > 1. \end{cases} \tag{9.13}$$

## 9.3 Estimation of Parameters

This section focusses on the development of two GQL equations to estimate the different model parameters.

### 9.3.1 GQL-I

The first GQL equation for estimation of the regression, serial- and cross-correlation coefficients is expressed as

$$\boldsymbol{D}_{\boldsymbol{\beta},\boldsymbol{\rho}}{}'\boldsymbol{\Sigma}_{\boldsymbol{\beta},\boldsymbol{\rho}}{}^{-1}(\boldsymbol{f}-\boldsymbol{\mu})=0, \tag{9.14}$$

where $\boldsymbol{f}=[\boldsymbol{f}'_1,\boldsymbol{f}'_2,\ldots,\boldsymbol{f}'_T]'$ with $\boldsymbol{f}_t=[Y_t^{[1]},Y_t^{[2]}]'$, and $\boldsymbol{\mu}=[\boldsymbol{\mu}'_1,\boldsymbol{\mu}'_2,\ldots,\boldsymbol{\mu}'_T]'$ with $\boldsymbol{\mu}_t=[\mu_t^{[1]},\mu_t^{[2]}]'$, are vectors with $2T$ elements. The matrix $\boldsymbol{D}_{\boldsymbol{\beta},\boldsymbol{\rho}}$ is the matrix of derivation which is given as

$$\boldsymbol{D}'_{\boldsymbol{\beta},\boldsymbol{\rho}}=\begin{pmatrix}
\frac{\partial\mu_1^{[1]}}{\partial\boldsymbol{\beta}^{[1]}} & \frac{\partial\mu_1^{[2]}}{\partial\boldsymbol{\beta}^{[1]}} & \cdots & \frac{\partial\mu_T^{[1]}}{\partial\boldsymbol{\beta}^{[1]}} & \frac{\partial\mu_T^{[2]}}{\partial\boldsymbol{\beta}^{[1]}}\\
\frac{\partial\mu_1^{[1]}}{\partial\boldsymbol{\beta}^{[2]}} & \frac{\partial\mu_1^{[2]}}{\partial\boldsymbol{\beta}^{[2]}} & \cdots & \frac{\partial\mu_T^{[1]}}{\partial\boldsymbol{\beta}^{[2]}} & \frac{\partial\mu_T^{[2]}}{\partial\boldsymbol{\beta}^{[2]}}\\
\frac{\partial\mu_1^{[1]}}{\partial\rho_{11}} & 0 & \cdots & \frac{\partial\mu_T^{[1]}}{\partial\rho_{11}} & 0\\
\frac{\partial\mu_1^{[1]}}{\partial\rho_{12}} & 0 & \cdots & \frac{\partial\mu_T^{[1]}}{\partial\rho_{12}} & 0\\
0 & \frac{\partial\mu_1^{[2]}}{\partial\rho_{21}} & \cdots & 0 & \frac{\partial\mu_T^{[2]}}{\partial\rho_{21}}\\
0 & \frac{\partial\mu_1^{[2]}}{\partial\rho_{22}} & \cdots & 0 & \frac{\partial\mu_T^{[2]}}{\partial\rho_{22}}
\end{pmatrix}_{(2p+4)\times 2T}. \tag{9.15}$$

The components of the matrix $\boldsymbol{D}_{\boldsymbol{\beta},\boldsymbol{\rho}}$ are given as

$$\begin{aligned}
\frac{\partial\mu_t^{[k]}}{\partial\boldsymbol{\beta}^{[l]}} &= \left[\frac{\partial\mu_t^{[k]}}{\partial\beta_1^{[l]}},\frac{\partial\mu_t^{[k]}}{\partial\beta_2^{[l]}},\ldots,\frac{\partial\mu_t^{[k]}}{\partial\beta_p^{[l]}}\right]',\ k,l\in\{1,2\},\\
\frac{\partial\mu_t^{[k]}}{\partial\beta_j^{[k]}} &= x_{tj}\lambda_t^{[k]}+\rho_{kk}x_{t-1,j}\lambda_{t-1}^{[k]},\ k\in\{1,2\},\\
\frac{\partial\mu_t^{[k]}}{\partial\beta_j^{[l]}} &= \rho_{kl}x_{t-1,j}\lambda_{t-1}^{[l]},\ k\neq l\in\{1,2\},\\
\frac{\partial\mu_t^{[k]}}{\partial\rho_{kl}} &= \lambda_{t-1}^{[l]},\ k,l\in\{1,2\}.
\end{aligned}$$

The covariance matrix $\boldsymbol{\Sigma}_{\boldsymbol{\beta},\boldsymbol{\rho}}$ consists of the covariances $\mathrm{Cov}(\boldsymbol{f}_i,\boldsymbol{f}_j)$ as $(i,j)$ elements, $i,j\in\{1,2,\ldots,T\}$. Using the expressions for the covariances $\boldsymbol{\Sigma}_{h,t}$ given in Equation (9.13), we obtain that $\boldsymbol{\Sigma}_{\boldsymbol{\beta},\boldsymbol{\rho}}$ can be expressed in tridiagonal form as

$$\boldsymbol{\Sigma}_{\boldsymbol{\beta},\boldsymbol{\rho}} = \begin{pmatrix} \boldsymbol{\Sigma}_{0,1} & \boldsymbol{\Sigma}'_{1,1} & 0 & 0 & \cdots & 0 & 0 \\ \boldsymbol{\Sigma}_{1,1} & \boldsymbol{\Sigma}_{0,2} & \boldsymbol{\Sigma}'_{1,2} & 0 & \cdots & 0 & 0 \\ 0 & \boldsymbol{\Sigma}_{1,2} & \boldsymbol{\Sigma}_{0,3} & \boldsymbol{\Sigma}'_{1,3} & \cdots & 0 & 0 \\ \vdots & \vdots & \vdots & \vdots & \ddots & \vdots & \vdots \\ 0 & 0 & 0 & 0 & \cdots & \boldsymbol{\Sigma}_{0,T-1} & \boldsymbol{\Sigma}'_{1,T-1} \\ 0 & 0 & 0 & 0 & \cdots & \boldsymbol{\Sigma}_{1,T-1} & \boldsymbol{\Sigma}_{0,T} \end{pmatrix}_{2T\times 2T} \tag{9.16}$$

Thus, we can conclude that the first GQL Equation (9.14) represents a system of $(2p+4)$ equations with the same number of unknown parameters.

The Newton-Raphson iterative scheme is used to solve Equation (9.14) yielding an iterative scheme of the form

$$\left(\hat{\boldsymbol{\psi}}_{r+1}\right) = \left(\hat{\boldsymbol{\psi}}_{r}\right) + [\boldsymbol{D}_{\boldsymbol{\psi}}'\boldsymbol{\Sigma}_{\boldsymbol{\psi}}{}^{-1}\boldsymbol{D}_{\boldsymbol{\psi}}]_r^{-1}[\boldsymbol{D}_{\boldsymbol{\psi}}'\boldsymbol{\Sigma}_{\boldsymbol{\psi}}{}^{-1}(\boldsymbol{f}-\boldsymbol{\mu})]_r, \tag{9.17}$$

where $\hat{\boldsymbol{\psi}} = [\hat{\beta}_1^{[1]}, \hat{\beta}_2^{[1]}, \hat{\beta}_1^{[2]}, \hat{\beta}_2^{[2]}, \hat{\rho}_{11}, \hat{\rho}_{12}, \hat{\rho}_{21}, \hat{\rho}_{22}]$ and $\hat{\boldsymbol{\psi}}_r$, are the estimates at the $r^{th}$ iteration and $[.]_r$ are the values of the expression at the $r^{th}$ iteration.

The working of this algorithm is as follows: For an initial $\hat{\boldsymbol{\psi}}$, the score and Hessian components are calculated and replaced in the Equation (9.17) to obtain an updated $\hat{\boldsymbol{\psi}}$. Under mild regulatory conditions, the regression and dependence estimators are consistent and $(\hat{\boldsymbol{\psi}}-\boldsymbol{\psi})'$ follows an asymptotic normal distribution with mean $\mathbf{0}$ and covariance matrix $[\boldsymbol{D}_{\boldsymbol{\psi}}{}'\boldsymbol{\Sigma}_{\boldsymbol{\psi}}{}^{-1}\boldsymbol{D}_{\boldsymbol{\psi}}]^{-1}$ (Refer to Subsection 2.3.1.2).

### 9.3.2 GQL-II

The second GQL equation is specified as

$$\boldsymbol{D_c}'\boldsymbol{\Sigma_c}^{-1}(\boldsymbol{Z}-\boldsymbol{m}) = 0, \tag{9.18}$$

where the score vector is $\boldsymbol{Z} = [Z_1, Z_2, \ldots, Z_T]'$ with $Z_t = \{Y_t^{[1]}Y_t^{[2]}\}$ and $\boldsymbol{m} = [m_1, m_2, \ldots, m_T]'$ with $m_t = E(Z_t) = E(Y_t^{[1]}Y_t^{[2]}) = \sum_{i=1}^{2} \rho_{1i}\rho_{2i}\lambda_{t-1}^{[i]}(1 + c_i\lambda_{t-1}^{[i]}) + \mu_t^{[1]}\mu_t^{[2]}$. The covariance matrix $\boldsymbol{\Sigma_c}$ consists of the entries $\text{Var}(Z_t)$ and $\text{Cov}(Z_t, Z_{t+1})$, that is,

$$\boldsymbol{\Sigma}_{\boldsymbol{\beta},\boldsymbol{\rho}} = \begin{pmatrix} \text{Var}(Z_1) & \text{Cov}(Z_1,Z_2) & 0 & 0 & \cdots & 0 & 0 \\ \text{Cov}(Z_2,Z_1) & \text{Var}(Z_2) & \text{Cov}(Z_2,Z_3) & 0 & \cdots & 0 & 0 \\ 0 & \text{Cov}(Z_3,Z_2) & \text{Var}(Z_3) & \text{Cov}(Z_3,Z_4) & \cdots & 0 & 0 \\ \vdots & \vdots & \vdots & \vdots & \ddots & \vdots & \vdots \\ 0 & 0 & 0 & 0 & \cdots & \text{Var}(Z_{T-1}) & \text{Cov}(Z_{T-1},Z_T) \\ 0 & 0 & 0 & 0 & \cdots & \text{Cov}(Z_T,Z_{T-1}) & \text{Var}(Z_T) \end{pmatrix}_{T\times T}. \tag{9.19}$$

To derive the above covariances, we use the following lemmas. By denoting $W_{jk,t} = \alpha_{jk,t+1} \circ R_t^{[k]}$ for $j, k \in \{1, 2\}$,

**Lemma 9.2.**

(a) $\text{Cov}(W_{kk,t-1}W_{jk,t-1}, R_{t+1}^{[k]}R_{t+1}^{[j]})$=0, $j, k \in \{1, 2\}$, $j \neq k$.

(b) $\text{Cov}(W_{kk,t-1}W_{jk,t-1}, W_{kj,t}W_{jj,t})$=0, $j, k \in \{1, 2\}$, $j \neq k$.

(c) $\text{Cov}(R_t^{[1]}R_t^{[2]}, R_{t+h}^{[1]}R_{t+h}^{[2]}) = 0$ for $h \neq 0, t, h \in \mathbb{Z}$.

(d) $\text{Cov}(W_{kk,t-1}W_{jk,t-1}, W_{kj,t}R_{t+1}^{[j]})$=0, $j, k \in \{1, 2\}$, $j \neq k$.

(e) $\text{Cov}(W_{kk,t-1}W_{jj,t-1}, W_{kj,t}W_{jk,t})$=0, $j, k \in \{1, 2\}$, $j \neq k$.

(f) $\text{Cov}(W_{kk,t-1}W_{jj,t-1}, W_{kj,t}R_{t+1}^{[j]})$=0, $j, k \in \{1, 2\}$, $j \neq k$.

(g) $\text{Cov}(W_{kk,t-1}W_{jj,t-1}, R_{t+1}^{[k]}R_{t+1}^{[j]})$=0, $j, k \in \{1, 2\}$, $j \neq k$.

(h) $\text{Cov}(W_{kk,t-1}R_t^{[j]}, W_{jk,t}R_{t+1}^{[k]})$=0, $j, k \in \{1, 2\}$, $j \neq k$.

(i) $\text{Cov}(W_{kk,t-1}R_t^{[j]}, R_{t+1}^{[k]}R_{t+1}^{[j]})$=0, $j, k \in \{1, 2\}$, $j \neq k$.

*Proof: Since the random variables $W_{kk,t-1}$, $W_{jk,t-1}$, $R_{t+1}^{[k]}$ and $R_{t+1}^{[j]}$ are independent the random variables for all $j, k \in \{1, 2\}$ and $t \in \mathbb{Z}$ and conditions (c) and (d) imply the independency of the random variables $W_{kk,t-1}W_{jk,t-1}$ and $R_{t+1}^{[k]}R_{t+1}^{[j]}$, this implies that the corresponding covariance is 0. The remaining Lemmas can be proved in a similar way.*

**Lemma 9.3.**

(a)
$$\begin{aligned}&\text{Cov}(W_{kk,t-1}W_{jk,t-1}, W_{kk,t}W_{jj,t})\\&= [\frac{\rho_{kk}(1-\rho_{kk})c_k}{1+c_k}]\rho_{jj}\rho_{jk}(\lambda_{t-1}^{[k]} + c_k{\lambda_{t-1}^{[k]}}^2 + {\lambda_{t-1}^{[k]}}^2)E(R_t^{[1]}R_t^{[2]}),\\&i, j, k \in \{1, 2\}.\end{aligned}$$

(b)
$$\begin{aligned}&\text{Cov}(W_{kk,t-1}W_{jk,t-1}, W_{kk,t}R_{t+1}^{[j]})\\&= [\frac{\rho_{kk}(1-\rho_{kk})c_k}{1+c_k}]\rho_{jk}(\lambda_{t-1}^{[k]} + c_k{\lambda_{t-1}^{[k]}}^2 + {\lambda_{t-1}^{[k]}}^2)\lambda_t^{[k]}\lambda_{t+1}^{[j]},\\&i, j, k \in \{1, 2\}.\end{aligned}$$

(c)
$$\begin{aligned}&\text{Cov}(R_t^{[k]}R_t^{[j]}, W_{kk,t}W_{jk,t})\\&= \rho_{kk}\rho_{jk}E({R_t^{[k]}}^3)\lambda_t^{[j]} - \rho_{kk}\rho_{jk}E({R_t^{[k]}}^2)\lambda_t^{[k]}\lambda_t^{[j]}, \quad j, k \in \{1, 2\},\ j \neq k.\end{aligned}$$

*Proof:*

(a) $\text{Cov}(W_{kk,t-1}W_{jk,t-1}, W_{kk,t}W_{jj,t})$

$$= E(W_{kk,t-1}W_{jk,t-1}W_{kk,t}W_{jj,t}) - E(W_{kk,t-1}W_{jk,t-1})E(W_{kk,t}W_{jj,t})$$
$$= E(\alpha_{kk,t}^2)E(\alpha_{jj,t+1})E(\alpha_{jk,t})[E(R_{t-1}^{[k]}{}^2)E(R_t^{[1]}R_t^{[2]})]$$
$$- E(\alpha_{kk,t})^2E(\alpha_{jj,t+1})E(\alpha_{jk,t})[E(R_{t-1}^{[k]}{}^2)E(R_t^{[1]}R_t^{[2]})]$$
$$= \text{Var}(\alpha_{kk,t})E(\alpha_{jj,t+1})E(\alpha_{jk,t})[E(R_{t-1}^{[k]}{}^2)E(R_t^{[1]}R_t^{[2]})]$$
$$= [\frac{\rho_{kk}(1-\rho_{kk})c_k}{1+c_k}]\rho_{jj}\rho_{jk}(\lambda_{t-1}^{[k]} + c_k\lambda_{t-1}^{[k]}{}^2 + \lambda_{t-1}^{[k]}{}^2)E(R_t^{[1]}R_t^{[2]}).$$

(b) $\text{Cov}(W_{kk,t-1}W_{jk,t-1}, W_{kk,t}R_{t+1}^{[j]})$

$$= E(W_{kk,t-1}W_{jk,t-1}W_{kk,t}R_{t+1}^{[j]}) - E(W_{kk,t-1}W_{jk,t-1})E(W_{kk,t}R_{t+1}^{[j]})$$
$$= E(\alpha_{kk,t}^2)E(\alpha_{jk,t})[(\lambda_{t-1}^{[k]} + c_k\lambda_{t-1}^{[k]}{}^2 + \lambda_{t-1}^{[k]}{}^2)\lambda_t^{[k]}\lambda_{t+1}^{[j]}]$$
$$- E(\alpha_{kk,t})^2E(\alpha_{jk,t})[(\lambda_{t-1}^{[k]} + c_k\lambda_{t-1}^{[k]}{}^2 + \lambda_{t-1}^{[k]}{}^2)\lambda_t^{[k]}\lambda_{t+1}^{[j]}]$$
$$= \text{Var}(\alpha_{kk,t})E(\alpha_{jk,t})[(\lambda_{t-1}^{[k]} + c_k\lambda_{t-1}^{[k]}{}^2 + \lambda_{t-1}^{[k]}{}^2)\lambda_t^{[k]}\lambda_{t+1}^{[j]}]$$
$$= [\frac{\rho_{kk}(1-\rho_{kk})c_k}{1+c_k}]\rho_{jk}(\lambda_{t-1}^{[k]} + c_k\lambda_{t-1}^{[k]}{}^2 + \lambda_{t-1}^{[k]}{}^2)\lambda_t^{[k]}\lambda_{t+1}^{[j]}.$$

(c) $\text{Cov}(R_t^{[k]}R_t^{[j]}, W_{kk,t}W_{jk,t})$

$$= E(R_t^{[k]}R_t^{[j]}W_{kk,t}W_{jk,t}) - E(R_t^{[k]}R_t^{[j]})E(W_{kk,t}W_{jk,t})$$
$$= E(\alpha_{kk,t+1})E(\alpha_{jk,t+1})[\lambda_t^{[j]}E(R_t^{[k]}{}^3)]$$
$$- E(\alpha_{kk,t+1})E(\alpha_{jk,t+1})[\lambda_t^{[j]}\lambda_t^{[k]}E(R_t^{[k]}{}^2)]$$
$$= \rho_{kk}\rho_{jk}E(R_t^{[k]}{}^3)\lambda_t^{[j]} - \rho_{kk}\rho_{jk}E(R_t^{[k]}{}^2)\lambda_t^{[k]}\lambda_t^{[j]}.$$

Hence,

$$\begin{aligned}
&\text{Cov}(Z_t, Z_{t+1}) \\
&\quad = \text{Cov}(Y_t^{[1]}Y_t^{[2]}, Y_{t+1}^{[1]}Y_{t+1}^{[2]}) \\
&\quad = \text{Cov}[(W_{11,t-1}W_{21,t-1} + W_{11,t-1}W_{22,t-1} + W_{11,t-1}R_t^{[2]} \\
&\qquad + W_{12,t-1}W_{21,t-1} + W_{12,t-1}W_{22,t-1} + W_{12,t-1}R_t^{[2]} + R_t^{[1]}W_{21,t-1} \\
&\qquad + R_t^{[1]}W_{22,t-1} + R_t^{[1]}R_t^{[2]}), \\
&(W_{11,t}W_{21,t} + W_{11,t}W_{22,t} + W_{11,t}R_{t+1}^{[2]} + W_{12,t}W_{21,t} + W_{12,t}W_{22,t} \\
&\qquad + W_{12,t}R_{t+1}^{[2]} + R_{t+1}^{[1]}W_{21,t} + R_{t+1}^{[1]}W_{22,t} + R_{t+1}^{[1]}R_{t+1}^{[2]})]. \qquad (9.20)
\end{aligned}$$

Expanding Equation (9.20) and using Lemmas 9.2 and 9.3 and the lemmas in Subsection 7.3.2 we obtain

$$
\begin{aligned}
&\mathrm{Cov}(Z_t, Z_{t+1}) \\
&\quad = \frac{\rho_{11}\rho_{21}c_1[c_1+\rho_{11}+\rho_{21}-2\rho_{11}\rho_{21}-c_1\rho_{11}\rho_{21}]}{(1+c_1)^2} \\
&\qquad \times [\lambda_{t-1}^{[1]}+c_1\lambda_{t-1}^{[1]^2}+\lambda_{t-1}^{[1]^2}][\lambda_t^{[1]}+c_1\lambda_t^{[1]^2}+\lambda_t^{[1]^2}] \\
&\qquad + (\frac{\rho_{11}(1-\rho_{11})c_1\rho_{21}}{1+c_1})(\lambda_{t-1}^{[1]}+c_1\lambda_{t-1}^{[1]^2}+\lambda_{t-1}^{[1]^2})(\lambda_t^{[1]})[\rho_{22}\lambda_t^{[2]}+\lambda_{t+1}^{[2]}] \\
&\qquad + (\frac{\rho_{21}(1-\rho_{21})c_1\rho_{11}}{1+c_1})(\lambda_{t-1}^{[1]}+c_1\lambda_{t-1}^{[1]^2}+\lambda_{t-1}^{[1]^2})(\lambda_t^{[1]})[\rho_{12}\lambda_t^{[2]}+\lambda_{t+1}^{[1]}] \\
&\qquad + (\frac{\rho_{11}(1-\rho_{11})c_1\rho_{21}}{1+c_1})(\lambda_t^{[1]}+c_1\lambda_t^{[1]^2}+\lambda_t^{[1]^2})(\lambda_{t-1}^{[1]})[\rho_{22}\lambda_{t-1}^{[2]}+\lambda_t^{[2]}] \\
&\qquad + \lambda_{t-1}^{[1]}\lambda_{t-1}^{[2]}\lambda_t^{[1]}\lambda_t^{[2]}[(\frac{\rho_{21}(1-\rho_{21})c_1}{1+c_1})(\frac{\rho_{12}(1-\rho_{12})c_2}{1+c_2})+(\frac{\rho_{11}(1-\rho_{11})c_1}{1+c_1}) \\
&\qquad \times (\frac{\rho_{22}(1-\rho_{22})c_2}{1+c_2})]+(\frac{\rho_{11}(1-\rho_{11})c_1}{1+c_1})(\lambda_{t-1}^{[1]}\lambda_t^{[1]}) \\
&\qquad \times [\rho_{22}\lambda_{t-1}^{[2]}\lambda_{t+1}^{[2]}+\rho_{22}(\lambda_t^{[2]}+c_2\lambda_t^{[2]^2}+\lambda_t^{[2]^2})+\lambda_t^{[2]}\lambda_{t+1}^{[2]}+\lambda_t^{[2]}\lambda_{t-1}^{[2]}] \\
&\qquad + (\frac{\rho_{22}(1-\rho_{22})c_2}{1+c_2})(\lambda_{t-1}^{[1]}\lambda_{t-1}^{[2]}) \\
&\qquad \times [\rho_{11}\lambda_{t+1}^{[1]}\lambda_t^{[2]}+\rho_{11}\rho_{12}(\lambda_t^{[2]}+c_2\lambda_t^{[2]^2}+\lambda_t^{[2]^2})+\lambda_t^{[1]}\lambda_t^{[2]}] \\
&\qquad + (\lambda_t^{[2]}+c_2\lambda_t^{[2]^2})(\rho_{11}\lambda_{t-1}^{[1]})[\rho_{12}\rho_{21}\lambda_t^{[1]}+\rho_{12}\lambda_{t+1}^{[2]}+\rho_{22}\lambda_{t+1}^{[1]}+\rho_{11}\rho_{22}\lambda_t^{[1]}] \\
&\qquad + (\frac{\rho_{21}(1-\rho_{21})c_1}{1+c_1})(\lambda_{t-1}^{[1]})(\lambda_t^{[1]}+c_1\lambda_t^{[1]^2}+\lambda_t^{[1]^2})[\rho_{12}\rho_{11}\lambda_{t-1}^{[2]}+\rho_{12}\lambda_t^{[2]}+\lambda_{t+1}^{[1]}] \\
&\qquad + (\frac{\rho_{12}(1-\rho_{12})c_2}{1+c_2})(\lambda_t^{[2]}+c_2\lambda_t^{[2]^2}+\lambda_t^{[2]^2})(\lambda_{t-1}^{[2]})[\rho_{21}\rho_{22}\lambda_{t-1}^{[1]}+\lambda_{t+1}^{[2]}+\lambda_t^{[1]}] \\
&\qquad + (\frac{\rho_{12}(1-\rho_{12})c_2\rho_{22}}{1+c_2})(\lambda_t^{[2]})(\lambda_{t-1}^{[2]}+c_2\lambda_{t-1}^{[2]^2}+\lambda_{t-1}^{[2]^2})[\rho_{21}\lambda_{t+1}^{[2]}+\lambda_t^{[1]}] \\
&\qquad + (\frac{\rho_{22}(1-\rho_{22})c_2\rho_{12}}{1+c_2})(\lambda_t^{[2]})(\lambda_{t-1}^{[2]}+c_2\lambda_{t-1}^{[2]^2}+\lambda_{t-1}^{[2]^2})[\rho_{11}\lambda_t^{[1]}+\lambda_{t+1}^{[1]}] \\
&\qquad + (\frac{\rho_{12}(1-\rho_{12})c_2}{1+c_2})\rho_{21}\lambda_{t-1}^{[1]}\lambda_{t-1}^{[2]}\lambda_t^{[2]}[\rho_{21}\lambda_t^{[1]}+\lambda_{t+1}^{[2]}] \\
&\qquad + (\frac{\rho_{21}(1-\rho_{21})c_1}{1+c_1})\rho_{12}\lambda_{t-1}^{[1]}\lambda_{t-1}^{[2]}\lambda_t^{[1]}[\rho_{12}\lambda_t^{[2]}+\lambda_{t+1}^{[1]}] \\
&\qquad + \frac{\rho_{12}\rho_{22}c_2[c_2+\rho_{12}+\rho_{22}-2\rho_{12}\rho_{22}-c_2\rho_{12}\rho_{22}]}{(1+c_2)^2} \\
&\qquad \times [\lambda_{t-1}^{[2]}+c_2\lambda_{t-1}^{[2]^2}+\lambda_{t-1}^{[2]^2}][\lambda_t^{[2]}+c_2\lambda_t^{[2]^2}+\lambda_t^{[2]^2}] \\
&\qquad + (\lambda_t^{[2]}+c_2\lambda_t^{[2]^2})[(\rho_{12}\lambda_{t-1}^{[2]})(\rho_{11}\rho_{22}\lambda_t^{[1]}+\rho_{12}\lambda_t^{[1]}+\rho_{12}\lambda_{t+1}^{[2]}+\rho_{22}\lambda_{t+1}^{[1]})
\end{aligned}
$$

$$
\begin{aligned}
&+\lambda_t^{[1]}(\rho_{12}\lambda_{t+1}^{[2]}+\rho_{22}\lambda_{t+1}^{[1]})]+(\lambda_t^{[1]}+c_1{\lambda_t^{[1]}}^2)\\
&\times[(\rho_{21}\lambda_{t-1}^{[1]}+\rho_{22}\lambda_{t-1}^{[2]})(\rho_{11}\rho_{22}\lambda_t^{[2]}+\rho_{11}\lambda_{t+1}^{[2]}+\rho_{12}\rho_{21}\lambda_t^{[2]})]\\
&+(\lambda_t^{[1]}+c_1{\lambda_t^{[1]}}^2)[\rho_{22}\lambda_{t-1}^{[2]}\rho_{21}\lambda_{t+1}^{[1]}+\rho_{12}^2\lambda_{t-1}^{[1]}\lambda_{t+1}^{[1]}+\rho_{11}\lambda_t^{[2]}\lambda_{t+1}^{[2]}+\rho_{21}\lambda_t^{[2]}\lambda_{t+1}^{[1]}]\\
&+(\frac{\rho_{22}(1-\rho_{22})c_2}{1+c_2})(\lambda_{t-1}^{[2]})[\rho_{11}\lambda_t^{[2]}(\lambda_t^{[1]}+c_1{\lambda_t^{[1]}}^2+{\lambda_t^{[1]}}^2)\\
&+\lambda_t^{[1]}(\lambda_t^{[2]}+c_2{\lambda_t^{[2]}}^2)+\lambda_{t+1}^{[1]}\lambda_t^{[1]}\lambda_t^{[2]}]+(\rho_{11}\rho_{22}+\rho_{12}\rho_{21})[\lambda_t^{[1]}+c_1{\lambda_t^{[1]}}^2+{\lambda_t^{[1]}}^2]\\
&\times[\lambda_t^{[2]}+c_2{\lambda_t^{[2]}}^2+{\lambda_t^{[2]}}^2]-(\rho_{11}\rho_{22}+\rho_{12}\rho_{21}){\lambda_t^{[1]}}^2{\lambda_t^{[2]}}^2+E({R_t^{[2]}}^3)\\
&\times[\rho_{11}\rho_{12}\rho_{22}\lambda_{t-1}^{[1]}+(\frac{\rho_{12}(1-\rho_{12})c_2}{1+c_2})\rho_{22}\lambda_{t-1}^{[2]}+\rho_{12}^2\rho_{22}\lambda_{t-1}^{[2]}+\rho_{12}\rho_{22}\lambda_t^{[1]}]\\
&-(\lambda_t^{[2]}+c_2{\lambda_t^{[2]}}^2+{\lambda_t^{[2]}}^2)[\rho_{11}\rho_{12}\rho_{22}\lambda_{t-1}^{[1]}\lambda_t^{[2]}+\rho_{12}^2\rho_{22}\lambda_{t-1}^{[2]}\lambda_t^{[2]}+\rho_{12}\rho_{22}\lambda_t^{[2]}\lambda_t^{[1]}]\\
&+E({R_t^{[1]}}^3)[(\frac{\rho_{21}(1-\rho_{21})c_1}{1+c_1})\rho_{11}\lambda_{t-1}^{[1]}+\rho_{11}\rho_{21}\rho_{22}\lambda_{t-1}^{[2]}+\rho_{11}\rho_{21}\lambda_t^{[2]}]\\
&-(\lambda_t^{[1]}+c_1{\lambda_t^{[1]}}^2+{\lambda_t^{[1]}}^2)[\rho_{21}^2\rho_{11}\lambda_{t-1}^{[1]}\lambda_t^{[1]}+\rho_{11}\rho_{21}\rho_{22}\lambda_{t-1}^{[2]}\lambda_t^{[1]}+\rho_{11}\rho_{21}\lambda_t^{[2]}\lambda_t^{[1]}].
\end{aligned}
\tag{9.21}
$$

**Lemma 9.4.**

(a) $\mathrm{Cov}(W_{kk,t-1}W_{jk,t-1}, R_t^{[k]}R_t^{[j]})=0,\ j,k\in\{1,2\},\ j\neq k.$

(b) $\mathrm{Cov}(W_{kk,t-1}W_{jk,t-1}, W_{kj,t-1}W_{jj,t-1})=0,\ j,k\in\{1,2\},\ j\neq k.$

(c) $\mathrm{Cov}(W_{kk,t-1}W_{jk,t-1}, W_{kj,t-1}R_t^{[j]})=0,\ j,k\in\{1,2\},\ j\neq k.$

*Proof: The random variables $W_{kk,t-1}$, $W_{jk,t-1}$, $R_t^{[k]}$ and $R_t^{[j]}$ are independent for all $j,k\in\{1,2\}$ and $t\in\mathbb{Z}$ and conditions (c) and (d) imply the independency of the random variables $W_{kk,t-1}W_{jk,t-1}$ and $R_t^{[k]}R_t^{[j]}$. Hence, this implies that its covariance is 0. The other statements can be proved in a similar way.*

**Lemma 9.5.**

(a)
$$
\begin{aligned}
&\mathrm{Cov}(W_{kk,t-1}W_{jk,t-1}, W_{kk,t-1}W_{jj,t-1})\\
&=[\frac{\rho_{kk}(1-\rho_{kk})c_k}{1+c_k}+\rho_{kk}^2]\rho_{jj}\rho_{jk}E({R_{t-1}^{[k]}}^3)\lambda_{t-1}^{[j]}\\
&\quad-\rho_{kk}^2\rho_{jj}\rho_{jk}(\lambda_{t-1}^{[k]}+c_k{\lambda_{t-1}^{[k]}}^2+{\lambda_{t-1}^{[k]}}^2)\lambda_{t-1}^{[j]}\lambda_{t-1}^{[k]},\quad j,k\in\{1,2\}.
\end{aligned}
$$

(b)
$$
\begin{aligned}
&\mathrm{Cov}(W_{kk,t-1}W_{jj,t-1}, W_{kk,t-1}R_t^{[j]})\\
&=[\frac{\rho_{kk}(1-\rho_{kk})c_k}{1+c_k}+\rho_{kk}^2]\rho_{jj}(\lambda_{t-1}^{[k]}+c_k{\lambda_{t-1}^{[k]}}^2+{\lambda_{t-1}^{[k]}}^2)\lambda_t^{[j]}\lambda_{t-1}^{[j]}\\
&\quad-\rho_{11}^2\rho_{22}{\lambda_{t-1}^{[k]}}^2\lambda_t^{[j]}\lambda_{t-1}^{[j]},\quad j,k\in\{1,2\}.
\end{aligned}
$$

(c) $$\begin{aligned}&\text{Cov}(W_{kk,t-1}W_{jj,t-1}, W_{kj,t-1}W_{jk,t-1})\\&= \rho_{kk}\rho_{jk}\rho_{kj}\rho_{jj}(\lambda_{t-1}^{[k]} + c_k\lambda_{t-1}^{[k]\,2} + \lambda_{t-1}^{[k]\,2})(\lambda_{t-1}^{[j]} + c_j\lambda_{t-1}^{[j]\,2} + \lambda_{t-1}^{[j]\,2})\\&\quad - \rho_{kk}\rho_{jk}\rho_{kj}\rho_{jj}\lambda_{t-1}^{[k]\,2}\lambda_{t-1}^{[j]\,2}, \quad j,k \in \{1,2\}, \quad j \neq k.\end{aligned}$$

*Proof:*

(a) $$\begin{aligned}&\text{Cov}(W_{kk,t-1}W_{jk,t-1}, W_{kk,t-1}W_{jj,t-1})\\&= E(W_{kk,t-1}W_{jk,t-1}W_{kk,t-1}W_{jj,t-1})\\&\quad - E(W_{kk,t-1}W_{jk,t-1})E(W_{kk,t-1}W_{jj,t-1})\\&= [\frac{\rho_{kk}(1-\rho_{kk})c_k}{1+c_k} + \rho_{kk}^2]\rho_{jj}\rho_{jk}E(R_{t-1}^{[k]\,3})\lambda_{t-1}^{[j]}\\&\quad - \rho_{kk}^2\rho_{jj}\rho_{jk}(\lambda_{t-1}^{[k]} + c_k\lambda_{t-1}^{[k]\,2} + \lambda_{t-1}^{[k]\,2})\lambda_{t-1}^{[j]}\lambda_{t-1}^{[k]}.\end{aligned}$$

(b) $$\begin{aligned}&\text{Cov}(W_{kk,t-1}W_{jj,t-1}, W_{kk,t-1}R_t^{[j]})\\&= E(W_{kk,t-1}W_{jj,t-1}W_{kk,t-1}R_t^{[j]}) - E(W_{kk,t-1}W_{jj,t-1})E(W_{kk,t-1}R_t^{[j]})\\&= [\frac{\rho_{kk}(1-\rho_{kk})c_k}{1+c_k} + \rho_{kk}^2]\rho_{jj}(\lambda_{t-1}^{[k]} + c_k\lambda_{t-1}^{[k]\,2} + \lambda_{t-1}^{[k]\,2})\lambda_t^{[j]}\lambda_{t-1}^{[j]}\\&\quad - \rho_{11}^2\rho_{22}\lambda_{t-1}^{[k]\,2}\lambda_t^{[j]}\lambda_{t-1}^{[j]}.\end{aligned}$$

(c) $$\begin{aligned}&\text{Cov}(W_{kk,t-1}W_{jj,t-1}, W_{kj,t-1}W_{jk,t-1})\\&= E(W_{kk,t-1}W_{jj,t-1}W_{kj,t-1}W_{jk,t-1})\\&\quad - E(W_{kk,t-1}W_{jj,t-1})E(W_{kj,t-1}W_{jk,t-1})\\&= \rho_{kk}\rho_{jk}\rho_{kj}\rho_{jj}(\lambda_{t-1}^{[k]} + c_k\lambda_{t-1}^{[k]\,2} + \lambda_{t-1}^{[k]\,2})(\lambda_{t-1}^{[j]} + c_j\lambda_{t-1}^{[j]\,2} + \lambda_{t-1}^{[j]\,2})\\&\quad - \rho_{kk}\rho_{jk}\rho_{kj}\rho_{jj}\lambda_{t-1}^{[k]\,2}\lambda_{t-1}^{[j]\,2}.\end{aligned}$$

Hence,

$$\begin{aligned}\text{Var}(Z_t) &= \text{Cov}(Z_t, Z_t) = \text{Cov}(Y_t^{[1]}Y_t^{[2]}, Y_t^{[1]}Y_t^{[2]})\\&= \text{Cov}[(W_{11,t-1}W_{21,t-1} + W_{11,t-1}W_{22,t-1} + W_{11,t-1}R_t^{[2]}\\&\quad + W_{12,t-1}W_{21,t-1} + W_{12,t-1}W_{22,t-1} + W_{12,t-1}R_t^{[2]} + R_t^{[1]}W_{21,t-1}\\&\quad + R_t^{[1]}W_{22,t-1} + R_t^{[1]}R_t^{[2]}),\\&\quad (W_{11,t-1}W_{21,t-1} + W_{11,t-1}W_{22,t-1} + W_{11,t-1}R_t^{[2]} + W_{12,t-1}W_{21,t-1}\\&\quad + W_{12,t-1}W_{22,t-1} + W_{12,t-1}R_t^{[2]} + R_t^{[1]}W_{21,t-1}\\&\quad + R_t^{[1]}W_{22,t-1} + R_t^{[1]}R_t^{[2]})]. \qquad (9.22)\end{aligned}$$

Expanding Equation (9.22) and using Lemmas 9.4 and 9.5 and the lemmas in Subsection 7.3.2, we obtain

$$
\begin{aligned}
&\mathrm{Cov}(Z_t, Z_t)\\
&\quad = \frac{\rho_{11}\rho_{21}c_1[c_1+\rho_{11}+\rho_{21}-2\rho_{11}\rho_{21}-c_1\rho_{11}\rho_{21}]}{(1+c_1)^2}E(R_{t-1}^{[1]\,4})\\
&\quad + \rho_{11}^2\rho_{21}^2E(R_{t-1}^{[1]\,4}) - \rho_{11}^2\rho_{21}^2[\lambda_{t-1}^{[1]} + c_1\lambda_{t-1}^{[1]\,2} + \lambda_{t-1}^{[1]\,2}]^2\\
&\quad + 2(\frac{\rho_{11}(\rho_{11}+c_1)}{1+c_1})\rho_{21}E(R_{t-1}^{[1]\,3})[\rho_{22}\lambda_{t-1}^{[2]}+\lambda_t^{[2]}] - 2(\lambda_{t-1}^{[1]}+c_1\lambda_{t-1}^{[1]\,2}+\lambda_{t-1}^{[1]\,2})\\
&\quad \times (\rho_{11}\lambda_{t-1}^{[1]})[\rho_{11}\rho_{22}\lambda_t^{[2]} + \rho_{21}^2\rho_{12}\lambda_{t-1}^{[2]} + \rho_{21}^2\lambda_t^{[1]} + \rho_{11}\rho_{21}\rho_{22}\lambda_{t-1}^{[2]}]\\
&\quad + 2(\frac{\rho_{21}(\rho_{21}+c_1)}{1+c_1})\rho_{11}E(R_{t-1}^{[1]\,3})[\rho_{12}\lambda_{t-1}^{[2]} + \lambda_t^{[1]}]\\
&\quad + 2\rho_{11}\rho_{21}\rho_{22}(\lambda_{t-1}^{[1]} + c_1\lambda_{t-1}^{[1]\,2})\lambda_{t-1}^{[2]}\lambda_t^{[1]}\\
&\quad + \frac{\rho_{11}\rho_{22}[-\rho_{11}\rho_{22}(c_1c_2+c_1+c_2)+\rho_{22}c_1+\rho_{11}c_2+c_1c_2]}{(1+c_1)(1+c_2)}E(R_{t-1}^{[1]\,3})\lambda_{t-1}^{[2]}\\
&\quad + \rho_{11}^2\rho_{22}^2E(R_{t-1}^{[1]\,3})\lambda_{t-1}^{[2]} + (\lambda_{t-1}^{[1]} + c_1\lambda_{t-1}^{[1]\,2} + \lambda_{t-1}^{[1]\,2})[\rho_{11}^2\rho_{22}^2\lambda_{t-1}^{[1]}\lambda_{t-1}^{[2]}\\
&\quad + 2(\frac{\rho_{11}(\rho_{11}+c_1)}{1+c_1})\rho_{22}\lambda_{t-1}^{[2]}\lambda_t^{[2]}]\\
&\quad - 2\rho_{11}\rho_{22}\lambda_{t-1}^{[1]}\lambda_{t-1}^{[2]}[\rho_{11}\lambda_{t-1}^{[1]}\lambda_t^{[2]} + \rho_{12}\rho_{21}\lambda_{t-1}^{[1]}\lambda_{t-1}^{[2]} + \rho_{22}\lambda_{t-1}^{[2]}\lambda_t^{[1]}]\\
&\quad + 2\rho_{11}\rho_{12}\rho_{21}\rho_{22}(\lambda_{t-1}^{[1]} + c_1\lambda_{t-1}^{[1]\,2} + \lambda_{t-1}^{[1]\,2})(\lambda_{t-1}^{[2]} + c_2\lambda_{t-1}^{[2]\,2} + \lambda_{t-1}^{[2]\,2})\\
&\quad + 2\rho_{11}\rho_{12}(\frac{\rho_{22}(\rho_{22}+c_2)}{1+c_2})E(R_{t-1}^{[2]\,3})\lambda_{t-1}^{[1]}\\
&\quad + 2\rho_{11}\rho_{12}\rho_{22}(\lambda_{t-1}^{[2]} + c_2\lambda_{t-1}^{[2]\,2})\lambda_t^{[2]}\lambda_{t-1}^{[1]}\\
&\quad + 2\rho_{11}\lambda_{t-1}^{[1]}(\lambda_{t-1}^{[2]} + c_2\lambda_{t-1}^{[2]\,2} + \lambda_{t-1}^{[2]\,2})[(\frac{\rho_{22}(\rho_{22}+c_2)}{1+c_2})\lambda_t^{[1]} - \rho_{12}\rho_{22}^2\lambda_{t-1}^{[2]}]\\
&\quad + (\frac{\rho_{11}(\rho_{11}+c_1)}{1+c_1})(\lambda_{t-1}^{[1]} + c_1\lambda_{t-1}^{[1]\,2} + \lambda_{t-1}^{[1]\,2})(\lambda_t^{[2]} + c_2\lambda_t^{[2]2} + \lambda_t^{[2]2})\\
&\quad - \rho_{11}^2\lambda_{t-1}^{[1]\,2}\lambda_t^{[2]2} + 2\rho_{11}\rho_{21}\lambda_t^{[2]}(\lambda_{t-1}^{[1]} + c_1\lambda_{t-1}^{[1]\,2})[\rho_{12}\lambda_{t-1}^{[2]} + \lambda_t^{[1]}]\\
&\quad + 2\rho_{11}\lambda_{t-1}^{[1]}(\lambda_t^{[2]} + c_2\lambda_t^{[2]2})[\rho_{12}\lambda_{t-1}^{[2]} + \lambda_t^{[1]}]\\
&\quad + \rho_{11}^2\rho_{21}^2(\lambda_{t-1}^{[1]} + c_1\lambda_{t-1}^{[1]\,2} + \lambda_{t-1}^{[1]\,2})(\lambda_{t-1}^{[2]} + c_2\lambda_{t-1}^{[2]\,2} + \lambda_{t-1}^{[2]\,2})\\
&\quad - \rho_{12}^2\rho_{21}^2\lambda_{t-1}^{[1]\,2}\lambda_{t-1}^{[2]\,2} + 2\rho_{12}(\lambda_{t-1}^{[2]} + c_2\lambda_{t-1}^{[2]\,2} + \lambda_{t-1}^{[2]\,2})[(\frac{(\rho_{12}+c_2)}{1+c_2})\lambda_{t-1}^{[1]}\lambda_t^{[2]}\\
&\quad - \rho_{22}\lambda_{t-1}^{[2]}(\rho_{12}\rho_{21}\lambda_{t-1}^{[1]} + \rho_{12}\lambda_t^{[2]} + \rho_{22}\lambda_t^{[1]})]\\
&\quad + 2\rho_{21}\rho_{22}(\frac{\rho_{12}(\rho_{12}+c_2)}{1+c_2})E(R_{t-1}^{[2]\,3})\lambda_{t-1}^{[1]} - \rho_{12}\lambda_{t-1}^{[2]}[2\rho_{12}\lambda_{t-1}^{[2]}\lambda_{t-1}^{[1]}\lambda_t^{[2]}\\
&\quad + 2\rho_{21}^2\lambda_{t-1}^{[1]\,2}\lambda_t^{[1]} + \rho_{12}\lambda_{t-1}^{[2]}\lambda_t^{[2]2}]
\end{aligned}
$$

$$
\begin{aligned}
&+ 2\rho_{12}(\lambda_{t-1}^{[1]} + c_1\lambda_{t-1}^{[1]\,2} + \lambda_{t-1}^{[1]\,2})[(\frac{\rho_{21}(\rho_{21}+c_1)}{1+c_1})\lambda_{t-1}^{[2]}\lambda_t^{[1]}] \\
&+ 2\rho_{22}(\lambda_{t-1}^{[2]} + c_2\lambda_{t-1}^{[2]\,2})[\rho_{21}\rho_{12}\lambda_{t-1}^{[1]} + \rho_{11}\lambda_t^{[1]}\lambda_t^{[2]}] \\
&+ 2(\frac{\rho_{22}(\rho_{22}+c_2)}{1+c_2})\rho_{12}E(R_{t-1}^{[2]\,3})\lambda_t^{[1]} \\
&+ \frac{\rho_{12}\rho_{22}c_2[c_2+\rho_{12}+\rho_{22}-2\rho_{12}\rho_{22}-c_2\rho_{12}\rho_{22}]}{(1+c_2)^2}E(R_{t-1}^{[2]\,4}) + \rho_{12}^2\rho_{22}^2E(R_{t-1}^{[2]\,4}) \\
&- \rho_{12}^2\rho_{22}^2(\lambda_{t-1}^{[2]} + c_2\lambda_{t-1}^{[2]\,2} + \lambda_{t-1}^{[2]\,2})^2 + 2(\frac{\rho_{12}(\rho_{12}+c_2)}{1+c_2})\rho_{22}E(R_{t-1}^{[2]\,3})\lambda_t^{[2]} \\
&+ 2(\frac{\rho_{12}(\rho_{12}+c_2)}{1+c_2})(\lambda_{t-1}^{[2]} + c_2\lambda_{t-1}^{[2]\,2} + \lambda_{t-1}^{[2]\,2})(\lambda_t^{[2]} + c_2\lambda_t^{[2]^2} + \lambda_t^{[2]^2}) \\
&+ 2\rho_{12}(\lambda_t^{[2]} + c_2\lambda_t^{[2]^2})\lambda_{t-1}^{[2]}\lambda_t^{[1]} + (\frac{\rho_{21}(\rho_{21}+c_1)}{1+c_1})(\lambda_{t-1}^{[1]} + c_1\lambda_{t-1}^{[1]\,2} + \lambda_{t-1}^{[1]\,2}) \\
&\times (\lambda_t^{[1]} + c_1\lambda_t^{[1]^2} + \lambda_t^{[1]^2}) - \lambda_t^{[1]^2}[\rho_{21}^2\lambda_{t-1}^{[1]\,2} + \rho_{22}^2\lambda_{t-1}^{[2]\,2}] + 2(\lambda_t^{[1]} + c_1\lambda_t^{[1]^2}) \\
&\times [\rho_{21}\rho_{22}\lambda_{t-1}^{[1]}\lambda_{t-1}^{[2]} + \rho_{21}\lambda_{t-1}^{[1]}\lambda_t^{[2]} + \rho_{22}\lambda_{t-1}^{[2]}\lambda_t^{[2]}] \\
&+ \frac{\rho_{12}\rho_{21}[-\rho_{12}\rho_{21}(c_1c_2+c_1+c_2) + \rho_{12}c_1 + \rho_{21}c_2 + c_1c_2]}{(1+c_1)(1+c_2)} \\
&\times (\lambda_{t-1}^{[1]} + c_1\lambda_{t-1}^{[1]\,2} + \lambda_{t-1}^{[1]\,2})(\lambda_{t-1}^{[2]} + c_2\lambda_{t-1}^{[2]\,2} + \lambda_{t-1}^{[2]\,2}) \\
&+ (\frac{\rho_{22}(\rho_{22}+c_2)}{1+c_2})(\lambda_t^{[1]} + c_1\lambda_t^{[1]^2} + \lambda_t^{[1]^2})(\lambda_{t-1}^{[2]} + c_2\lambda_{t-1}^{[2]\,2} + \lambda_{t-1}^{[2]\,2}),
\end{aligned}
\tag{9.23}
$$

where from Subsection 8.3.2, we obtain

$$
E(R_t^{[k]^3}) = \lambda_t^{[k]} + 3\lambda_t^{[k]^2} + 3c_k\lambda_t^{[k]^2} + 3c_k\lambda_t^{[k]^3} + 2c_k^2\lambda_t^{[k]^3} + \lambda_t^{[k]^3} \tag{9.24}
$$

and

$$
\begin{aligned}
E(R_t^{[k]^4}) = {} & \lambda_t^{[k]} + 7\lambda_t^{[k]^2} + 7c_k\lambda_t^{[k]^2} + 6\lambda_t^{[k]^3} + 18c_k\lambda_t^{[k]^3} \\
& + 12c_k^2\lambda_t^{[k]^3} + 6c_k\lambda_t^{[k]^4} + 11c_k^2\lambda_t^{[k]^4} + 6c_k^3\lambda_t^{[k]^4} + \lambda_t^{[k]^4}.
\end{aligned}
\tag{9.25}
$$

The entries of the derivative matrix $\boldsymbol{D_c}$ are expressed as

$$
\boldsymbol{D}_c' = \begin{pmatrix} \frac{\partial m_1}{\partial c_1} & \frac{\partial m_2}{\partial c_1} & \cdots & \frac{\partial m_T}{\partial c_1} \\ \frac{\partial m_1}{\partial c_2} & \frac{\partial m_2}{\partial c_2} & \cdots & \frac{\partial m_T}{\partial c_2} \end{pmatrix}_{2\times T} \tag{9.26}
$$

with

$$
m_t = \rho_{11}\rho_{21}\lambda_{t-1}^{[1]}(1 + c_1\lambda_{t-1}^{[1]}) + \rho_{12}\rho_{22}\lambda_{t-1}^{[2]}(1 + c_2\lambda_{t-1}^{[2]}) + \mu_t^{[1]}\mu_t^{[2]}. \tag{9.27}
$$

The entries of the derivative matrix are given as:

1. $\frac{\partial m_t}{\partial c_1} = \rho_{11}\rho_{21}{\lambda_{t-1}^{[1]}}^2$.

2. $\frac{\partial m_t}{\partial c_2} = \rho_{12}\rho_{22}{\lambda_{t-1}^{[2]}}^2$.

Equation (9.18) is solved using the Newton-Raphson iteration procedure that yields

$$\begin{pmatrix} \hat{c}_{1r+1} \\ \hat{c}_{2r+1} \end{pmatrix} = \begin{pmatrix} \hat{c}_{1r} \\ \hat{c}_{2r} \end{pmatrix} + [\boldsymbol{D_c}' \boldsymbol{\Sigma_c}^{-1} \boldsymbol{D_c}]_r^{-1} [\boldsymbol{D_c}' \boldsymbol{\Sigma_c}^{-1} (\boldsymbol{Z} - \boldsymbol{m})]_r, \tag{9.28}$$

where $\hat{c}_{1r}$ and $\hat{c}_{2r}$ are the estimates at the $r^{th}$ iteration and $[.]_r$ are the values of the expression at the $r^{th}$ iteration.

The algorithm works as follows: using the updated estimates of the regression and dependence parameters obtained from the iterative Equation (9.17) and an initial value of $c_1$ and $c_2$, the iterative Equation (9.28) is solved to obtain an updated over-dispersion parameters which are then in turn used to obtain a new set of regression and dependence using Equation (9.17) and this cycle of iteration continues until convergence. These estimates are consistent and under mild regularity conditions, it can be shown that $(\hat{\boldsymbol{c}} - \boldsymbol{c})'$ is asymptotic normal with mean $\boldsymbol{0}$ and covariance matrix $[\boldsymbol{D_c}' \boldsymbol{\Sigma_c}^{-1} \boldsymbol{D_c}]^{-1}$ (Refer to Subsection 2.3.1.2).

Using the converged estimates from the above GQL equations, we derive the forecasting equations for the BINMA(1) model with NB innovations. The conditional expectation and variance of the one-step ahead forecast $Y_{t+1}^{[k]}$ given $r_t^{[k]}$ are as follows:

$$\begin{aligned} E(Y_{t+1}^{[1]}|r_t^{[1]}, r_t^{[2]}) &= E(\alpha_{11,t} \circ r_t^{[1]}|r_t^{[1]}) + E(\alpha_{12,t} \circ r_t^{[2]}|r_t^{[2]}) + E(R_{t+1}^{[1]}) \\ &= \rho_{11} r_t^{[1]} + \rho_{12} r_t^{[2]} + \lambda_{t+1}^{[1]}, \end{aligned} \tag{9.29}$$

$$\begin{aligned} E(Y_{t+1}^{[2]}|r_t^{[1]}, r_t^{[2]}) &= E(\alpha_{21,t} \circ r_t^{[1]}|r_t^{[1]}) + E(\alpha_{22,t} \circ r_t^{[2]}|r_t^{[2]}) + E(R_{t+1}^{[2]}) \\ &= \rho_{21} r_t^{[1]} + \rho_{22} r_t^{[2]} + \lambda_{t+1}^{[2]}, \end{aligned} \tag{9.30}$$

with the corresponding variances:

$$\begin{aligned} &\text{Var}(Y_{t+1}^{[1]}|r_t^{[1]}, r_t^{[2]}) \\ &= \text{Var}(\alpha_{11,t} \circ R_t^{[1]}|r_1^{[1]}) + \text{Var}(\alpha_{12,t} \circ R_t^{[2]}|r_1^{[2]}) + \text{Var}(R_{t+1}^{[1]}) \\ &= E_{\alpha_{11,t}}[\text{Var}(\alpha_{11,t} \circ R_t^{[1]}|r_t^{[1]}, \alpha_{11,t})] + \text{Var}_{\alpha_{11,t}}[E(\alpha_{11,t} \circ R_t^{[1]}|r_t^{[1]}, \alpha_{11,t})] \\ &\quad + E_{\alpha_{12,t}}[\text{Var}(\alpha_{12,t} \circ R_t^{[2]}|r_t^{[2]}, \alpha_{12,t})] + \text{Var}_{\alpha_{12,t}}[E(\alpha_{12,t} \circ R_t^{[2]}|r_t^{[2]}, \alpha_{12,t})] \\ &\quad + \text{Var}(R_{t+1}^{[1]}) \end{aligned}$$

$$
\begin{aligned}
&= E_{\alpha_{11,t}}[\alpha_{11,t}(1-\alpha_{11,t})r_t^{[1]}]+\text{Var}_{\alpha_{11,t}}[\alpha_{11,t}r_t^{[1]}]+E_{\alpha_{12,t}}[\alpha_{12,t}(1-\alpha_{12,t})r_t^{[2]}] \\
&\quad + \text{Var}_{\alpha_{12,t}}[\alpha_{12,t}r_t^{[2]}] + \text{Var}(R_{t+1}^{[1]}) \\
&= \left[\rho_{11} - (\frac{\rho_{11}(1-\rho_{11})c_1}{1+c_1} + \rho_{11}^2)\right] r_t^{[1]} + \frac{\rho_{11}(1-\rho_{11})c_1}{1+c_1}(r_t^{[1]})^2 \\
&\quad + \left[\rho_{12} - (\frac{\rho_{12}(1-\rho_{12})c_2}{1+c_2} + \rho_{12}^2)\right] r_t^{[2]} + \frac{\rho_{12}(1-\rho_{12})c_2}{1+c_2}(r_t^{[2]})^2 \\
&\quad + [\lambda_{t+1}^{[1]} + c_1(\lambda_{t+1}^{[1]})^2] \\
&= \frac{\rho_{11}(1-\rho_{11})}{1+c_1}r_t^{[1]} + \frac{\rho_{11}(1-\rho_{11})c_1}{1+c_1}(r_t^{[1]})^2 + \frac{\rho_{12}(1-\rho_{12})}{1+c_2}r_t^{[2]} \\
&\quad + \frac{\rho_{12}(1-\rho_{12})c_2}{1+c_2}(r_t^{[2]})^2 + [\lambda_{t+1}^{[1]} + c_1(\lambda_{t+1}^{[1]})^2] \\
&= \frac{\rho_{11}(1-\rho_{11})}{1+c_1}r_t^{[1]}[1+c_1r_t^{[1]}] + \frac{\rho_{12}(1-\rho_{12})}{1+c_2}r_t^{[2]}[1+c_2r_t^{[2]}] \\
&\quad + \lambda_{t+1}^{[1]} + c_1(\lambda_{t+1}^{[1]})^2. \qquad (9.31)
\end{aligned}
$$

Similarly,

$$
\begin{aligned}
\text{Var}(Y_{t+1}^{[2]}|r_t^{[1]}, r_t^{[2]}) &= \frac{\rho_{21}(1-\rho_{21})}{1+c_1}r_t^{[1]}[1+c_1r_t^{[1]}] + \frac{\rho_{22}(1-\rho_{22})}{1+c_2}r_t^{[2]}[1+c_2r_t^{[2]}] \\
&\quad + \lambda_{t+1}^{[2]} + c_2(\lambda_{t+1}^{[2]})^2, \qquad (9.32)
\end{aligned}
$$

where $r_t^{[k]}$ are approximated by their respective means.

## 9.4 Simulation Study

In this section, we present a simulation study to simulate non-stationary BINMA(1) series with NB innovations. The covariate design is assumed as follows:

$$
x_{t1} = \begin{cases} -\cos(2\pi t) + 0.15 & (t = 1, \ldots, T/4) \\ \sin(2\pi t) + 0.25 & (t = (T/4) + 1, \ldots, 3T/4) \\ \cos(2\pi t) + 0.30 & (t = (3T/4) + 1, \ldots, T) \end{cases}
$$

$$
x_{t2} = \begin{cases} (1/2t) & (t = 1, \ldots, T/4) \\ (-1/2t) & (t = (T/4) + 1, \ldots, 3T/4) \\ -2t & (t = (3T/4) + 1, \ldots, T) \end{cases}
$$

where $[\beta_1^{[1]}, \beta_2^{[1]}, \beta_1^{[2]}, \beta_2^{[2]}] = [0.2, 0.2, 0.2, 0.2]$, $\rho_{11}, \rho_{22}, \rho_{21}, \rho_{12}$ are assumed combination of values 0.3 and 0.9 and $c_1$ and $c_2$ take combination of values

[0.2, 0.6], [0.9, 1.5] and [0.4,0.8] for $T = 60, 100, 600$. $y_t^{[1]}$ and $y_t^{[2]}$ are generated using the OD model (9.3), where $R_t^{[k]} \sim \text{NB}(\frac{1}{c_k}, c_k\lambda_t^{[k]})$. Five thousand Monte Carlo replications are implemented and the results are shown in Tables 9.1–9.2.

**TABLE 9.1**
Estimates of the Parameters and Standard Errors under Non-Stationary BINMA(1) Process for $[c_1, c_2] = [0.9, 1.5], [0.4, 0.8]$

| T | $\hat{\beta}_1^{[1]}$ | $\hat{\beta}_2^{[1]}$ | $\hat{\beta}_1^{[2]}$ | $\hat{\beta}_2^{[2]}$ | $\hat{c}_1 = 0.9$ | $\hat{c}_2 = 1.5$ | $\hat{\rho}_{11} = 0.9$ | $\hat{\rho}_{22} = 0.9$ | $\hat{\rho}_{12} = 0.3$ | $\hat{\rho}_{21} = 0.3$ |
|---|---|---|---|---|---|---|---|---|---|---|
| 60 | 0.1842 | 0.1864 | 0.1816 | 0.1879 | 0.8853 | 1.4874 | 0.8801 | 0.8811 | 0.2887 | 0.2879 |
| | (0.0921) | (0.0914) | (0.0923) | (0.0927) | (0.1610) | (0.1314) | (0.1297) | (0.1058) | (0.1173) | (0.1028) |
| 100 | 0.1912 | 0.1906 | 0.1945 | 0.2089 | 0.8917 | 1.4924 | 0.8908 | 0.8944 | 0.3070 | 0.3062 |
| | (0.0830) | (0.0777) | (0.0611) | (0.0848) | (0.1160) | (0.1285) | (0.0766) | (0.0917) | (0.0839) | (0.0980) |
| 600 | 0.1964 | 0.1945 | 0.1967 | 0.1984 | 0.8970 | 1.4985 | 0.8997 | 0.8962 | 0.2972 | 0.3042 |
| | (0.0559) | (0.0426) | (0.0316) | (0.0552) | (0.0891) | (0.0942) | (0.0597) | (0.0649) | (0.0483) | (0.0352) |
| T | $\hat{\beta}_1^{[1]}$ | $\hat{\beta}_2^{[1]}$ | $\hat{\beta}_1^{[2]}$ | $\hat{\beta}_2^{[2]}$ | $\hat{c}_1 = 0.9$ | $\hat{c}_2 = 1.5$ | $\hat{\rho}_{11} = 0.3$ | $\hat{\rho}_{22} = 0.3$ | $\hat{\rho}_{12} = 0.9$ | $\hat{\rho}_{21} = 0.9$ |
| 60 | 0.1859 | 0.1869 | 0.2182 | 0.2153 | 0.8815 | 1.4817 | 0.2823 | 0.2808 | 0.8892 | 0.8875 |
| | (0.0911) | (0.0930) | (0.1053) | (0.1061) | (0.1590) | (0.1380) | (0.1018) | (0.1269) | (0.1012) | (0.1181) |
| 100 | 0.1902 | 0.1922 | 0.2092 | 0.1928 | 0.8911 | 1.4917 | 0.2923 | 0.2908 | 0.9070 | 0.9065 |
| | (0.0849) | (0.0679) | (0.0745) | (0.0819) | (0.1388) | (0.1204) | (0.0845) | (0.0926) | (0.0731) | (0.0957) |
| 600 | 0.1943 | 0.1966 | 0.1965 | 0.2018 | 0.8989 | 1.4967 | 0.2944 | 0.2958 | 0.8980 | 0.9015 |
| | (0.0436) | (0.0340) | (0.0545) | (0.0204) | (0.0721) | (0.0909) | (0.0452) | (0.0638) | (0.0563) | (0.0676) |
| T | $\hat{\beta}_1^{[1]}$ | $\hat{\beta}_2^{[1]}$ | $\hat{\beta}_1^{[2]}$ | $\hat{\beta}_2^{[2]}$ | $\hat{c}_1 = 0.9$ | $\hat{c}_2 = 1.5$ | $\hat{\rho}_{11} = 0.3$ | $\hat{\rho}_{22} = 0.3$ | $\hat{\rho}_{12} = 0.3$ | $\hat{\rho}_{21} = 0.3$ |
| 60 | 0.2120 | 0.1840 | 0.2180 | 0.1810 | 0.8897 | 1.4889 | 0.2811 | 0.2838 | 0.2887 | 0.2891 |
| | (0.1034) | (0.0911) | (0.1002) | (0.0903) | (0.1432) | (0.1310) | (0.1146) | (0.1097) | (0.1226) | (0.1030) |
| 100 | 0.1916 | 0.1914 | 0.1945 | 0.2073 | 0.8920 | 1.4913 | 0.2921 | 0.2914 | 0.3066 | 0.2968 |
| | (0.0837) | (0.0642) | (0.0839) | (0.0763) | (0.1172) | (0.1270) | (0.0798) | (0.0970) | (0.0813) | (0.0939) |
| 600 | 0.1945 | 0.1975 | 0.1957 | 0.1972 | 0.8969 | 1.4977 | 0.2975 | 0.2998 | 0.3016 | 0.2985 |
| | (0.0512) | (0.0302) | (0.0567) | (0.0263) | (0.0790) | (0.0991) | (0.0477) | (0.0618) | (0.0375) | (0.0633) |
| T | $\hat{\beta}_1^{[1]}$ | $\hat{\beta}_2^{[1]}$ | $\hat{\beta}_1^{[2]}$ | $\hat{\beta}_2^{[2]}$ | $\hat{c}_1 = 0.4$ | $\hat{c}_2 = 0.8$ | $\hat{\rho}_{11} = 0.9$ | $\hat{\rho}_{22} = 0.9$ | $\hat{\rho}_{12} = 0.3$ | $\hat{\rho}_{21} = 0.3$ |
| 60 | 0.2181 | 0.1869 | 0.2183 | 0.1886 | 0.3824 | 0.7879 | 0.8861 | 0.8864 | 0.3110 | 0.3115 |
| | (0.1073) | (0.0915) | (0.1033) | (0.0940) | (0.1673) | (0.1381) | (0.1129) | (0.1061) | (0.1269) | (0.1037) |
| 100 | 0.1923 | 0.1909 | 0.1910 | 0.1914 | 0.3925 | 0.7909 | 0.8924 | 0.8911 | 0.3090 | 0.3085 |
| | (0.0818) | (0.0630) | (0.0843) | (0.0746) | (0.1424) | (0.1220) | (0.0782) | (0.0952) | (0.0871) | (0.0946) |
| 600 | 0.2026 | 0.1967 | 0.1942 | 0.1948 | 0.3976 | 0.7990 | 0.8984 | 0.8942 | 0.3046 | 0.3036 |
| | (0.0509) | (0.0374) | (0.0508) | (0.0272) | (0.0925) | (0.0869) | (0.0540) | (0.0625) | (0.0407) | (0.0671) |
| T | $\hat{\beta}_1^{[1]}$ | $\hat{\beta}_2^{[1]}$ | $\hat{\beta}_1^{[2]}$ | $\hat{\beta}_2^{[2]}$ | $\hat{c}_1 = 0.4$ | $\hat{c}_2 = 0.8$ | $\hat{\rho}_{11} = 0.3$ | $\hat{\rho}_{22} = 0.3$ | $\hat{\rho}_{12} = 0.9$ | $\hat{\rho}_{21} = 0.9$ |
| 60 | 0.1812 | 0.1854 | 0.1847 | 0.2112 | 0.3893 | 0.7810 | 0.2863 | 0.2832 | 0.8895 | 0.8898 |
| | (0.0901) | (0.0921) | (0.0911) | (0.1025) | (0.1308) | (0.1387) | (0.1112) | (0.1051) | (0.1279) | (0.1018) |
| 100 | 0.1914 | 0.1920 | 0.1943 | 0.2076 | 0.3915 | 0.7916 | 0.2917 | 0.2905 | 0.9063 | 0.9077 |
| | (0.0858) | (0.0677) | (0.0876) | (0.0796) | (0.1194) | (0.1273) | (0.0815) | (0.0944) | (0.0736) | (0.0910) |
| 600 | 0.1982 | 0.1952 | 0.1950 | 0.1976 | 0.3971 | 0.7996 | 0.2961 | 0.2968 | 0.9020 | 0.9031 |
| | (0.0361) | (0.0551) | (0.0255) | (0.0525) | (0.0844) | (0.0956) | (0.0459) | (0.0682) | (0.0587) | (0.0697) |
| T | $\hat{\beta}_1^{[1]}$ | $\hat{\beta}_2^{[1]}$ | $\hat{\beta}_1^{[2]}$ | $\hat{\beta}_2^{[2]}$ | $\hat{c}_1 = 0.4$ | $\hat{c}_2 = 0.8$ | $\hat{\rho}_{11} = 0.3$ | $\hat{\rho}_{22} = 0.3$ | $\hat{\rho}_{12} = 0.3$ | $\hat{\rho}_{21} = 0.3$ |
| 60 | 0.1828 | 0.2123 | 0.1852 | 0.1885 | 0.3870 | 0.7830 | 0.2835 | 0.2884 | 0.2821 | 0.2898 |
| | (0.0912) | (0.1011) | (0.0864) | (0.0942) | (0.1319) | (0.1710) | (0.1167) | (0.1089) | (0.1219) | (0.1008) |
| 100 | 0.0919 | 0.0938 | 0.0901 | 0.0947 | 0.3910 | 0.7907 | 0.2935 | 0.2913 | 0.3091 | 0.3082 |
| | (0.0644) | (0.0759) | (0.0822) | (0.0816) | (0.1129) | (0.1249) | (0.0820) | (0.0975) | (0.0778) | (0.0961) |
| 600 | 0.0972 | 0.0962 | 0.0962 | 0.0979 | 0.3993 | 0.7980 | 0.2988 | 0.2979 | 0.3035 | 0.3023 |
| | (0.0148) | (0.0522) | (0.0243) | (0.0569) | (0.0851) | (0.0970) | (0.0431) | (0.0626) | (0.0423) | (0.0687) |

**TABLE 9.2**
Estimates of the Parameters and Standard Errors under Non-Stationary BINMA(1) Process for $[c_1, c_2] = [0.3, 0.6]$

| T | $\hat{\beta}_1^{[1]}$ | $\hat{\beta}_2^{[1]}$ | $\hat{\beta}_1^{[2]}$ | $\hat{\beta}_2^{[2]}$ | $\hat{c}_1 = 0.3$ | $\hat{c}_2$=0.6 | $\hat{\rho}_{11} = 0.9$ | $\hat{\rho}_{22} = 0.9$ | $\hat{\rho}_{12} = 0.3$ | $\hat{\rho}_{21} = 0.3$ |
|---|---|---|---|---|---|---|---|---|---|---|
| 60 | 0.2176 | 0.1839 | 0.2122 | 0.1813 | 0.2870 | 0.5885 | 0.8889 | 0.8890 | 0.2891 | 0.2881 |
| | (0.1060) | (0.1009) | (0.0903) | (0.0901) | (0.1230) | (0.1343) | (0.0902) | (0.1032) | (0.1178) | (0.1066) |
| 100 | 0.1913 | 0.1911 | 0.1902 | 0.2093 | 0.2929 | 0.5927 | 0.8926 | 0.8901 | 0.3051 | 0.3070 |
| | (0.0878) | (0.0642) | (0.0854) | (0.0776) | (0.1122) | (0.1279) | (0.0754) | (0.0907) | (0.0891) | (0.0977) |
| 600 | 0.1944 | 0.1986 | 0.1965 | 0.2955 | 0.2943 | 0.5969 | 0.8969 | 0.8978 | 0.3037 | 0.3020 |
| | (0.0332) | (0.0577) | (0.0214) | (0.0579) | (0.0704) | (0.0917) | (0.0471) | (0.0395) | (0.0661) | (0.0553) |
| T | $\hat{\beta}_1^{[1]}$ | $\hat{\beta}_2^{[1]}$ | $\hat{\beta}_1^{[2]}$ | $\hat{\beta}_2^{[2]}$ | $\hat{c}_1 = 0.3$ | $\hat{c}_2$=0.6 | $\hat{\rho}_{11} = 0.3$ | $\hat{\rho}_{22} = 0.3$ | $\hat{\rho}_{12} = 0.9$ | $\hat{\rho}_{21} = 0.9$ |
| 60 | 0.1874 | 0.1821 | 0.1889 | 0.2120 | 0.2816 | 0.5885 | 0.2898 | 0.2872 | 0.8881 | 0.8890 |
| | (0.0903) | (0.0904) | (0.0923) | (0.0919) | (0.1227) | (0.1361) | (0.1126) | (0.1033) | (0.0940) | (0.1094) |
| 100 | 0.1918 | 0.1942 | 0.1935 | 0.1929 | 0.3096 | 0.5919 | 0.2932 | 0.2914 | 0.9060 | 0.9067 |
| | (0.0816) | (0.0619) | (0.0870) | (0.0729) | (0.1238) | (0.1112) | (0.0849) | (0.0977) | (0.0858) | (0.0939) |
| 600 | 0.1989 | 0.1966 | 0.2015 | 0.1960 | 0.2968 | 0.5944 | 0.2959 | 0.2980 | 0.9014 | 0.9046 |
| | (0.0536) | (0.0408) | (0.0552) | (0.0312) | (0.0905) | (0.0722) | (0.0428) | (0.0652) | (0.0312) | (0.0678) |
| T | $\hat{\beta}_1^{[1]}$ | $\hat{\beta}_2^{[1]}$ | $\hat{\beta}_1^{[2]}$ | $\hat{\beta}_2^{[2]}$ | $\hat{c}_1 = 0.3$ | $\hat{c}_2$=0.6 | $\hat{\rho}_{11} = 0.3$ | $\hat{\rho}_{22} = 0.3$ | $\hat{\rho}_{12} = 0.3$ | $\hat{\rho}_{21} = 0.3$ |
| 60 | 0.1821 | 0.1873 | 0.1838 | 0.2169 | 0.2867 | 0.5860 | 0.2806 | 0.2885 | 0.3144 | 0.3189 |
| | (0.0908) | (0.0935) | (0.0918) | (0.1015) | (0.1353) | (0.1290) | (0.1119) | (0.1071) | (0.0938) | (0.1069) |
| 100 | 0.1936 | 0.1908 | 0.1939 | 0.1916 | 0.2926 | 0.5929 | 0.2941 | 0.2931 | 0.3091 | 0.3061 |
| | (0.0856) | (0.0720) | (0.0873) | (0.0643) | (0.1203) | (0.1019) | (0.0708) | (0.0954) | (0.0847) | (0.0912) |
| 600 | 0.1979 | 0.2019 | 0.1980 | 0.1958 | 0.2962 | 0.5974 | 0.2955 | 0.2969 | 0.3010 | 0.2985 |
| | (0.0512) | (0.0315) | (0.0598) | (0.0230) | (0.0720) | (0.0972) | (0.0409) | (0.0656) | (0.0354) | (0.0617) |

Tables 9.1–9.2 show the estimates of the parameters of interest are close to their corresponding population values and as expected, the standard errors of the different estimates decrease as the sample size increases. We also note a higher standard errors for the over-dispersion parameters as compared to the other parameters. However, some non-convergent simulations are noted during the simulation: For $\rho_{11} = \rho_{22} = 0.9$, $\rho_{12} = \rho_{21} = 0.3$, $c_1 = 0.2$ and $c_2 = 0.6$, the number of non-convergent simulations is 280 for $T = 60$, 250 for $T = 100$ and 180 for $T = 600$. For same values of over-dispersion and cross-correlation but lower values of serial-correlation parameters, that is $\rho_{11} = \rho_{22} = 0.3$, we observe a slight decrease in the number of non-convergent simulations. In addition, for a higher over-dispersion values, there is an increase in the number of failures. For $\rho_{11} = \rho_{22} = 0.9$, $\rho_{12} = \rho_{21} = 0.3$, $c_1 = 0.9$ and $c_2 = 1.5$, the number of failures in the GQL algorithm is 290 for $T = 60$, 265 for $T = 100$ and 195 for $T = 600$. The non-convergent simulations were due to the ill-conditioned covariance matrix (9.19) or the singularity of the Hessian matrix (9.28).

## 9.5 Summary

In this chapter, we have developed a novel BINMA(1) model with NB innovations. It is remarked that the marginal distribution of the counting series is rather complicated to identify. Using the multivariate binomial thinning properties, the serial and cross auto-covariances have been derived for a general lag-$h$. Since the likelihood function is difficult to construct, we have considered the GQL approach to estimate the different parameters of interest. The simulation results have illustrated that GQL provides consistent estimates with relatively fewer non-convergent experiments. In the next chapter, we develop a constrained BINAR(1) model with COM-Poisson innovations that can analyze counting series that exhibit equi-, over- or under-dispersion.

# 10

# *Constrained BINAR(1) Model with Correlated COM-Poisson Innovations*

## 10.1 Introduction

The BINAR(1) models introduced in earlier chapters can only model counting series that exhibit equi- or over-dispersion. However, count observations can be under-dispersed in some cases. COM-Poisson distribution (Shmueli et al. (2005)) can model counts that are equi-, over- or under-dispersed. COM-Poisson belongs to the exponential family of distribution and thus inherits the statistical properties attributed to the moment expressions of these distributions. The COM-Poisson model has achieved some popularity in the recent years due to its flexibility in handling all forms of dispersion in discrete data analysis. It has found applications in marketing (Boatwright et al. (2003), Borle et al. (2005, 2007, 2006), Kalyanam et al. (2007), Singh et al. (2009)), linguistic (Best (2001), Wimmer et al. (1994)) and transportation (Lord et al. (2010, 2008)) at cross-sectional level. Hence, in this chapter we develop a constrained BINAR(1) model with COM-Poisson innovations that can handle different levels of equi-, over- or under-dispersion.

A number of real-life applications in simple univariate setting has made use of the COM-Poisson as the underlying model for their analysis wherein reliable estimates and fits were obtained (Mamodekhan & Jowaheer (2010), Mamode Khan & Jowaheer (2013), Sellers et al. (2012)). Researchers have further extended its purpose to model time series processes under the autoregressive structures mainly (Mamodekhan & Jowaheer (2010), Mamode Khan & Jowaheer (2013), Mamode Khan (2016)). However, so far there exists no model that accommodates non-stationarity along with different levels of over-, equi- or in particular under-dispersion in bivariate time series of counts. Hence, this chapter bridges this gap by proposing a constrained BINAR(1) model where the innovations are marginally COM-Poisson (Minka et al. (2003), Shmueli et al. (2005)) and are correlated, under time-dependent covariates. Inferential analysis is based on the GQL estimation method which is used to estimate the regression and dispersion parameters, while the dependence parameters are estimated using the moment estimating equations, as in Chapter 2.

DOI: 10.1201/9781003600435-10 

The main objectives of this chapter are to construct a constrained BINAR(1) model with COM-Poisson innovations and to develop a GQL approach to estimate the different sets of unknown parameters.

The organization of the chapter is as follows: In Section 10.2, we provide an outline of the COM-Poisson model. In Section 10.3, a constrained BINAR(1) model under COM-Poisson innovations is derived. In Section 10.4, the GQL equation is developed for estimating the regression and dispersion parameters, while the serial- and cross-correlation parameters are estimated using the moment estimating equations. In Section 10.5, we present some numerical experiments where the GQL approach is assessed by simulating non-stationary BINAR(1) series, where the COM-Poisson innovations are simulated through an algorithm. The conclusion is presented in the last section.

## 10.2 COM-Poisson Model

The COM-Poisson probability model (Shmueli et al. (2005)) is specified as

$$P(Y=y) = \frac{\theta^y}{(y!)^\nu}\frac{1}{Z(\theta,\nu)}, \quad y=0,1,2,3,\ldots, \quad \text{for} \quad \theta>0, \nu\geq 0, \quad (10.1)$$

where the normalizing constant $Z(\theta,\nu) = \sum_{j=0}^{\infty}\frac{\theta^j}{(j!)^\nu}$. The COM-Poisson model (10.1) generalizes several discrete distributions. Likewise, under $\nu=0$ and $\theta<1$, the Geometric distribution becomes a special case of the COM-Poisson. Under $\nu=1$, it generalizes the Poisson distribution with mean parameter $\theta$ and becomes Bernoulli when $\nu\rightarrow\infty$ with probability of success given by $\left(\frac{\theta}{1+\theta}\right)$. Under any parametrization of $\theta$ and $\nu$, the COM-Poisson is shown to be a member of the exponential family with link predictor function $\ln\theta = (\boldsymbol{x}'\boldsymbol{\beta})$ for $\boldsymbol{x} = [x_1,x_2,\ldots,x_j,\ldots,x_p]'$ and $\boldsymbol{\beta} = [\beta_1,\beta_2,\ldots,\beta_j,\ldots,\beta_p]'$. Furthermore, $\nu$ is regarded as the dispersion parameter such that $\nu>1$ represents under-dispersion and $\nu<1$ over-dispersion.

The mgf and pgf are expressed as

$$E[e^{Ys}] = \frac{Z(\theta e^s,\nu)}{Z(\theta,\nu)} \quad \text{and} \quad E[s^Y] = \frac{Z(\theta s,\nu)}{Z(\theta,\nu)} \quad (10.2)$$

while the moments of the COM-Poisson model are derived iteratively using

$$E(Y^{s+1}) = \begin{cases} \theta E(Y+1)^{1-\nu} & s=0, \\ \theta\frac{\partial}{\partial\theta}E(Y^s) + E(Y)E(Y^s) & s>0. \end{cases} \quad (10.3)$$

In particular, the mean and variance are given as follows:

$$E(Y)\equiv\mu = \theta\frac{\partial}{\partial\theta}\log Z(\theta,\nu) \quad \text{and} \quad \mathrm{Var}(Y)\equiv\sigma^2 = \theta\frac{\partial E(Y)}{\partial\theta}. \quad (10.4)$$

However, these moments can be approximated in closed forms using the asymptotic approximation by Minka et al. (2003)

$$Z(\theta,\nu)=\frac{\exp(\nu\theta^{1/\nu})}{(2\pi\theta^{1/\nu})^{\frac{\nu-1}{2}}\sqrt{\nu}}[1+O(\theta^{-1/\nu})] \tag{10.5}$$

that yields

$$\mu\simeq\theta^{1/\nu}-\frac{\nu-1}{2\nu}\quad\text{and}\quad\sigma^2\simeq\theta\frac{\partial}{\partial\theta}(\theta^{1/\nu}-\frac{\nu-1}{2\nu})=\frac{\theta^{1/\nu}}{\nu} \tag{10.6}$$

or alternatively, from Equation (10.6), $\sigma^2$ can be re-written as $\sigma^2\simeq\frac{\mu}{\nu}+\frac{\nu-1}{2\nu^2}$ (See Mamode Khan & Jowaheer (2013) and Mamode Khan (2016)). Hence, under this parametrization, $(\mu,\nu)$ is the pair of parameters to describe the COM-Poisson variable $Y$, such that $Y\sim$ COM-Poisson$(\mu,\nu)$.

In the context of the above approximated moments, Minka et al. (2003), Shmueli et al. (2005) and Zhu (2012) presented some trial experiments that reveal that these approximations are more accurate for large $\theta$. For large $\theta$ and $\nu<1$, Var(Y) outweighs the E(Y) while for large $\theta$ and $\nu>1$, E(Y) $>$ Var(Y) indicating under-dispersion, but in fact for $\nu>1$, Nadarajah (2009) provides an exact expression for the mean and variance based on generalized hypergeometric function. However, the approximation by Minka et al. (2003) is rather more suitable as it explores the full range of $\nu\geq 0$.

## 10.3 The Constrained BINAR(1) Model

The constrained BINAR(1) model is specified as

$$\begin{pmatrix} Y_t^{[1]} \\ Y_t^{[2]} \end{pmatrix}=\begin{pmatrix} \rho_{11} & 0 \\ 0 & \rho_{22} \end{pmatrix}\circ\begin{pmatrix} Y_{t-1}^{[1]} \\ Y_{t-1}^{[2]} \end{pmatrix}+\begin{pmatrix} R_t^{[1]} \\ R_t^{[2]} \end{pmatrix}, \tag{10.7}$$

where $\rho_{kk}\in[0,1]$ and $R_t^{[k]}$ is the innovation term at the $t^{th}$ time point for $k\in\{1,2\}$. In the above, the pair $(Y_{t-1}^{[k]},R_t^{[k]})$ is mutually independent.

'$\circ$' indicates the binomial thinning operator (Steutel & Van Harn (1979), Wei$\beta$ (2008b)) such that $\{\rho_{kk}\circ Y_{t-1}^{[k]}|Y_{t-1}^{[k]}\}\sim$ Binomial$(Y_{t-1}^{[k]},\rho_{kk})$ with the following properties:

1. $E(\rho_{kk}\circ Y_{t-1}^{[k]})=\rho_{kk}E(Y_{t-1}^{[k]})$.
2. $\text{Var}(\rho_{kk}\circ Y_{t-1}^{[k]})=\rho_{kk}(1-\rho_{kk})E(Y_{t-1}^{[k]})+\rho_{kk}^2\text{Var}(Y_{t-1}^{[k]})$.
3. $\text{Cov}(\rho_{11}\circ Y_{t-1}^{[1]},\rho_{22}\circ Y_{t-1}^{[2]})=\rho_{11}\rho_{22}\text{Cov}(Y_{t-1}^{[1]},Y_{t-1}^{[2]})$.

In the above model (10.7), $R_t^{[k]}$ follows marginally the COM-Poisson distribution. Since we consider the model in a more general case we do not suppose any assumption about the joint distribution of the innovation terms $R_t^{[1]}$ and $R_t^{[2]}$. We only suppose that the correlation between these terms is $\kappa_{12,t}$, such that

$$\text{Corr}(R_t^{[1]}, R_{t'}^{[2]}) = \begin{cases} \kappa_{12,t} & t = t' \\ 0 & t \neq t'. \end{cases} \tag{10.8}$$

We further assume that $Y_{t-1}^{[k]} \sim$ COM-Poisson$(\mu_{t-1}^{[k]}, \nu_k)$ and the innovation term $R_t^{[k]} \sim$ COM-Poisson$(\lambda_t^{[k]}, \tilde{\nu}_k)$, where $\lambda_t^{[k]} = \mu_t^{[k]} - \rho_{kk}\mu_{t-1}^{[k]}$ and

$$\tilde{\nu}_k = \frac{(2q_0^{[k]} + 1) + \sqrt{(2q_0^{[k]} + 1)^2 - 8q_1^{[k]}}}{4q_1^{[k]}} > 0 \tag{10.9}$$

with $q_0^{[k]} = [\mu_t^{[k]} - \rho_{kk}\mu_{t-1}^{[k]}]$ and $q_1^{[k]} = [(1 - \rho_{kk}^2)(\frac{\nu_k - 1}{2\nu_k^2}) + \frac{\mu_t^{[k]}}{\nu_k} - \frac{\rho_{kk}\mu_{t-1}^{[k]}}{\nu_k}(\rho_{kk} + \nu_k(1 - \rho_{kk}))]$ (See Mamode Khan & Jowaheer (2013) and Mamode Khan (2016) for more details) and $\mu_t^{[k]} = (\theta_t^{[k]})^{1/\nu_k} - (\frac{\nu_k - 1}{2\nu_k})$, where $\theta_t^{[k]} = \exp(\boldsymbol{x}_t'\boldsymbol{\beta}^{[k]})$ with $\boldsymbol{x}_t = [x_{t1}, x_{t2}, \ldots, x_{tj}, \ldots, x_{tp}]'$ and $\boldsymbol{\beta}^{[k]} = [\beta_1^{[k]}, \beta_2^{[k]}, \ldots, \beta_j^{[k]}, \ldots, \beta_p^{[k]}]'$ for $k \in \{1, 2\}$.

Based on the above conditions, it is proved that

$$\begin{aligned} E(Y_t^{[k]}) &= E(\rho_{kk} \circ Y_{t-1}^{[k]}) + E(R_t^{[k]}) \\ &= \rho_{kk}E(Y_{t-1}^{[k]}) + E(R_t^{[k]}) \\ &= \mu_t^{[k]} \end{aligned} \tag{10.10}$$

while

$$\begin{aligned} \text{Var}(Y_t^{[k]}) &= \text{Var}(\rho_{kk} \circ Y_{t-1}^{[k]}) + \text{Var}(R_t^{[k]}) \\ &= \rho_{kk}(1 - \rho_{kk})E(Y_{t-1}^{[k]}) + \rho_{kk}^2\text{Var}(Y_{t-1}^{[k]}) + \text{Var}(R_t^{[k]}) \\ &= \rho_{kk}(1 - \rho_{kk})\mu_{t-1}^{[k]} + \rho_{kk}^2\left[\frac{\mu_{t-1}^{[k]}}{\nu_k} + \frac{\nu_k - 1}{2\nu_k^2}\right] + \frac{\mu_t^{[k]} - \rho_{kk}\mu_{t-1}^{[k]}}{\tilde{\nu}_k} + \frac{\tilde{\nu}_k - 1}{2\tilde{\nu}_k^2} \\ &= \frac{\mu_t^{[k]}}{\nu_k} + \frac{\nu_k - 1}{2\nu_k^2} \end{aligned} \tag{10.11}$$

when replacing $\tilde{\nu}_k$ by Equation (10.9).
Hence, $Y_t^{[k]} \sim$ COM-Poisson$(\mu_t^{[k]}, \nu_k)$.

The serial lag-$h$ correlation for the $k^{th}$ series is

$$\begin{aligned}
\text{Cov}(Y_t^{[k]}, Y_{t+h}^{[k]}) &= \text{Cov}[Y_t^{[k]}, (\rho_{kk} \circ Y_{t+h-1}^{[k]} + R_{t+h}^{[k]})] \\
&= \rho_{kk}\text{Cov}[Y_t^{[k]}, Y_{t+h-1}^{[k]}] \\
&= \rho_{kk}\text{Cov}[Y_t^{[k]}, (\rho_{kk} \circ Y_{t+h-2}^{[k]} + R_{t+h-1}^{[k]})] \\
&= \rho_{kk}^2\text{Cov}[Y_t^{[k]}, Y_{t+h-2}^{[k]}] \\
&\vdots \\
&= \rho_{kk}^h\text{Cov}[Y_t^{[k]}, Y_t^{[k]}] \\
&= \rho_{kk}^h \left[\frac{\mu_t^{[k]}}{\nu_k} + \frac{\nu_k - 1}{2\nu_k^2}\right]
\end{aligned} \tag{10.12}$$

and hence

$$\text{Corr}(Y_t^{[k]}, Y_{t+h}^{[k]}) = \frac{\rho_{kk}^h\sqrt{\left(\frac{\mu_t^{[k]}}{\nu_k} + \frac{\nu_k-1}{2\nu_k^2}\right)}}{\sqrt{\left(\frac{\mu_{t+h}^{[k]}}{\nu_k} + \frac{\nu_k-1}{2\nu_k^2}\right)}}. \tag{10.13}$$

The cross-covariance expressions are given as follows:

$$\begin{aligned}
\text{Cov}(Y_t^{[1]}, Y_t^{[2]}) &= \text{Cov}(\rho_{11} \circ Y_{t-1}^{[1]} + R_t^{[1]}, \rho_{22} \circ Y_{t-1}^{[2]} + R_t^{[2]}) \\
&= \rho_{11}\rho_{22}\text{Cov}(Y_{t-1}^{[1]}, Y_{t-1}^{[2]}) + \text{Cov}(R_t^{[1]}, R_t^{[2]}) \\
&= \rho_{11}\rho_{22}\text{Cov}(Y_{t-1}^{[1]}, Y_{t-1}^{[2]}) \\
&\quad + \left[\kappa_{12,t}\sqrt{\frac{\mu_t^{[1]} - \rho_{11}\mu_{t-1}^{[1]}}{\tilde{\nu}_1} + \frac{\tilde{\nu}_1 - 1}{2\tilde{\nu}_1^2}}\sqrt{\frac{\mu_t^{[2]} - \rho_{22}\mu_{t-1}^{[2]}}{\tilde{\nu}_2} + \frac{\tilde{\nu}_2 - 1}{2\tilde{\nu}_2^2}}\right].
\end{aligned} \tag{10.14}$$

As for the other off-diagonal cross-covariances

$$\begin{aligned}
\text{Cov}(Y_t^{[1]}, Y_{t+h}^{[2]}) &= \text{Cov}(Y_t^{[1]}, \rho_{22} \circ Y_{t+h-1}^{[2]} + R_{t+h}^{[2]}) \\
&= \rho_{22}\text{Cov}(Y_t^{[1]} Y_{t+h-1}^{[2]}) \\
&= \rho_{22}\text{Cov}(Y_t^{[1]}, \rho_{22} \circ Y_{t+h-2}^{[2]} + R_{t+h-1}^{[2]}) \\
&= \rho_{22}^2\text{Cov}(Y_t^{[1]}, Y_{t+h-2}^{[2]}) \\
&\vdots \\
&= \rho_{22}^h\text{Cov}(Y_t^{[1]}, Y_t^{[2]}).
\end{aligned} \tag{10.15}$$

Similarly,

$$\begin{aligned}
\mathrm{Cov}(Y_{t+h}^{[1]}, Y_t^{[2]}) &= \mathrm{Cov}(\rho_{11} \circ Y_{t+h-1}^{[1]} + R_{t+h}^{[1]}, Y_t^{[2]}) \\
&= \rho_{11}\mathrm{Cov}(Y_{t+h-1}^{[1]}, Y_t^{[2]}) \\
&= \rho_{11}\mathrm{Cov}(\rho_{11} \circ Y_{t+h-2}^{[1]} + R_{t+h-1}^{[1]}, Y_t^{[2]}) \\
&= \rho_{11}^2\mathrm{Cov}(Y_{t+h-2}^{[1]}, Y_t^{[2]}) \\
&\vdots \\
&= \rho_{11}^h\mathrm{Cov}(Y_t^{[1]}, Y_t^{[2]}). \qquad (10.16)
\end{aligned}$$

**Remark 10.1.** Under stationary moment conditions, we have

$$E(Y_t^{[k]}) = \mu_1^{[k]} \quad , \quad \mathrm{Var}(Y_t^{[k]}) = \frac{\mu_1^{[k]}}{\nu_k} + \frac{\nu_k - 1}{2\nu_k^2} \qquad (10.17)$$

and

$$\mathrm{Cov}(Y_t^{[1]}, Y_t^{[2]}) = \frac{\kappa_{12,1}\sqrt{\frac{(1-\rho_{11})\mu_1^{[1]}}{\tilde{\nu}_1} + \frac{\tilde{\nu}_1 - 1}{2\tilde{\nu}_1^2}}\sqrt{\frac{(1-\rho_{22})\mu_1^{[2]}}{\tilde{\nu}_2} + \frac{\tilde{\nu}_2 - 1}{2\tilde{\nu}_2^2}}}{(1-\rho_{11}\rho_{22})} \qquad (10.18)$$

**Remark 10.2.** It can also be easily verified that under $\nu_k = 1$, we obtain

$$\begin{aligned}
\tilde{\nu}_k &= \frac{(2\mu_t^{[k]} - 2\rho_{kk}\mu_{t-1}^{[k]} + 1) + \sqrt{(2\mu_t^{[k]} - 2\rho_{kk}\mu_{t-1}^{[k]} + 1)^2 - 8[\mu_t^{[k]} - \rho_{kk}\mu_{t-1}^{[k]}]}}{4[\mu_t^{[k]} - \rho_{kk}\mu_{t-1}^{[k]}]} \\
&= \frac{(2\mu_t^{[k]} - 2\rho_{kk}\mu_{t-1}^{[k]} + 1) + \sqrt{(2\mu_t^{[k]} - 2\rho_{kk}\mu_{t-1}^{[k]} - 1)^2}}{4[\mu_t^{[k]} - \rho_{kk}\mu_{t-1}^{[k]}]} \\
&= \frac{4[\mu_t^{[k]} - \rho_{kk}\mu_{t-1}^{[k]}]}{4[\mu_t^{[k]} - \rho_{kk}\mu_{t-1}^{[k]}]} \\
&= 1 \qquad (10.19)
\end{aligned}$$

and

$$E(Y_t^{[k]}) = \mathrm{Var}(Y_t^{[k]}) = \mu_t^{[k]}. \qquad (10.20)$$

Hence, $Y_t^{[k]} \sim \mathrm{Poisson}(\mu_t^{[k]})$. Thus, the constrained BINAR(1) model with Poisson innovations introduced in Chapter 2 is a special case of the constrained BINAR(1) model with COM-Poisson innovations.

## 10.4 Estimation of Parameters

The GQL equation is expressed as:

$$\boldsymbol{D}_{\boldsymbol{\beta},\boldsymbol{\nu}}{}' \boldsymbol{\Sigma}_{\boldsymbol{\beta},\boldsymbol{\nu}}{}^{-1}(\boldsymbol{y} - \boldsymbol{\mu}) = 0, \tag{10.21}$$

where

$$\boldsymbol{y} = \begin{pmatrix} \boldsymbol{y}^{[1]} \\ \boldsymbol{y}^{[2]} \end{pmatrix}_{2T\times 1}, \boldsymbol{y}^{[1]} = [y_1^{[1]}, y_2^{[1]}, \ldots, y_T^{[1]}]', \boldsymbol{y}^{[2]} = [y_1^{[2]}, y_2^{[2]}, \ldots, y_T^{[2]}]' \tag{10.22}$$

and

$$\boldsymbol{\mu} = \begin{pmatrix} \boldsymbol{\mu}^{[1]} \\ \boldsymbol{\mu}^{[2]} \end{pmatrix}_{2T\times 1}, \boldsymbol{\mu}^{[1]} = [\mu_1^{[1]}, \mu_2^{[1]}, \ldots, \mu_T^{[1]}]', \boldsymbol{\mu}^{[2]} = [\mu_1^{[2]}, \mu_2^{[2]}, \ldots, \mu_T^{[2]}]' \tag{10.23}$$

where $\mu_t^{[k]} = (\theta_t^{[k]})^{1/\nu_k} - (\frac{\nu_k - 1}{2\nu_k})$.

The derivative matrix is a block diagonal $2T \times (2p+2)$ matrix with $[\boldsymbol{D}^{[1]}, \boldsymbol{D}^{[2]}]$ on the diagonals where

$$\boldsymbol{D}^{[k]} = \begin{pmatrix} \frac{\partial \mu_1^{[k]}}{\partial \beta_1^{[k]}} & \frac{\partial \mu_1^{[k]}}{\partial \beta_2^{[k]}} & \cdots & \frac{\partial \mu_1^{[k]}}{\partial \beta_p^{[k]}} & \frac{\partial \mu_1^{[k]}}{\partial \nu_k} \\ \frac{\partial \mu_2^{[k]}}{\partial \beta_1^{[k]}} & \frac{\partial \mu_2^{[k]}}{\partial \beta_2^{[k]}} & \cdots & \frac{\partial \mu_2^{[k]}}{\partial \beta_p^{[k]}} & \frac{\partial \mu_2^{[k]}}{\partial \nu_k} \\ \vdots & \vdots & \ddots & \vdots & \vdots \\ \frac{\partial \mu_T^{[k]}}{\partial \beta_1^{[k]}} & \frac{\partial \mu_T^{[k]}}{\partial \beta_2^{[k]}} & \cdots & \frac{\partial \mu_T^{[k]}}{\partial \beta_p^{[k]}} & \frac{\partial \mu_T^{[k]}}{\partial \nu_k} \end{pmatrix}_{T\times(p+1)} \tag{10.24}$$

with $\frac{\partial \mu_t^{[k]}}{\partial \beta_j^{[k]}} = \frac{(\theta_t^{[k]})^{\frac{1}{\nu_k}} x_{tj}}{\nu_k}$ and $\frac{\partial \mu_t^{[k]}}{\partial \nu_k} = -\frac{(\theta_t^{[k]})^{\frac{1}{\nu_k}} ln(\theta_t^{[k]})}{\nu_k^2} - \frac{1}{2\nu_k} + \frac{\nu_k - 1}{2\nu_k^2}$.

The covariance structure $\boldsymbol{\Sigma}_{\boldsymbol{\beta},\boldsymbol{\nu}}$ is

$$
\begin{pmatrix}
V(Y_1^{[1]}) & Cov(Y_1^{[1]}, Y_2^{[1]}) & \cdots & Cov(Y_1^{[1]}, Y_T^{[1]}) & \vdots & Cov(Y_1^{[1]}, Y_1^{[2]}) & Cov(Y_1^{[1]}, Y_2^{[2]}) & \cdots & Cov(Y_1^{[1]}, Y_T^{[2]}) \\
Cov(Y_2^{[1]}, Y_1^{[1]}) & V(Y_2^{[1]}) & \cdots & Cov(Y_2^{[1]}, Y_T^{[1]}) & \vdots & Cov(Y_2^{[1]}, Y_1^{[2]}) & Cov(Y_2^{[1]}, Y_2^{[2]}) & \cdots & Cov(Y_2^{[1]}, Y_T^{[2]}) \\
\vdots & \vdots & \ddots & \vdots & \vdots & \vdots & \vdots & \ddots & \vdots \\
Cov(Y_T^{[1]}, Y_1^{[1]}) & Cov(Y_T^{[1]}, Y_2^{[1]}) & \cdots & V(Y_T^{[1]}) & \vdots & Cov(Y_T^{[1]}, Y_1^{[2]}) & Cov(Y_T^{[1]}, Y_2^{[2]}) & \cdots & Cov(Y_T^{[1]}, Y_T^{[2]}) \\
\cdots & \cdots & \cdots & \cdots & \cdots & \cdots & \cdots & \cdots & \cdots \\
Cov(Y_1^{[2]}, Y_1^{[1]}) & Cov(Y_1^{[2]}, Y_2^{[1]}) & \cdots & Cov(Y_1^{[2]}, Y_T^{[1]}) & \vdots & V(Y_1^{[2]}) & Cov(Y_1^{[2]}, Y_2^{[2]}) & \cdots & Cov(Y_1^{[2]}, Y_T^{[2]}) \\
Cov(Y_2^{[2]}, Y_1^{[1]}) & Cov(Y_2^{[2]}, Y_2^{[1]}) & \cdots & Cov(Y_2^{[2]}, Y_T^{[1]}) & \vdots & Cov(Y_2^{[2]}, Y_1^{[2]}) & V(Y_2^{[2]}) & \cdots & Cov(Y_2^{[2]}, Y_T^{[2]}) \\
\vdots & \vdots & \ddots & \vdots & \vdots & \vdots & \vdots & \ddots & \vdots \\
Cov(Y_T^{[2]}, Y_1^{[1]}) & Cov(Y_T^{[2]}, Y_2^{[1]}) & \cdots & Cov(Y_T^{[2]}, Y_T^{[1]}) & \vdots & Cov(Y_T^{[2]}, Y_1^{[2]}) & Cov(Y_T^{[2]}, Y_2^{[2]}) & \cdots & V(Y_T^{[2]})
\end{pmatrix}_{2T \times 2T}
\quad (10.25)
$$

where the entries are computed using Equation (10.11)–(10.16). However, for $t = 1$, $\text{Cov}(Y_t^{[1]}, Y_t^{[2]})$ is computed using Equation (10.18), while for $t = 2, \ldots, T$, $\text{Cov}(Y_t^{[1]}, Y_t^{[2]})$ is computed iteratively using Equation (10.14).

The serial-correlation coefficients for the two series $\rho_{11}$ and $\rho_{22}$ in Equation (10.13) are estimated using Sutradhar & Das (1999) and Jowaheer & Sutradhar (2005) moment estimation approaches

$$\hat{\rho}_{kk} = \frac{T \sum_{t=1}^{T-1} \tilde{y}_t^{[k]} \tilde{y}_{t+1}^{[k]}}{\left[\sum_{t=1}^{T} (\tilde{y}_t^{[k]})^2\right] \left[\sum_{t=2}^{T} \sqrt{\frac{\left(\frac{\mu_{t-1}^{[k]}}{\nu_k} + \frac{\nu_k - 1}{2\nu_k^2}\right)}{\left(\frac{\mu_t^{[k]}}{\nu_k} + \frac{\nu_k - 1}{2\nu_k^2}\right)}}\right]} \tag{10.26}$$

where $\tilde{y_t}^{[k]} = \frac{y_t^{[k]} - \mu_t^{[k]}}{\sqrt{\left(\frac{\mu_t^{[k]}}{\nu_k} + \frac{\nu_k - 1}{2\nu_k^2}\right)}}$. As for the cross-correlation $\kappa_{12,t}$, it is estimated using Equation (10.14), that is,

$$\hat{\kappa}_{12,t} = \frac{\tilde{\text{Cov}}(Y_t^{[1]}, Y_t^{[2]}) - \hat{\rho_{11}}\hat{\rho_{22}}\tilde{\text{Cov}}(Y_{t-1}^{[1]}, Y_{t-1}^{[2]})}{\sqrt{\frac{(\hat{\mu}_t^{[1]} - \hat{\rho_{11}}\hat{\mu}_{t-1}^{[1]})}{\hat{\nu}_1} + \frac{\hat{\nu}_1 - 1}{2\hat{\nu}_1^2}} \sqrt{\frac{(\hat{\mu}_t^{[2]} - \hat{\rho_{22}}\hat{\mu}_{t-1}^{[2]})}{\hat{\nu}_2} + \frac{\hat{\nu}_2 - 1}{2\hat{\nu}_2^2}}} \tag{10.27}$$

where $\tilde{\text{Cov}}(Y_t^{[1]}, Y_t^{[2]}) = \frac{1}{T} \sum_{t=1}^{T} (y_t^{[1]} - \hat{\mu}_t^{[1]})(y_t^{[2]} - \hat{\mu}_t^{[2]})$, $\tilde{\text{Cov}}(Y_{t-1}^{[1]}, Y_{t-1}^{[2]}) = \frac{1}{T-1} \sum_{t=2}^{T} (y_{t-1}^{[1]} - \hat{\mu}_{t-1}^{[1]})(y_{t-1}^{[2]} - \hat{\mu}_{t-1}^{[2]})$, based on the assumption that $\hat{\mu}_0^{[k]} = \hat{\mu}_1^{[k]}$.

The Newton-Raphson iterative technique is used to solve the GQL in Equation (10.21) and this yields

$$(\ \hat{\boldsymbol{\psi}}_{r+1}\ ) = (\ \hat{\boldsymbol{\psi}}_r\ ) + [\boldsymbol{D_\psi}' \boldsymbol{\Sigma_\psi}^{-1} \boldsymbol{D_\psi}]_r^{-1} [\boldsymbol{D_\psi}' \boldsymbol{\Sigma_\psi}^{-1} (\boldsymbol{y} - \boldsymbol{\mu})]_r, \tag{10.28}$$

where $\hat{\boldsymbol{\psi}} = [\hat{\beta}_1^{[1]}, \hat{\beta}_2^{[1]}, \hat{\beta}_1^{[2]}, \hat{\beta}_2^{[2]}, \hat{\nu_1}, \hat{\nu_2}]$ and $\hat{\boldsymbol{\psi}}_r$ are the estimates at the $r^{th}$ iteration and $[.]_r$ are the values of the expression at the $r^{th}$ iteration.

The working of this algorithm is as follows: For an initial $\hat{\boldsymbol{\psi}}$, the serial-correlations (10.26) and cross-correlations (10.27) are computed. These estimates are then used to estimate the covariance structure (10.25). Then, using Equation (10.28), an updated estimate of $\hat{\boldsymbol{\psi}}$ is obtained. The new set of regression and dispersion parameters is then used to re-calculate components (10.26) and (10.27). This iterative process continues until the absolute difference between the updated and previous regression estimates is less than a tolerance $10^{-5}$. Under mild regulatory conditions, the regression and over-dispersion estimators are consistent and $(\hat{\boldsymbol{\psi}} - \boldsymbol{\psi})'$ follows an asymptotic normal distribution with mean $\boldsymbol{0}$ and covariance matrix $[\boldsymbol{D_\psi}' \boldsymbol{\Sigma_\psi}^{-1} \boldsymbol{D_\psi}]^{-1}$ (Refer to Subsection 2.3.1.2).

Using the converged estimates from the GQL method, we now derive the forecasting equations of the constrained BINAR(1) with COM-Poisson innovations. The conditional expectation and variance of the one-step ahead forecasting equations $Y_{t+1}^{[k]}$ given $y_t^{[k]}$ are as follows:

$$\begin{aligned}E(Y_{t+1}^{[k]}|y_t^{[k]}) &= E(\rho_{kk}\circ y_t^{[k]}|y_t^{[k]}) + E(R_{t+1}^{[k]})\\ &= \rho_{kk}y_t^{[k]} + \mu_{t+1}^{[k]} - \rho_{kk}\mu_t^{[k]}\\ &= \mu_{t+1}^{[k]} + \rho_{kk}(y_t^{[k]} - \mu_t^{[k]}) \end{aligned} \tag{10.29}$$

and

$$\begin{aligned}\mathrm{Var}(Y_{t+1}^{[k]}|y_t^{[k]}) &= \mathrm{Var}(\rho_{kk}\circ y_t^{[k]}|y_t^{[k]}) + \mathrm{Var}(R_{t+1}^{[k]})\\ &= \rho_{kk}(1-\rho_{kk})y_t^{[k]} + \frac{\mu_{t+1}^{[k]} - \rho_{kk}\mu_t^{[k]}}{\tilde{\nu}_k} + \frac{\tilde{\nu}_k - 1}{2\tilde{\nu}_k^2}. \end{aligned} \tag{10.30}$$

## 10.5 Simulation Study

This section focusses on simulating non-stationary BINAR(1) time series data of COM-Poisson marginals using the parameters of Equation (10.7). The counting series at the time point $T = 1$ is generated under specified values of $\boldsymbol{\beta}^{[1]}, \boldsymbol{\beta}^{[2]}, \rho_{11}, \rho_{22}, \nu_1$ and $\nu_2$, using the COM-Poisson package in R (https://cran.r-project.org/web/packages/compoisson/index.html). The simulated $Y_1^{[1]}$ and $Y_1^{[2]}$ are then substituted in Equation (10.7) to generate subsequent observations for the respective counting series. Based on an iterative OD technique in the generation of multivariate COM-Poisson data, we develop an algorithm in this section to generate correlated $R_t^{[1]}$ and $R_t^{[2]}$. The procedure for generating the correlated innovations is explained as follows:
From the assumptions in Section 10.3,

$$R_t^{[k]} \sim \text{COM-Poisson}(\mu_t^{[k]} - \rho_{kk}\mu_{t-1}^{[k]}, \tilde{\nu}_k) \quad \text{and} \tag{10.31}$$

$$R_t^{[2]} = \gamma \circ R_t^{[1]} + \varepsilon_t, \tag{10.32}$$

where $\varepsilon_t \in (0,1)$ and and $\varepsilon_t \sim$ COM-Poisson$(\mu^*, \nu^*)$, where $\mu^* = \mu_t^{[2]} - \rho_{22}\mu_{t-1}^{[2]} - \gamma(\mu_t^{[1]} - \rho_{11}\mu_{t-1}^{[1]})$ and

$$\nu^* = \frac{2q_0 + 1 + \sqrt{(2q_0+1)^2 - 8q_1}}{4q_1} > 0 \tag{10.33}$$

with $q_0 = [\mu_t^{[2]} - \rho_{22}\mu_{t-1}^{[2]} - \gamma(\mu_t^{[1]} - \rho_{11}\mu_{t-1}^{[1]})]$, $q_1 = [\frac{\mu_t^{[2]} - \rho_{22}\mu_{t-1}^{[2]}}{\tilde{\nu}_2} + \frac{\tilde{\nu}_2 - 1}{2\tilde{\nu}_2^2} - \gamma(1-\gamma)(\mu_t^{[1]} - \rho_{11}\mu_{t-1}^{[1]}) - \gamma^2(\frac{\mu_t^{[1]} - \rho_{11}\mu_{t-1}^{[1]}}{\tilde{\nu}_1} + \frac{\tilde{\nu}_1 - 1}{2\tilde{\nu}_1^2})]$ and also under this assumption $\text{Cov}(R_t^{[1]}, R_t^{[2]}) = \gamma(\frac{\mu_t^{[1]} - \rho_{11}\mu_{t-1}^{[1]}}{\tilde{\nu}_1} + \frac{\tilde{\nu}_1 - 1}{2\tilde{\nu}_1^2}) \neq 0$. Note, the purpose here is to simply generate a correlated pair $(R_t^{[1]}, R_t^{[2]})$ that are marginally COM-Poisson with a covariance structure. In this simulation set-up, irrespective of $\varepsilon_t$, we may assume $Y_t^{[1]}$ is COM-Poisson with $\nu_1 < 1$ and $Y_t^{[2]}$ is COM-Poisson with $\nu_2 \geq 1$ or vice-versa.

The sample sizes are assumed to be $T = 60, 100, 600$ and the value of $\kappa_{12,1} = 0.5$, $\gamma = 0.5$ and $(\rho_{11}, \rho_{22}) = (0.9, 0.9), (0.3, 0.3)$. We set values of $\beta_k^{[1]} = 0.4$ and $\beta_k^{[2]} = 0.8$ for $k \in \{1, 2\}$. The covariate design is assumed to be a $(2 \times 1)$ vector for each time point where

$$x_{t1} = \begin{cases} -1 + t & (t = 1, \ldots, T/4), \\ rnorm(1, 0, 1) & (t = (T/4) + 1, \ldots, 3T/4), \\ 1 + t & (t = (3T/4) + 1, \ldots, T), \end{cases}$$

and $x_{t2}$=$runif$(1,0,1) is a randomly generated number following the uniform (0,1) distribution in R for each $t = 1, \ldots, T$.

As for the values of $\nu_1$ and $\nu_2$, the following combinations are considered (Table 10.1).

The number of Monte Carlo replications for each situation was 5000. Tables 10.2–10.3 present the simulated mean, dispersion, correlation estimates along with their standard errors in brackets under non-stationary set-up.

The above set of results refers to the simulated mean estimates for the regression, dependence and dispersion parameters under the GQL approach under the non-stationary set-up. The first remark to be made is the estimates for the different parameters are consistent with the population values under all combinations taken irrespective of the number of time points. As expected, for large sample size, the standard errors decrease and the bias in the estimates also decreases. It should be noted that the cross-dependence $\kappa_{12,t}$ is given only for $t = 1$ because of high dimension. However, a number of non-convergent simulations occurring under the different combination is noted. For instance, under $\rho_{11} = \rho_{22} = 0.3$ and $\nu_1 = \nu_2 = 1$, the number of non-convergent simulations is approximately 260 for $T = 60$, 240 for $T = 100$ and 190 for $T = 600$. However, when $\rho_{11}, \rho_{22}$ increase to 0.9, the number of non-convergent

**TABLE 10.1**
Different Scenarios of Dispersion

| | | | | |
|---|---|---|---|---|
| Scenario | :1 | $\nu_1$=1, | $\nu_2$=1 | Equi- and equi-dispersed |
| | :2 | $\nu_1$=0.7, | $\nu_2$=0.7 | Over- and over-dispersed |
| | :3 | $\nu_1$=1.5, | $\nu_2$=1.5 | Under- and under-dispersed |
| | :4 | $\nu_1$=0.7, | $\nu_2$=1.5 | Over- and under-dispersed |
| | :5 | $\nu_1$=1.5, | $\nu_2$=0.7 | Under- and over-dispersed |

**TABLE 10.2**
Estimates of the Parameters and Standard Errors under Non-Stationary BINAR(1) Process for $[\nu_1, \nu_2] = [1, 1], [0.7, 0.7], [1.5, 1.5]$

| T | $\hat{\beta}_1^{[1]}$ | $\hat{\beta}_2^{[1]}$ | $\hat{\beta}_1^{[2]}$ | $\hat{\beta}_2^{[2]}$ | $\hat{\nu}_1 = 1$ | $\hat{\nu}_2=1$ | $\hat{\rho}_{11} = 0.9$ | $\hat{\rho}_{22} = 0.9$ | $\hat{\kappa}_{12,1}$ |
|---|---|---|---|---|---|---|---|---|---|
| 60 | 0.3847 | 0.3802 | 0.7815 | 0.7862 | 0.9877 | 0.9894 | 0.8850 | 0.8895 | 0.4880 |
| | (0.0773) | (0.0820) | (0.0778) | (0.0961) | (0.1134) | (0.1038) | | | |
| 100 | 0.3918 | 0.3941 | 0.7914 | 0.7922 | 0.9912 | 0.9953 | 0.8964 | 0.8910 | 0.4930 |
| | (0.0595) | (0.0638) | (0.0582) | (0.0660) | (0.0923) | (0.0873) | | | |
| 600 | 0.3974 | 0.3971 | 0.7977 | 0.7987 | 0.9999 | 0.9970 | 0.8917 | 0.8976 | 0.4969 |
| | (0.0259) | (0.0211) | (0.0141) | (0.0243) | (0.0453) | (0.0581) | | | |
| T | $\hat{\beta}_1^{[1]}$ | $\hat{\beta}_2^{[1]}$ | $\hat{\beta}_1^{[2]}$ | $\hat{\beta}_2^{[2]}$ | $\hat{\nu}_1 = 1$ | $\hat{\nu}_2=1$ | $\hat{\rho}_{11}=0.3$ | $\hat{\rho}_{22} = 0.3$ | $\hat{\kappa}_{12,1}$ |
| 60 | 0.3877 | 0.4160 | 0.7862 | 0.7817 | 0.9866 | 0.9826 | 0.3105 | 0.3111 | 0.5184 |
| | (0.0794) | (0.0851) | (0.0763) | (0.0803) | (0.1285) | (0.1073) | | | |
| 100 | 0.3923 | 0.3934 | 0.7909 | 0.7945 | 0.9915 | 0.9901 | 0.2909 | 0.2980 | 0.4950 |
| | (0.0525) | (0.0529) | (0.0655) | (0.0622) | (0.0914) | (0.0895) | | | |
| 600 | 0.3925 | 0.3916 | 0.7914 | 0.7993 | 0.9907 | 0.9977 | 0.2972 | 0.2915 | 0.4953 |
| | (0.0237) | (0.0294) | (0.0118) | (0.0193) | (0.0468) | (0.0526) | | | |
| T | $\hat{\beta}_1^{[1]}$ | $\hat{\beta}_2^{[1]}$ | $\hat{\beta}_1^{[2]}$ | $\hat{\beta}_2^{[2]}$ | $\hat{\nu}_1=0.7$ | $\hat{\nu}_2=0.7$ | $\hat{\rho}_{11}=0.9$ | $\hat{\rho}_{22}=0.9$ | $\hat{\kappa}_{12,1}$ |
| 60 | 0.3888 | 0.3869 | 0.7883 | 0.7818 | 0.6810 | 0.6855 | 0.8643 | 0.8830 | 0.4857 |
| | (0.0812) | (0.0716) | (0.0846) | (0.0753) | (0.1202) | (0.1007) | | | |
| 100 | 0.3925 | 0.3999 | 0.7982 | 0.7964 | 0.6977 | 0.6936 | 0.8992 | 0.8980 | 0.4958 |
| | (0.0541) | (0.0615) | (0.0573) | (0.0625) | (0.1038) | (0.0850) | | | |
| 600 | 0.3962 | 0.3964 | 0.7909 | 0.7920 | 0.7026 | 0.7008 | 0.8937 | 0.8910 | 0.4984 |
| | (0.0184) | (0.0261) | (0.0215) | (0.0163) | (0.0573) | (0.0516) | | | |
| T | $\hat{\beta}_1^{[1]}$ | $\hat{\beta}_2^{[1]}$ | $\hat{\beta}_1^{[2]}$ | $\hat{\beta}_2^{[2]}$ | $\hat{\nu}_1=0.7$ | $\hat{\nu}_2=0.7$ | $\hat{\rho}_{11} = 0.3$ | $\hat{\rho}_{22} = 0.3$ | $\hat{\kappa}_{12,1}$ |
| 60 | 0.4111 | 0.3854 | 0.7823 | 0.7834 | 0.6871 | 0.6813 | 0.2870 | 0.2856 | 0.5168 |
| | (0.0840) | (0.0867) | (0.0716) | (0.0732) | (0.1107) | (0.1235) | | | |
| 100 | 0.3945 | 0.3937 | 0.7996 | 0.7970 | 0.6918 | 0.6963 | 0.2945 | 0.2989 | 0.4931 |
| | (0.0542) | (0.0635) | (0.0556) | (0.0571) | (0.0910) | (0.0935) | | | |
| 600 | 0.3962 | 0.3951 | 0.7910 | 0.7942 | 0.6996 | 0.6920 | 0.2910 | 0.2970 | 0.4994 |
| | (0.0184) | (0.0222) | (0.0211) | (0.0140) | (0.0486) | (0.0570) | | | |
| T | $\hat{\beta}_1^{[1]}$ | $\hat{\beta}_2^{[1]}$ | $\hat{\beta}_1^{[2]}$ | $\hat{\beta}_2^{[2]}$ | $\hat{\nu}_1=1.5$ | $\hat{\nu}_2 = 1.5$ | $\hat{\rho}_{11}=0.9$ | $\hat{\rho}_{22}=0.9$ | $\hat{\kappa}_{12,1}$ |
| 60 | 0.4131 | 0.3898 | 0.8143 | 0.7811 | 1.4840 | 1.4825 | 0.8882 | 0.8834 | 0.4885 |
| | (0.0713) | (0.0828) | (0.0724) | (0.0953) | (0.1251) | (0.1096) | | | |
| 100 | 0.3927 | 0.3910 | 0.7943 | 0.7942 | 1.4955 | 1.4949 | 0.8962 | 0.8953 | 0.4991 |
| | (0.0570) | (0.0629) | (0.0615) | (0.0590) | (0.1022) | (0.0870) | | | |
| 600 | 0.3326 | 0.3989 | 0.7913 | 0.7948 | 1.5046 | 1.5027 | 0.8943 | 0.8967 | 0.4916 |
| | (0.0212) | (0.0137) | (0.0271) | (0.0201) | (0.0592) | (0.0574) | | | |
| T | $\hat{\beta}_1^{[1]}$ | $\hat{\beta}_2^{[1]}$ | $\hat{\beta}_1^{[2]}$ | $\hat{\beta}_2^{[2]}$ | $\hat{\nu}_1=1.5$ | $\hat{\nu}_2 = 1.5$ | $\hat{\rho}_{11}=0.3$ | $\hat{\rho}_{22}=0.3$ | $\hat{\kappa}_{12,1}$ |
| 60 | 0.3818 | 0.3812 | 0.8133 | 0.7825 | 1.5117 | 1.4820 | 0.3193 | 0.3159 | 0.5144 |
| | (0.0950) | (0.0716) | (0.0949) | (0.0776) | (0.1196) | (0.1020) | | | |
| 100 | 0.4071 | 0.3925 | 0.8082 | 0.7970 | 1.4986 | 1.4942 | 0.2931 | 0.2934 | 0.4992 |
| | (0.0677) | (0.0572) | (0.0510) | (0.0639) | (0.0917) | (0.0861) | | | |
| 600 | 0.3986 | 0.3980 | 0.7910 | 0.7950 | 1.4998 | 1.4954 | 0.2924 | 0.2913 | 0.4987 |
| | (0.0117) | (0.0163) | (0.0281) | (0.0247) | (0.0571) | (0.0596) | | | |

simulations is approximately 275 for $T = 60$, 250 for $T = 100$ and 200 for $T = 600$. Secondly, for values of $\nu_1, \nu_2 > 1$, the non-convergent simulations slightly increase than under the equi- or over-dispersed cases. For instance, under $\rho_{11} = \rho_{22} = 0.3$ and $\nu_1 = \nu_2 = 1.5$, the number of non-convergent simulations is approximately 300 for $T = 60$, 275 for $T = 100$ and 225 for $T = 600$.

**TABLE 10.3**
Estimates of the Parameters and Standard Errors under Non-Stationary BINAR(1) Process for $[\nu_1, \nu_2] = [0.7, 1.5], [1.5, 0.7]$

| T | $\hat{\beta}_1^{[1]}$ | $\hat{\beta}_2^{[1]}$ | $\hat{\beta}_1^{[2]}$ | $\hat{\beta}_2^{[2]}$ | $\hat{\nu}_1$=0.7 | $\hat{\nu}_2 = 1.5$ | $\hat{\rho}_{11} = 0.9$ | $\hat{\rho}_{22} = 0.9$ | $\hat{\kappa}_{12,1}$ |
|---|---|---|---|---|---|---|---|---|---|
| 60 | 0.3846 | 0.3833 | 0.7892 | 0.7884 | 0.6868 | 1.4804 | 0.8868 | 0.8801 | 0.4825 |
| | (0.0990) | (0.0712) | (0.1059) | (0.0737) | (0.1072) | (0.1177) | | | |
| 100 | 0.3917 | 0.3931 | 0.7946 | 0.7947 | 0.6913 | 1.4907 | 0.8972 | 0.8968 | 0.4950 |
| | (0.0643) | (0.0555) | (0.0610) | (0.0591) | (0.0987) | (0.0806) | | | |
| 600 | 0.3950 | 0.3976 | 0.7904 | 0.7962 | 0.6921 | 1.4966 | 0.8951 | 0.8935 | 0.4929 |
| | (0.0165) | (0.0212) | (0.0118) | (0.0261) | (0.0434) | (0.0544) | | | |
| T | $\hat{\beta}_1^{[1]}$ | $\hat{\beta}_2^{[1]}$ | $\hat{\beta}_1^{[2]}$ | $\hat{\beta}_2^{[2]}$ | $\hat{\nu}_1$=0.7 | $\hat{\nu}_2 = 1.5$ | $\hat{\rho}_{11}$=0.3 | $\hat{\rho}_{22}$=0.3 | $\hat{\kappa}_{12,1}$ |
| 60 | 0.4152 | 0.3834 | 0.7832 | 0.7823 | 0.7126 | 1.4853 | 0.2856 | 0.3106 | 0.4820 |
| | (0.1035) | (0.0775) | (0.0745) | (0.0731) | (0.1112) | (0.1057) | | | |
| 100 | 0.3985 | 0.3991 | 0.7905 | 0.7943 | 0.6938 | 1.4914 | 0.2911 | 0.2920 | 0.4993 |
| | (0.0719) | (0.0534) | (0.0543) | (0.0524) | (0.0991) | (0.0865) | | | |
| 600 | 0.3988 | 0.3915 | 0.7981 | 0.7928 | 0.6947 | 1.4970 | 0.2924 | 0.2990 | 0.4905 |
| | (0.0217) | (0.0211) | (0.0157) | (0.0126) | (0.0440) | (0.0507) | | | |
| T | $\hat{\beta}_1^{[1]}$ | $\hat{\beta}_2^{[1]}$ | $\hat{\beta}_1^{[2]}$ | $\hat{\beta}_2^{[2]}$ | $\hat{\nu}_1$=1.5 | $\hat{\nu}_2 = 0.7$ | $\hat{\rho}_{11}$=0.9 | $\hat{\rho}_{22}$=0.9 | $\hat{\kappa}_{12,1}$ |
| 60 | 0.3813 | 0.3811 | 0.7899 | 0.7848 | 1.4822 | 0.6807 | 0.8866 | 0.8826 | 0.4871 |
| | (0.0920) | (0.0945) | (0.0765) | (0.0703) | (0.1168) | (0.1083) | | | |
| 100 | 0.4092 | 0.4072 | 0.7931 | 0.7929 | 1.4905 | 0.6949 | 0.8978 | 0.8921 | 0.4976 |
| | (0.0630) | (0.0593) | (0.0558) | (0.0581) | (0.0853) | (0.0856) | | | |
| 600 | 0.3993 | 0.3960 | 0.7911 | 0.7919 | 1.4980 | 0.6988 | 0.8920 | 0.8990 | 0.4914 |
| | (0.0216) | (0.0154) | (0.0297) | (0.0267) | (0.0570) | (0.0586) | | | |
| T | $\hat{\beta}_1^{[1]}$ | $\hat{\beta}_2^{[1]}$ | $\hat{\beta}_1^{[2]}$ | $\hat{\beta}_2^{[2]}$ | $\hat{\nu}_1$=1.5 | $\hat{\nu}_2 = 0.7$ | $\hat{\rho}_{11}$=0.3 | $\hat{\rho}_{22}$=0.3 | $\hat{\kappa}_{12,1}$ |
| 60 | 0.4120 | 0.3849 | 0.7806 | 0.7868 | 1.5186 | 0.6843 | 0.2834 | 0.3154 | 0.5163 |
| | (0.0930) | (0.0721) | (0.0974) | (0.0785) | (0.1048) | (0.1020) | | | |
| 100 | 0.3982 | 0.3997 | 0.7919 | 0.7970 | 1.4987 | 0.6939 | 0.2932 | 0.2997 | 0.4984 |
| | (0.0576) | (0.0628) | (0.0541) | (0.0503) | (0.0958) | (0.0876) | | | |
| 600 | 0.3980 | 0.3914 | 0.7994 | 0.7999 | 1.4919 | 0.6921 | 0.2935 | 0.2983 | 0.4964 |
| | (0.0279) | (0.0261) | (0.0182) | (0.0223) | (0.0560) | (0.0550) | | | |

## 10.6 Summary

This chapter has focussed on the modeling of non-stationary bivariate time series of counts using a novel class of constrained BINAR(1) model with cross-related COM-Poisson innovations. The CBINAR(1)CMP model is rather flexible as compared to other BINAR(1) models introduced in earlier chapters since it facilitates the modeling of time series with different levels of dispersion as shown in the simulation section. However, the estimation of the regression effects and dispersion parameters has been conducted using the GQL equation based on a simple score function while the serial- and cross-correlation parameters are estimated using robust moment estimation equations. The performance of the GQL method has been assessed via a simulation study wherein a simulation algorithm to generate bivariate COM-Poisson data has been developed. A similar analysis is conducted for the constrained BINMA(1) model with COM-Poisson innovations in the next chapter.

# 11

# *Constrained BINMA(1) Model with Correlated COM-Poisson Innovations*

## 11.1 Introduction

In this chapter, we develop a constrained BINMA(1) model where the cross-correlation between the series is induced by the correlated COM-Poisson innovations and formulate a GQL estimation method for the estimation of the unknown model parameters.

The organization of the chapter is as follows: In Section 11.2, we introduce a BINMA(1) model with COM-Poisson innovations under the non-stationary moment assumptions. In Section 11.3, we propose a trial experiment under the stationary set-up that depicts the different combinations of over- and under-dispersion under the different parameter assumptions. Section 11.4 focuses on the extended version of the GQL approach to estimate the parameters in the BINMA(1) process along with the forecasting equations and the asymptotic properties of the GQL estimators. In Section 11.5, the Monte Carlo experiments are conducted to assess the performance of the GQL approach. The conclusion is presented in Section 11.7.

## 11.2 The Constrained BINMA(1) Model

The constrained BINMA(1) model is specified as

$$\begin{pmatrix} Y_t^{[1]} \\ Y_t^{[2]} \end{pmatrix} = \begin{pmatrix} \rho_{11} & 0 \\ 0 & \rho_{22} \end{pmatrix} \circ \begin{pmatrix} R_{t-1}^{[1]} \\ R_{t-1}^{[2]} \end{pmatrix} + \begin{pmatrix} R_t^{[1]} \\ R_t^{[2]} \end{pmatrix}, \tag{11.1}$$

where $R_t^{[k]}$ is a sequence of identically and independently distributed COM-Poisson innovation terms, that is, $R_t^{[k]} \sim \text{COM-Poisson}(\lambda_t^{[k]}, \nu_k)$ and $\text{Corr}(R_i^{[k]}, R_j^{[k]}) = 0$ for $i \neq j$ and $k \in \{1, 2\}$. In the event there exists some covariates $\boldsymbol{x}_t = [x_{t1}, x_{t2}, \ldots, x_{tj}, \ldots, x_{tp}]'$ with regression coefficients $\boldsymbol{\beta}^{[k]} = [\beta_1^{[k]}, \beta_2^{[k]}, \ldots, \beta_j^{[k]}, \ldots, \beta_p^{[k]}]'$ affecting the counting series $Y_t^{[k]}$, it then implies

 DOI: 10.1201/9781003600435-11

that its corresponding link function $\theta_t^{[k]}$ is given by $\theta_t^{[k]} = \exp(\boldsymbol{x}_t'\boldsymbol{\beta}^{[k]})$ and $E(R_t^{[k]}) = \lambda_t^{[k]} = (\theta_t^{[k]})^{1/\nu_k} - (\frac{\nu_k - 1}{2\nu_k})$ for $k \in \{1, 2\}$.

In Equation (11.1),

$$\rho_{kk} \circ R_{t-1}^{[k]} = \begin{cases} \sum_{l=1}^{R_{t-1}^{[k]}} b_l(\rho_{kk}), & R_{t-1}^{[k]} > 0, \\ 0 & R_{t-1}^{[k]} = 0 \end{cases} \tag{11.2}$$

with the following properties:

1. $E(\rho_{kk} \circ R_{t-1}^{[k]}) = \rho_{kk} E(R_{t-1}^{[k]})$.
2. $\text{Var}(\rho_{kk} \circ R_{t-1}^{[k]}) = \rho_{kk}(1 - \rho_{kk}) E(R_{t-1}^{[k]}) + \rho_{kk}^2 \text{Var}(R_{t-1}^{[k]})$.
3. $\text{Cov}(\rho_{11} \circ R_{t-1}^{[1]}, \rho_{22} \circ R_{t-1}^{[2]}) = \rho_{11}\rho_{22} \text{Cov}(R_{t-1}^{[1]}, R_{t-1}^{[2]})$.

Note, as well, there is no dependency between the counting observation $Y_t^{[k]}$ and $R_u^{[k]}$, that is $\text{Corr}(Y_t^{[k]}, R_u^{[k]}) = 0$ for all $k \in \{1, 2\}$, $u, t \in \mathbb{Z}$, $u \neq t$.

In addition, we assume $R_t^{[1]}$ and $R_t^{[2]}$ are pairwise correlated such that

$$\text{Corr}(R_t^{[1]}, R_{t'}^{[2]}) = \begin{cases} \kappa_{12,t} & t = t' \\ 0 & t \neq t' \end{cases} \tag{11.3}$$

where $\kappa_{12,t}$ is to be estimated.

Based on the above conditions, it is proved that

$$\begin{aligned} E(Y_t^{[k]}) &= E(\rho_{kk} \circ R_{t-1}^{[k]}) + E(R_t^{[k]}) \\ &= \rho_{kk} E(R_{t-1}^{[k]}) + E(R_t^{[k]}) \\ &= \rho_{kk} \lambda_{t-1}^{[k]} + \lambda_t^{[k]} \end{aligned} \tag{11.4}$$

and

$$\begin{aligned} \text{Var}(Y_t^{[k]}) &= \text{Var}(\rho_{kk} \circ R_{t-1}^{[k]}) + \text{Var}(R_t^{[k]}) \\ &= \rho_{kk}(1 - \rho_{kk}) E(R_{t-1}^{[k]}) + \rho_{kk}^2 \text{Var}(R_{t-1}^{[k]}) + \text{Var}(R_t^{[k]}) \\ &= \rho_{kk}(1 - \rho_{kk}) \lambda_{t-1}^{[k]} + \rho_{kk}^2 \left[ \frac{\lambda_{t-1}^{[k]}}{\nu_k} + \frac{\nu_k - 1}{2\nu_k^2} \right] + \frac{\lambda_t^{[k]}}{\nu_k} + \frac{\nu_k - 1}{2\nu_k^2}. \end{aligned} \tag{11.5}$$

The serial-correlation at lag-$h$ is

$$\begin{aligned} \text{Cov}(Y_t^{[k]}, Y_{t+h}^{[k]}) &= \text{Cov}(Y_t^{[k]}, \rho_{kk} \circ R_{t+h-1}^{[k]} + R_{t+h}^{[k]}) \\ &= \rho_{kk} \text{Cov}(Y_t^{[k]}, R_{t+h-1}^{[k]}) \\ &= \rho_{kk} \text{Cov}(R_t^{[k]}, R_{t+h-1}^{[k]}) \\ &= \begin{cases} \rho_{kk} \left[ \frac{\lambda_t^{[k]}}{\nu_k} + \frac{\nu_k - 1}{2\nu_k^2} \right] & h = 1 \\ 0 & h > 1. \end{cases} \end{aligned} \tag{11.6}$$

As for the covariance between $Y_t^{[1]}$ and $Y_t^{[2]}$,

$$
\begin{aligned}
\text{Cov}(Y_t^{[1]}, Y_t^{[2]}) &= \text{Cov}(\rho_{11} \circ R_{t-1}^{[1]} + R_t^{[1]}, \rho_{22} \circ R_{t-1}^{[2]} + R_t^{[2]}) \\
&= \text{Cov}(\rho_{11} \circ R_{t-1}^{[1]}, \rho_{22} \circ R_{t-1}^{[2]}) + \text{Cov}(\rho_{11} \circ R_{t-1}^{[1]}, R_t^{[2]}) \\
&+ \text{Cov}(R_t^{[1]}, \rho_{22} \circ R_{t-1}^{[2]}) + \text{Cov}(R_t^{[1]}, R_t^{[2]}) \\
&= \rho_{11}\rho_{22}\text{Cov}(R_{t-1}^{[1]} R_{t-1}^{[2]}) + \text{Cov}(R_t^{[1]} R_t^{[2]}) \\
&= \rho_{11}\rho_{22}\kappa_{12,t-1}\sqrt{\frac{\lambda_{t-1}^{[1]}}{\nu_1} + \frac{\nu_1 - 1}{2\nu_1^2}}\sqrt{\frac{\lambda_{t-1}^{[2]}}{\nu_2} + \frac{\nu_2 - 1}{2\nu_2^2}} \\
&+ \kappa_{12,t}\sqrt{\frac{\lambda_t^{[1]}}{\nu_1} + \frac{\nu_1 - 1}{2\nu_1^2}}\sqrt{\frac{\lambda_t^{[2]}}{\nu_2} + \frac{\nu_2 - 1}{2\nu_2^2}},
\end{aligned}
\tag{11.7}
$$

As for the other off-diagonal cross-covariances,

$$
\begin{aligned}
\text{Cov}(Y_t^{[1]}, Y_{t+h}^{[2]}) &= \text{Cov}(Y_t^{[1]}, \rho_{22} \circ R_{t+h-1}^{[2]} + R_{t+h}^{[2]}) \\
&= \rho_{22}\text{Cov}(Y_t^{[1]}, R_{t+h-1}^{[2]}) \\
&= \rho_{22}\text{Cov}(R_t^{[1]}, R_{t+h-1}^{[2]}) \\
&= \begin{cases} \rho_{22}\kappa_{12,t}\sqrt{\frac{\lambda_t^{[1]}}{\nu_1} + \frac{\nu_1-1}{2\nu_1^2}}\sqrt{\frac{\lambda_t^{[2]}}{\nu_2} + \frac{\nu_2-1}{2\nu_2^2}} & h = 1 \\ 0 & h > 1 \end{cases}
\end{aligned}
\tag{11.8}
$$

and similarly

$$
\text{Cov}(Y_{t+h}^{[1]}, Y_t^{[2]}) = \begin{cases} \rho_{11}\kappa_{12,t}\sqrt{\frac{\lambda_t^{[1]}}{\nu_1} + \frac{\nu_1-1}{2\nu_1^2}}\sqrt{\frac{\lambda_t^{[2]}}{\nu_2} + \frac{\nu_2-1}{2\nu_2^2}} & h = 1, \\ 0 & h > 1. \end{cases}
\tag{11.9}
$$

**Remark 11.1.** Under stationary moment conditions, we have

$$
E(Y_t^{[k]}) = \mu_1^{[k]} = (1 + \rho_{kk})\lambda_1^{[k]},
\tag{11.10}
$$

$$
\text{Var}(Y_t^{[k]}) = \frac{\lambda_1^{[k]}}{\nu_k}[1 + \rho_{kk}^2(1 - \nu_k) + \rho_{kk}\nu_k] + (1 + \rho_{kk}^2)\left(\frac{\nu_k - 1}{2\nu_k^2}\right)
\tag{11.11}
$$

and

$$
\text{Cov}(Y_t^{[1]}, Y_t^{[2]}) = (1 + \rho_{11}\rho_{22})\left[\kappa_{12,1}\sqrt{\frac{\lambda_1^{[1]}}{\nu_1} + \frac{\nu_1 - 1}{2\nu_1^2}}\sqrt{\frac{\lambda_1^{[2]}}{\nu_2} + \frac{\nu_2 - 1}{2\nu_2^2}}\right]
\tag{11.12}
$$

**Remark 11.2.** For $\nu_1 = 1$, $\nu_2 = 1$, Equations (11.4) and (11.5) reduce to

$$
E(Y_t^{[k]}) = \text{Var}(Y_t^{[k]}) = \rho_{kk}\lambda_{t-1}^{[k]} + \lambda_t^{[k]}
\tag{11.13}
$$

and hence, $Y_t^{[k]} \sim \text{Poisson}(\rho_{kk}\lambda_{t-1}^{[k]} + \lambda_t^{[k]})$. Thus, the constrained BINMA(1) model with Poisson innovations introduced in Chapter 3 may be viewed as a special case of the constrained BINMA(1) model with COM-Poisson innovations.

## 11.3 Trial Experiment

This section explores the behavior of the counting series $Y_t^{[k]}$ via an ad-hoc trial experiment under different low and high values of $\rho_{kk}$, $[\nu_1, \nu_2]$ and $[\theta_1^{[1]}, \theta_1^{[2]}]$ for $k \in \{1, 2\}$. In particular, we are interested in investigating the level of dispersion in $Y_t^{[1]}$ and $Y_t^{[2]}$ for the different parameter values $[\rho_{11}, \rho_{22}] = [0.1, 0.2], [0.5, 0.6]$ and $[0.9, 0.9]$ and $[\theta_1^{[1]}, \theta_1^{[2]}] = [0.1, 0.2]$, $[0.9, 0.95], [1.1, 1.2]$ and $[2, 2.1]$. Four scenarios for $\nu_1$ and $\nu_2$ are considered: $\nu_1 = 1, \nu_2 = 1$ (equi-dispersed), $\nu_1 < 1, \nu_2 < 1$ (both over-dispersed), $\nu_1 > 1, \nu_2 > 1$ (under-dispersed), $\nu_1 < 1, \nu_2 > 1$ (over- and under-dispersed). The experiment is conducted under stationary condition for simplicity. The expected mean $\text{E}(Y_t^{[1]})$, $\text{E}(Y_t^{[2]})$ and variance $\text{Var}(Y_t^{[1]})$, $\text{Var}(Y_t^{[2]})$ are computed from Equations (11.10) and (11.11) and the results are presented in Tables 11.1–11.4.

We make the following observations from Tables 11.1–11.4:

(a) As proved in Equation (11.13), both counting series are equi-dispersed at different levels from Table 11.1.

**TABLE 11.1**
Experimental Results under Different Serial and Dispersion Combinations for $[\nu_1 = 1, \nu_2 = 1]$

| $\rho_{11}$ | $\rho_{22}$ | $\nu_1$ | $\nu_2$ | $\theta_1^{[1]}$ | $\theta_1^{[2]}$ | $\text{E}(Y_t^{[1]})$ | $\text{Var}(Y_t^{[1]})$ | $\text{E}(Y_t^{[2]})$ | $\text{Var}(Y_t^{[2]})$ |
|---|---|---|---|---|---|---|---|---|---|
| 0.1 | 0.2 | 1 | 1 | 0.1 | 0.2 | 0.11 | 0.11 | 0.24 | 0.24 |
| | | | | 0.9 | 0.95 | 0.99 | 0.99 | 1.14 | 1.14 |
| | | | | 1.1 | 1.2 | 1.21 | 1.21 | 1.44 | 1.44 |
| | | | | 2 | 2.1 | 2.2 | 2.2 | 2.52 | 2.52 |
| 0.5 | 0.6 | 1 | 1 | 0.1 | 0.2 | 0.15 | 0.15 | 0.32 | 0.32 |
| | | | | 0.9 | 0.95 | 1.35 | 1.35 | 1.52 | 1.52 |
| | | | | 1.1 | 1.2 | 1.65 | 1.65 | 1.92 | 1.92 |
| | | | | 2 | 2.1 | 3 | 3 | 3.36 | 3.36 |
| 0.8 | 0.9 | 1 | 1 | 0.1 | 0.2 | 0.18 | 0.18 | 0.38 | 0.38 |
| | | | | 0.9 | 0.95 | 1.62 | 1.62 | 1.805 | 1.805 |
| | | | | 1.1 | 1.2 | 1.96 | 1.96 | 2.28 | 2.28 |
| | | | | 2 | 2.1 | 3.6 | 3.6 | 3.99 | 3.99 |

**TABLE 11.2**
Experimental Results under Different Serial and Dispersion Combinations for $[\nu_1 < 1, \nu_2 < 1]$

| $\rho_{11}$ | $\rho_{22}$ | $\nu_1$ | $\nu_2$ | $\theta_1^{[1]}$ | $\theta_1^{[2]}$ | $E(Y_t^{[1]})$ | $Var(Y_t^{[1]})$ | $E(Y_t^{[2]})$ | $Var(Y_t^{[2]})$ |
|---|---|---|---|---|---|---|---|---|---|
| 0.1 | 0.2 | 0.1 | 0.15 | 0.1 | 0.2 | 4.95 | 0.405 | 3.4 | 0.453 |
| | | | | 0.9 | 0.95 | 5.333 | 3.958 | 4.252 | 5.492 |
| | | | | 1.1 | 1.2 | 7.803 | 26.83 | 7.45 | 24.37 |
| | | | | 2 | 2.1 | 1131 | 10435 | 172.2 | 998.10 |
| 0.5 | 0.6 | 0.1 | 0.15 | 0.1 | 0.2 | 6.75 | 1.125 | 4.533 | 0.680 |
| | | | | 0.9 | 0.95 | 7.273 | 5.57 | 5.67 | 7.291 |
| | | | | 1.1 | 1.2 | 10.641 | 34.20 | 9.93 | 32.06 |
| | | | | 2 | 2.1 | 1543 | 13057 | 229.6 | 1309.6 |
| 0.8 | 0.9 | 0.1 | 0.15 | 0.1 | 0.2 | 8.1 | 0.72 | 5.38 | 0.255 |
| | | | | 0.9 | 0.95 | 8.73 | 6.49 | 6.733 | 8.891 |
| | | | | 1.1 | 1.2 | 12.77 | 43.67 | 11.79 | 41.24 |
| | | | | 2 | 2.1 | 1851 | 16958 | 272 | 1710 |
| 0.1 | 0.2 | 0.8 | 0.9 | 0.1 | 0.2 | 0.1994 | 0.0873 | 0.2674 | 0.2289 |
| | | | | 0.9 | 0.95 | 1.102 | 1.197 | 1.200 | 1.252 |
| | | | | 1.1 | 1.2 | 1.377 | 1.535 | 1.536 | 1.620 |
| | | | | 2 | 2.1 | 2.754 | 3.228 | 2.803 | 3.009 |
| 0.5 | 0.6 | 0.8 | 0.9 | 0.1 | 0.2 | 0.2719 | 0.1332 | 0.3565 | 0.3062 |
| | | | | 0.9 | 0.95 | 1.502 | 1.62 | 1.600 | 1.667 |
| | | | | 1.1 | 1.2 | 1.887 | 2.073 | 2.048 | 2.158 |
| | | | | 2 | 2.1 | 3.755 | 4.342 | 3.738 | 4.007 |
| 0.8 | 0.9 | 0.8 | 0.9 | 0.1 | 0.2 | 0.3262 | 0.1443 | 0.4233 | 0.3564 |
| | | | | 0.9 | 0.95 | 1.803 | 1.957 | 1.900 | 1.9900 |
| | | | | 1.1 | 1.2 | 2.253 | 2.510 | 2.432 | 2.578 |
| | | | | 2 | 2.1 | 4.506 | 5.276 | 4.438 | 4.796 |

(b) From Table 11.2, for $\theta_1^{[k]} > 1$ and irrespective of $\{\rho_{jk}\}$, both counting series $Y_t^{[1]}$ and $Y_t^{[2]}$ exhibit over-dispersion but at different levels for $j, k \in \{1, 2\}$.

(c) From Table 11.3, we note that for $\theta_1^{[k]} > 0.9$ and for different combinations of $\{\rho_{jk}\}$, both counting series show various levels of under-dispersion, for $j, k \in \{1, 2\}$.

(d) From Table 11.4, we observe that for $\theta_1^{[k]} > 0.9$ and for any combinations of $\{\rho_{jk}\}$, the first series is over-dispersed while the second series is under-dispersed.

Hence, we conclude that the BINMA(1) model given by Equation (11.1) is suitable for modeling bivariate time series of various forms and levels of dispersion.

**TABLE 11.3**
Experimental Results under Different Serial and Dispersion Combinations for $[\nu_1 > 1, \nu_2 > 1]$

| $\rho_{11}$ | $\rho_{22}$ | $\nu_1$ | $\nu_2$ | $\theta_1^{[1]}$ | $\theta_1^{[2]}$ | $E(Y_t^{[1]})$ | $Var(Y_t^{[1]})$ | $E(Y_t^{[2]})$ | $Var(Y_t^{[2]})$ |
|---|---|---|---|---|---|---|---|---|---|
| 0.1 | 0.2 | 1.1 | 1.2 | 0.1 | 0.2 | 0.0856 | 0.1202 | 0.2139 | 0.2552 |
| | | | | 0.9 | 0.95 | 0.9495 | 0.912 | 1.050 | 0.9704 |
| | | | | 1.1 | 1.2 | 1.150 | 1.095 | 1.297 | 1.182 |
| | | | | 2 | 2.1 | 2.016 | 1.889 | 2.127 | 1.892 |
| 0.5 | 0.6 | 1.1 | 1.2 | 0.1 | 0.2 | 0.1167 | 0.1596 | 0.2851 | 0.3392 |
| | | | | 0.9 | 0.95 | 1.295 | 1.248 | 1.400 | 1.296 |
| | | | | 1.1 | 1.2 | 1.568 | 1.500 | 1.729 | 1.579 |
| | | | | 2 | 2.1 | 2.749 | 2.592 | 2.836 | 2.529 |
| 0.8 | 0.9 | 1.1 | 1.2 | 0.1 | 0.2 | 0.14 | 0.196 | 0.3386 | 0.4105 |
| | | | | 0.9 | 0.95 | 1.554 | 1.493 | 1.662 | 1.524 |
| | | | | 1.1 | 1.2 | 1.8811 | 1.793 | 2.053 | 1.853 |
| | | | | 2 | 2.1 | 3.298 | 3.093 | 3.368 | 2.959 |
| 0.1 | 0.2 | 2 | 2.5 | 0.1 | 0.2 | 0.0729 | 0.1657 | 0.2704 | 0.2546 |
| | | | | 0.9 | 0.95 | 0.769 | 0.542 | 0.816 | 0.516 |
| | | | | 1.1 | 1.2 | 0.879 | 0.6015 | 0.9308 | 0.5716 |
| | | | | 2 | 2.1 | 1.281 | 0.819 | 1.255 | 0.727 |
| 0.5 | 0.6 | 2 | 2.5 | 0.1 | 0.2 | 0.0993 | 0.2142 | 0.3605 | 0.3398 |
| | | | | 0.9 | 0.95 | 1.0481 | 0.7676 | 1.0876 | 0.6961 |
| | | | | 1.1 | 1.2 | 1.198 | 0.855 | 1.241 | 0.7713 |
| | | | | 2 | 2.1 | 1.746 | 1.175 | 1.673 | 0.983 |
| 0.8 | 0.9 | 2 | 2.5 | 0.1 | 0.2 | 0.1192 | 0.2699 | 0.4281 | 0.4006 |
| | | | | 0.9 | 0.95 | 1.2576 | 0.8897 | 1.2914 | 0.7705 |
| | | | | 1.1 | 1.2 | 1.4379 | 0.9878 | 1.4737 | 0.8486 |
| | | | | 2 | 2.1 | 2.096 | 1.346 | 1.986 | 1.068 |

**TABLE 11.4**
Experimental Results under Different Serial and Dispersion Combinations for $[\nu_1 < 1, \nu_2 > 1]$

| $\rho_{11}$ | $\rho_{22}$ | $\nu_1$ | $\nu_2$ | $\theta_1^{[1]}$ | $\theta_1^{[2]}$ | $E(Y_t^{[1]})$ | $Var(Y_t^{[1]})$ | $E(Y_t^{[2]})$ | $Var(Y_t^{[2]})$ |
|---|---|---|---|---|---|---|---|---|---|
| 0.1 | 0.2 | 0.8 | 1.5 | 0.1 | 0.2 | 0.1994 | 0.0873 | 0.2104 | 0.2652 |
| | | | | 0.9 | 0.95 | 1.102 | 1.197 | 0.960 | 0.798 |
| | | | | 1.1 | 1.2 | 1.377 | 1.535 | 1.155 | 0.937 |
| | | | | 2 | 2.1 | 2.754 | 3.228 | 1.768 | 1.373 |
| 0.5 | 0.6 | 0.8 | 1.5 | 0.1 | 0.2 | 0.2719 | 0.1332 | 0.2805 | 0.3522 |
| | | | | 0.9 | 0.95 | 1.502 | 1.62 | 1.2796 | 1.068 |
| | | | | 1.1 | 1.2 | 1.887 | 2.073 | 1.540 | 1.255 |
| | | | | 2 | 2.1 | 3.755 | 4.342 | 2.357 | 1.8404 |
| 0.8 | 0.9 | 0.8 | 1.5 | 0.1 | 0.2 | 0.3262 | 0.1443 | 0.3331 | 0.4285 |
| | | | | 0.9 | 0.95 | 1.803 | 1.957 | 1.519 | 1.238 |
| | | | | 1.1 | 1.2 | 2.253 | 2.5096 | 1.829 | 1.449 |
| | | | | 2 | 2.1 | 4.506 | 5.276 | 2.799 | 2.111 |

## 11.4 Estimation of Parameters

The GQL equation is formulated as

$$\left(\frac{\partial\boldsymbol{\mu}}{\partial\boldsymbol{\beta},\partial\boldsymbol{\rho},\partial\boldsymbol{\nu}}\right)'\boldsymbol{\Sigma}^{-1}(\boldsymbol{y}-\boldsymbol{\mu})=0, \tag{11.14}$$

where $\boldsymbol{y}=[\boldsymbol{y}^{[1]},\boldsymbol{y}^{[2]}]'$, with $\boldsymbol{y}^{[k]}=[y_1^{[k]},y_2^{[k]},\ldots,y_T^{[k]}]'$ and $\boldsymbol{\mu}=[\boldsymbol{\mu}^{[1]},\boldsymbol{\mu}^{[2]}]'$, with $\boldsymbol{\mu}^{[k]}=[\mu_1^{[k]},\mu_2^{[k]},\ldots,\mu_T^{[k]}]'$, $\mu_t^{[k]}=\rho_{kk}[(\theta_{t-1}^{[k]})^{1/\nu_k}-(\frac{\nu_k-1}{2\nu_k})]+(\theta_t^{[k]})^{1/\nu_k}-(\frac{\nu_k-1}{2\nu_k})$ for $k\in\{1,2\}$.

$\left(\frac{\partial\boldsymbol{\mu}}{\partial\boldsymbol{\beta},\partial\boldsymbol{\rho},\partial\boldsymbol{\nu}}\right)$ is the block diagonal with $\left[\left(\frac{\partial\boldsymbol{\mu}}{\partial\boldsymbol{\beta},\partial\boldsymbol{\rho},\partial\boldsymbol{\nu}}\right)^{[1]},\left(\frac{\partial\boldsymbol{\mu}}{\partial\boldsymbol{\beta},\partial\boldsymbol{\rho},\partial\boldsymbol{\nu}}\right)^{[2]}\right]$ with the entries in $\left(\frac{\partial\boldsymbol{\mu}}{\partial\boldsymbol{\beta},\partial\boldsymbol{\rho},\partial\boldsymbol{\nu}}\right)^{[k]}$ expressed as derivatives with respect to the unknown parameters:

$$\left(\frac{\partial\boldsymbol{\mu}}{\partial\boldsymbol{\beta},\partial\boldsymbol{\rho},\partial\boldsymbol{\nu}}\right)^{[k]}=\begin{pmatrix}\frac{\partial\mu_1^{[k]}}{\partial\beta_1^{[k]}} & \frac{\partial\mu_1^{[k]}}{\partial\beta_2^{[k]}} & \cdots & \frac{\partial\mu_1^{[k]}}{\partial\beta_p^{[k]}} & \frac{\partial\mu_1^{[k]}}{\partial\rho_{kk}} & \frac{\partial\mu_1^{[k]}}{\partial\nu_k}\\ \frac{\partial\mu_2^{[k]}}{\partial\beta_1^{[k]}} & \frac{\partial\mu_2^{[k]}}{\partial\beta_2^{[k]}} & \cdots & \frac{\partial\mu_2^{[k]}}{\partial\beta_p^{[k]}} & \frac{\partial\mu_2^{[k]}}{\partial\rho_{kk}} & \frac{\partial\mu_2^{[k]}}{\partial\nu_k}\\ \vdots & \vdots & \ddots & \vdots & \vdots & \vdots\\ \frac{\partial\mu_T^{[k]}}{\partial\beta_1^{[k]}} & \frac{\partial\mu_T^{[k]}}{\partial\beta_2^{[k]}} & \cdots & \frac{\partial\mu_T^{[k]}}{\partial\beta_p^{[k]}} & \frac{\partial\mu_T^{[k]}}{\partial\rho_{kk}} & \frac{\partial\mu_T^{[k]}}{\partial\nu_k}\end{pmatrix}_{T\times(p+2)} \tag{11.15}$$

where

1. $\frac{\partial\mu_t^{[k]}}{\partial\beta_j^{[k]}}=\rho_{kk}\frac{(\theta_{t-1}^{[k]})^{\frac{1}{\nu_k}}x_{t-1,j}}{\nu_k}+\frac{(\theta_t^{[k]})^{\frac{1}{\nu_k}}x_{tj}}{\nu_k}$.

2. $\frac{\partial\mu_t^{[k]}}{\partial\rho_{kk}}=\lambda_{t-1}^{[k]}$.

3. $\frac{\partial\mu_t^{[k]}}{\partial\nu_k}=\rho_{kk}\left[-\frac{(\theta_{t-1}^{[k]})^{\frac{1}{\nu_k}}ln(\theta_{t-1}^{[k]})}{\nu_k^2}-\frac{1}{2\nu_k}+\frac{\nu_k-1}{2\nu_k^2}\right]-\frac{(\theta_t^{[k]})^{\frac{1}{\nu_k}}ln(\theta_t^{[k]})}{\nu_k^2}-\frac{1}{2\nu_k}+\frac{\nu_k-1}{2\nu_k^2}$.

The covariance structure $\boldsymbol{\Sigma}_\psi$ consists of

$$\begin{pmatrix}\mathrm{Var}(\boldsymbol{Y}^{[1]}) & \mathrm{Cov}(\boldsymbol{Y}^{[1]},\boldsymbol{Y}^{[2]})\\ \mathrm{Cov}(\boldsymbol{Y}^{[2]},\boldsymbol{Y}^{[1]}) & \mathrm{Var}(\boldsymbol{Y}^{[2]})\end{pmatrix} \tag{11.16}$$

with the entries of $\text{Var}(\boldsymbol{Y}^{[k]})$ specified as

$$\text{Cov}(Y_t^{[k]}, Y_{t+h}^{[k]}) = \begin{cases} \rho_{kk}(1-\rho_{kk})\lambda_{t-1}^{[k]} + \rho_{kk}^2\left[\frac{\lambda_{t-1}^{[k]}}{\nu_k} + \frac{\nu_k - 1}{2\nu_k^2}\right] + \frac{\lambda_t^{[k]}}{\nu_k} + \frac{\nu_k-1}{2\nu_k^2} & h = 0, \\ \rho_{kk}\left[\frac{\lambda_t^{[k]}}{\nu_k} + \frac{\nu_k-1}{2\nu_k^2}\right] & h = 1, \\ 0 & h > 1. \end{cases} \tag{11.17}$$

The entries of the sub-matrix $\text{Cov}(\boldsymbol{Y}^{[1]}, \boldsymbol{Y}^{[2]})$ are:
$\text{Cov}(Y_t^{[1]}, Y_{t+h}^{[2]})$

$$= \begin{cases} \rho_{11}\rho_{22}\kappa_{12,t-1}\sqrt{\frac{\lambda_{t-1}^{[1]}}{\nu_1} + \frac{\nu_1-1}{2\nu_1^2}}\sqrt{\frac{\lambda_{t-1}^{[2]}}{\nu_2} + \frac{\nu_2-1}{2\nu_2^2}} & h = 0, \\ \quad +\kappa_{12,t}\sqrt{\frac{\lambda_t^{[1]}}{\nu_1} + \frac{\nu_1-1}{2\nu_1^2}}\sqrt{\frac{\lambda_t^{[2]}}{\nu_2} + \frac{\nu_2-1}{2\nu_2^2}} & \\ \rho_{22}\kappa_{12,t}\sqrt{\frac{\lambda_t^{[1]}}{\nu_1} + \frac{\nu_1-1}{2\nu_1^2}}\sqrt{\frac{\lambda_t^{[2]}}{\nu_2} + \frac{\nu_2-1}{2\nu_2^2}} & h = 1, \\ 0 & h > 1 \end{cases} \tag{11.18}$$

and similarly for $\text{Cov}(\boldsymbol{Y}^{[2]}, \boldsymbol{Y}^{[1]})$,
$\text{Cov}(Y_t^{[2]}, Y_{t+h}^{[1]})$

$$= \begin{cases} \rho_{11}\rho_{22}\kappa_{12,t-1}\sqrt{\frac{\lambda_{t-1}^{[1]}}{\nu_1} + \frac{\nu_1-1}{2\nu_1^2}}\sqrt{\frac{\lambda_{t-1}^{[2]}}{\nu_2} + \frac{\nu_2-1}{2\nu_2^2}} & h = 0, \\ \quad +\kappa_{12,t}\sqrt{\frac{\lambda_t^{[1]}}{\nu_1} + \frac{\nu_1-1}{2\nu_1^2}}\sqrt{\frac{\lambda_t^{[2]}}{\nu_2} + \frac{\nu_2-1}{2\nu_2^2}} & \\ \rho_{11}\kappa_{12,t}\sqrt{\frac{\lambda_t^{[1]}}{\nu_1} + \frac{\nu_1-1}{2\nu_1^2}}\sqrt{\frac{\lambda_t^{[2]}}{\nu_2} + \frac{\nu_2-1}{2\nu_2^2}} & h = 1, \\ 0 & h > 1. \end{cases} \tag{11.19}$$

The cross-correlation parameter $\kappa_{12,t}$ for $t = 1$ is estimated using Equation (11.12), while for $t = 2, \ldots, T$, $\kappa_{12,t}$ is estimated iteratively using equation (11.7) as follows:

$$\hat{\kappa}_{12,t} = \begin{cases} \frac{\tilde{\text{Cov}}(Y_t^{[1]}, Y_t^{[2]})}{(1+\hat{\rho}_{11}\hat{\rho}_{22})\sqrt{\frac{\hat{\lambda}_1^{[1]}}{\hat{\nu}_1} + \frac{\hat{\nu}_1-1}{2\hat{\nu}_1^2}}\sqrt{\frac{\hat{\lambda}_1^{[2]}}{\hat{\nu}_2} + \frac{\hat{\nu}_2-1}{2\hat{\nu}_2^2}}} & t = 1, \\ \frac{\tilde{\text{Cov}}(Y_t^{[1]}, Y_t^{[2]}) - \hat{\rho}_{11}\hat{\rho}_{22}\hat{\kappa}_{12,t-1}\sqrt{\frac{\hat{\lambda}_{t-1}^{[1]}}{\hat{\nu}_1} + \frac{\hat{\nu}_1-1}{2\hat{\nu}_1^2}}\sqrt{\frac{\hat{\lambda}_{t-1}^{[2]}}{\hat{\nu}_2} + \frac{\hat{\nu}_2-1}{2\hat{\nu}_2^2}}}{\sqrt{\frac{\hat{\lambda}_t^{[1]}}{\hat{\nu}_1} + \frac{\hat{\nu}_1-1}{2\hat{\nu}_1^2}}\sqrt{\frac{\hat{\lambda}_t^{[2]}}{\hat{\nu}_2} + \frac{\hat{\nu}_2-1}{2\hat{\nu}_2^2}}} & t = 2, \ldots, T. \end{cases} \tag{11.20}$$

where $\tilde{\text{Cov}}(Y_t^{[1]}, Y_t^{[2]}) = \frac{1}{T}\sum_{t=1}^{T}(y_t^{[1]} - \hat{\mu}_t^{[1]})(y_t^{[2]} - \hat{\mu}_t^{[2]})$.

Equation (11.14) is solved iteratively by the Newton-Raphson technique as:

$$\left( \hat{\boldsymbol{\psi}}_{r+1} \right) = \left( \hat{\boldsymbol{\psi}}_{r} \right) + \left[ \left(\frac{\partial \boldsymbol{\mu}}{\partial \boldsymbol{\psi}}\right)' \boldsymbol{\Sigma}_{\boldsymbol{\psi}}^{-1} \left(\frac{\partial \boldsymbol{\mu}}{\partial \boldsymbol{\psi}}\right) \right]_r^{-1} \left[ \left(\frac{\partial \boldsymbol{\mu}}{\partial \boldsymbol{\psi}}\right)' \boldsymbol{\Sigma}_{\boldsymbol{\psi}}^{-1} (\boldsymbol{y} - \boldsymbol{\mu}) \right]_r \tag{11.21}$$

where $\hat{\boldsymbol{\psi}} = [\hat{\beta}_1^{[2]}, \hat{\beta}_2^{[1]}, \hat{\beta}_1^{[2]}, \hat{\beta}_2^{[2]}, \hat{\rho_{11}}, \hat{\rho_{22}}, \hat{\nu_1}, \hat{\nu_2}]$ and $\hat{\boldsymbol{\psi}}_r$ are the values at the $r^{th}$ iteration and $[.]_r$ are the values of the expression at the $r^{th}$ iteration.

The algorithm works as follows: For an initial $\hat{\boldsymbol{\psi}}$, we compute $\kappa_{12,t}$, the score and the Hessian structures. These estimates are used to solve the iterative Equation (11.21) to obtain an updated $\hat{\boldsymbol{\psi}}$, which are in turn used to re-compute $\kappa_{12,t}$, the score and Hessian structures. Using these updated values, the iterative Equation (11.21) is solved again and this cycle of iteration continues until convergence of all the parameters. Under mild regulatory conditions, the asymptotic normality and consistency properties of $\hat{\boldsymbol{\psi}}$ follows the normal distribution with mean $\mathbf{0}$ and covariance $\left[ \left(\frac{\partial \boldsymbol{\mu}}{\partial \boldsymbol{\psi}}\right)' \boldsymbol{\Sigma}_{\boldsymbol{\psi}}^{-1} \left(\frac{\partial \boldsymbol{\mu}}{\partial \boldsymbol{\psi}}\right) \right]^{-1}$ (Refer to Subsection 2.3.1.2).

Using the converged estimates from the GQL method, the forecasting values of $Y_{t+1}^{[k]}$ are computed using the conditional expectation and variance of the one step-ahead forecast $Y_{t+1}^{[k]}$ given $r_t^{[k]}$ as follows:

$$\begin{aligned} E(Y_{t+1}^{[k]} | r_t^{[k]}) &= E(\rho_{kk} \circ r_t^{[k]} | r_t^{[k]}) + E(R_{t+1}^{[k]}) \\ &= \rho_{kk} r_t^{[k]} + \lambda_{t+1}^{[k]} \end{aligned} \tag{11.22}$$

and

$$\begin{aligned} \text{Var}(Y_{t+1}^{[k]} | r_t^{[k]}) &= \text{Var}(\rho_{kk} \circ r_t^{[k]} | r_t^{[k]}) + \text{Var}(R_{t+1}^{[k]}) \\ &= \rho_{kk}(1 - \rho_{kk}) r_t^{[k]} + \frac{\lambda_{t+1}^{[k]}}{\nu_k} + \frac{\nu_k - 1}{2\nu_k^2} \end{aligned} \tag{11.23}$$

where $r_t^{[k]}$ is approximated by $\text{E}(R_t^{[k]})$.

## 11.5 Numerical Experiment

Monte Carlo experiments are performed in this section based on non-stationary BINMA(1) simulated data from Equation (11.1) using a generating algorithm similar to the one developed in Section 10.5 to obtain simulated pair

$(R_t^{[1]}, R_t^{[2]})$ with $\theta_t^{[k]} = \exp(\boldsymbol{x}_t^{'}\boldsymbol{\beta}^{[k]})$ where

$$x_{t1} = \begin{cases} -t/T & (t = 1, \ldots, T/4), \\ rnorm(1,0,1) & (t = (T/4)+1, \ldots, 3T/4), \\ t/T & (t = (3T/4)+1, \ldots, T), \end{cases}$$

and $x_{t2}$ is the random uniform generated number in the interval (0,1), that is, $x_{t2} = runif(1,0,1)$ from the *runif* function in R for each $t = 1, \ldots, T$.

In particular, $R_t^{[1]} \sim$ COM-Poisson$(\lambda_t^{[1]}, \nu_1)$ and $R_t^{[2]} \sim$ COM-Poisson $(\lambda_t^{[2]}, \nu_2)$, with

$$R_t^{[2]} = \gamma * R_t^{[1]} + \epsilon_t \tag{11.24}$$

where $\gamma \in (0,1)$ and the parameter of $\epsilon_t$ is determined as $\epsilon_t \sim$ COM-Poisson$(\lambda_t^*, \nu^*)$, where $\lambda_t^* = \lambda_t^{[2]} - \gamma\lambda_t^{[1]}$ and

$$\nu^* = \frac{2q_0 + 1 + \sqrt{(2q_0+1)^2 - 8q_1}}{4q_1} > 0 \tag{11.25}$$

with $q_0 = [\lambda_t^{[2]} - \gamma\lambda_t^{[1]}]$, $q_1 = \left[\frac{\lambda_t^{[2]}}{\nu_2} + \frac{\nu_2-1}{2\nu_2^2} - \gamma(1-\gamma)\lambda_t^{[1]} - \gamma^2(\frac{\lambda_t^{[1]}}{\nu_1} + \frac{\nu_1-1}{2\nu_1^2})\right]$. $R_t^{[1]}$ and $R_t^{[2]}$ are autocorrelated through Equation (11.24) and it is proved $\text{Cov}(R_t^{[1]}, R_t^{[2]}) = \gamma\left[\frac{\lambda_t^{[1]}}{\nu_1} + \frac{\nu_1-1}{2\nu_1^2}\right] \neq 0$. The *rcom* function in R is used to simulate the COM-Poisson variables $R_t^{[1]}$, $R_t^{[2]}$ and $\epsilon_t$ with their given parameters. Assuming $\gamma = 0.5$, with the pair $[\nu_1, \nu_2] = [1,1], [0.6, 0.4], [1.7, 1.9]$, $[0.6, 1.7]$ and $[1.7, 0.6]$, $[\rho_{11}, \rho_{22}] = [0.3, 0.3], [0.9, 0.9]$ and $\kappa_{12,1} = 0.5$, with $\beta_1^{[1]} = \beta_2^{[1]} = 0.3$, $\beta_1^{[2]} = \beta_2^{[2]} = 0.5$, 5000 Monte-Carlo replications are made for each of the above combination and the results are displayed in Tables 11.5–11.6.

As shown in Tables 11.5–11.6, the mean estimates of all the parameters are close to the population values under all combinations and time points. With increasing time points, the standard errors decrease. However, some non-convergent simulations are noted. When $\nu_1 = 1$, $\nu_2 = 1$, $\rho_{11} = \rho_{22} = 0.9$ and $\kappa_{12,1} = 0.5$, around 325 simulations did not converge for $T = 60$, 270 for $T = 100$ and 165 for $T = 600$. For same values of dependence parameters, but $\nu_1 = 0.6$, $\nu_2 = 0.6$, it is remarked that the GQL algorithms fail in 350 simulations for $T = 60$, 280 for $T = 100$ and 180 for T=600. For $\nu_1, \nu_2 > 1$, we note an increase in the number of non-convergent simulations compared to $\nu_1, \nu_2 = 1$ and $\nu_1, \nu_2 < 1$. For instance, when $\nu_1 = 1.7$ and $\nu_2 = 1.9$, the number of failures is 435 for $T = 60$, 380 for $T = 100$ and 245 for $T = 600$. As for $\nu_1 = 0.6$, $\nu_2 = 1.7$ and $\nu_1 = 1.7$, $\nu_2 = 0.6$, GQL algorithm did not converge in approximately 380 simulations for $T = 60$, 315 for $T = 100$ and 210 for $T = 600$. The failures were mainly due to the singularity of the Hessian matrix in Equation (11.21). Note that the cross-correlation between the innovation

**TABLE 11.5**

Estimates of the Parameters and Standard Errors under Non-Stationary BINMA(1) Process for $[\nu_1, \nu_2] = [1.7, 0.6], [0.6, 1.7]$

| T | $\hat{\beta}_1^{[1]}$ | $\hat{\beta}_2^{[1]}$ | $\hat{\beta}_1^{[2]}$ | $\hat{\beta}_2^{[2]}$ | $\hat{\nu}_1 = 1.7$ | $\hat{\nu}_2 = 0.6$ | $\hat{\rho}_{11} = 0.9$ | $\hat{\rho}_{21} = 0.9$ | $\hat{\kappa}_{12,1}$ |
|---|---|---|---|---|---|---|---|---|---|
| 60 | 0.3126 | 0.3187 | 0.4894 | 0.4815 | 1.6829 | 0.5890 | 0.8811 | 0.8877 | 0.4830 |
| | (0.1256) | (0.0814) | (0.0909) | (0.1088) | (0.1460) | (0.1474) | (0.0889) | (0.0921) | |
| 100 | 0.3067 | 0.3022 | 0.5007 | 0.5057 | 1.6940 | 0.5944 | 0.8911 | 0.8929 | 0.4936 |
| | (0.0866) | (0.0663) | (0.0641) | (0.0720) | (0.1179) | (0.1266) | (0.0602) | (0.0628) | |
| 600 | 0.2980 | 0.2987 | 0.4996 | 0.5027 | 1.6967 | 0.5970 | 0.8944 | 0.8956 | 0.4959 |
| | (0.0386) | (0.0191) | (0.0273) | (0.0321) | (0.0658) | (0.0661) | (0.0041) | (0.0027) | |
| T | $\hat{\beta}_1^{[1]}$ | $\hat{\beta}_2^{[1]}$ | $\hat{\beta}_1^{[2]}$ | $\hat{\beta}_2^{[2]}$ | $\hat{\nu}_1 = 1.7$ | $\hat{\nu}_2 = 0.6$ | $\hat{\rho}_{11} = 0.3$ | $\hat{\rho}_{22} = 0.3$ | $\hat{\kappa}_{12,1}$ |
| 60 | 0.3184 | 0.3158 | 0.4895 | 0.4817 | 1.6893 | 0.5870 | 0.2831 | 0.2857 | 0.4812 |
| | (0.1015) | (0.0891) | (0.0940) | (0.0853) | (0.1259) | (0.1116) | (0.0739) | (0.0725) | |
| 100 | 0.3030 | 0.3019 | 0.5021 | 0.5050 | 1.7084 | 0.6076 | 0.2923 | 0.2978 | 0.4924 |
| | (0.0781) | (0.0614) | (0.0795) | (0.0690) | (0.1147) | (0.1087) | (0.0628) | (0.0690) | |
| 600 | 0.2989 | 0.2993 | 0.4987 | 0.5018 | 1.7015 | 0.6021 | 0.2955 | 0.2988 | 0.4936 |
| | (0.0316) | (0.0353) | (0.0261) | (0.0125) | (0.0507) | (0.0660) | (0.0095) | (0.0060) | |
| T | $\hat{\beta}_1^{[1]}$ | $\hat{\beta}_2^{[1]}$ | $\hat{\beta}_1^{[2]}$ | $\hat{\beta}_2^{[2]}$ | $\hat{\nu}_1 = 0.6$ | $\hat{\nu}_2 = 1.7$ | $\hat{\rho}_{11} = 0.9$ | $\hat{\rho}_{22} = 0.9$ | $\hat{\kappa}_{12,1}$ |
| 60 | 0.3127 | 0.3194 | 0.4858 | 0.4832 | 0.5870 | 1.6849 | 0.8888 | 0.8872 | 0.4896 |
| | (0.1170) | (0.0850) | (0.1215) | (0.0831) | (0.1459) | (0.1438) | (0.0747) | (0.0706) | |
| 100 | 0.3094 | 0.3011 | 0.5071 | 0.5021 | 0.5934 | 1.6977 | 0.8941 | 0.8973 | 0.4950 |
| | (0.0792) | (0.0635) | (0.0854) | (0.0604) | (0.1265) | (0.1070) | (0.0550) | (0.0622) | |
| 600 | 0.2927 | 0.2995 | 0.4950 | 0.5011 | 0.5951 | 1.6989 | 0.8963 | 0.8979 | 0.4974 |
| | (0.0345) | (0.0232) | (0.0309) | (0.0179) | (0.0630) | (0.0667) | (0.0087) | (0.0091) | |
| T | $\hat{\beta}_1^{[1]}$ | $\hat{\beta}_2^{[1]}$ | $\hat{\beta}_1^{[2]}$ | $\hat{\beta}_2^{[2]}$ | $\hat{\nu}_1 = 0.6$ | $\hat{\nu}_2 = 1.7$ | $\hat{\rho}_{11} = 0.3$ | $\hat{\rho}_{22} = 0.3$ | $\hat{\kappa}_{12,1}$ |
| 60 | 0.3161 | 0.3130 | 0.4839 | 0.4854 | 0.5805 | 1.6866 | 0.2801 | 0.2836 | 0.4812 |
| | (0.0878) | (0.0852) | (0.0810) | (0.0805) | (0.0462) | (0.0457) | (0.0312) | (0.0364) | |
| 100 | 0.3065 | 0.3058 | 0.5025 | 0.5010 | 0.5960 | 1.6924 | 0.2923 | 0.2987 | 0.4911 |
| | (0.0692) | (0.0614) | (0.0648) | (0.0627) | (0.0339) | (0.0391) | (0.0226) | (0.0259) | |
| 600 | 0.2956 | 0.2978 | 0.4988 | 0.5005 | 0.5980 | 1.6945 | 0.2944 | 0.2994 | 0.4930 |
| | (0.0384) | (0.0397) | (0.0373) | (0.0316) | (0.0054) | (0.0015) | (0.0073) | (0.0034) | |

**TABLE 11.6**

Estimates of the Parameters and Standard Errors under Non-Stationary BINMA(1) Process for $[\nu_1, \nu_2] = [1, 1], [0.6, 0.4], [1.7, 1.9]$

| T | $\hat{\beta}_1^{[1]}$ | $\hat{\beta}_2^{[1]}$ | $\hat{\beta}_1^{[2]}$ | $\hat{\beta}_2^{[2]}$ | $\hat{\nu}_1 = 1$ | $\hat{\nu}_2 = 1$ | $\hat{\rho}_{11} = 0.9$ | $\hat{\rho}_{22} = 0.9$ | $\hat{\kappa}_{12,1}$ |
|---|---|---|---|---|---|---|---|---|---|
| 60 | 0.3115 | 0.3154 | 0.4881 | 0.4810 | 0.9881 | 0.9892 | 0.8863 | 0.8895 | 0.4878 |
| | (0.1060) | (0.0809) | (0.0997) | (0.0883) | (0.1477) | (0.1435) | (0.0715) | (0.0732) | |
| 100 | 0.3072 | 0.3012 | 0.5021 | 0.5067 | 0.9993 | 0.9979 | 0.8907 | 0.8932 | 0.4995 |
| | (0.0838) | (0.0650) | (0.0698) | (0.0718) | (0.1250) | (0.1224) | (0.0683) | (0.0539) | |
| 600 | 0.2988 | 0.2990 | 0.4987 | 0.5040 | 0.9996 | 0.9989 | 0.8951 | 0.8990 | 0.4997 |
| | (0.0318) | (0.0270) | (0.0382) | (0.0121) | (0.0841) | (0.0698) | (0.0331) | (0.0207) | |
| T | $\hat{\beta}_1^{[1]}$ | $\hat{\beta}_2^{[1]}$ | $\hat{\beta}_1^{[2]}$ | $\hat{\beta}_2^{[2]}$ | $\hat{\nu}_1 = 1$ | $\hat{\nu}_2 = 1$ | $\hat{\rho}_{11}$=0.3 | $\hat{\rho}_{22} = 0.3$ | $\hat{\kappa}_{12,1}$ |
| 60 | 0.3170 | 0.3130 | 0.4895 | 0.4815 | 0.9887 | 0.9895 | 0.2831 | 0.2888 | 0.5175 |
| | (0.0841) | (0.0847) | (0.0870) | (0.0846) | (0.0454) | (0.0430) | (0.0399) | (0.0387) | |
| 100 | 0.3039 | 0.3080 | 0.5013 | 0.5091 | 0.9959 | 0.9917 | 0.2994 | 0.2931 | 0.4970 |
| | (0.0688) | (0.0611) | (0.0664) | (0.0621) | (0.0374) | (0.0352) | (0.0239) | (0.0209) | |
| 600 | 0.2978 | 0.2976 | 0.4990 | 0.5025 | 0.9974 | 0.9959 | 0.2997 | 0.2948 | 0.4988 |
| | (0.0313) | (0.0380) | (0.0303) | (0.0338) | (0.0025) | (0.0028) | (0.0051) | (0.0090) | |

*(Continued)*

**TABLE 11.6**
*(Continued)*

| T | $\hat{\beta}_1^{[1]}$ | $\hat{\beta}_2^{[1]}$ | $\hat{\beta}_1^{[2]}$ | $\hat{\beta}_2^{[2]}$ | $\hat{\nu}_1 = 0.6$ | $\hat{\nu}_2 = 0.4$ | $\hat{\rho}_{11} = 0.9$ | $\hat{\rho}_{22} = 0.9$ | $\hat{\kappa}_{12,1}$ |
|---|---|---|---|---|---|---|---|---|---|
| 60 | 0.3188 | 0.3140 | 0.4875 | 0.4881 | 0.5829 | 0.3810 | 0.8882 | 0.8851 | 0.4867 |
| | (0.1167) | (0.0991) | (0.0862) | (0.0893) | (0.1460) | (0.1204) | (0.0814) | (0.0764) | |
| 100 | 0.3081 | 0.3093 | 0.5037 | 0.5074 | 0.5915 | 0.3990 | 0.8924 | 0.8997 | 0.4966 |
| | (0.0883) | (0.0710) | (0.0696) | (0.0628) | (0.1146) | (0.1032) | (0.0518) | (0.0611) | |
| 600 | 0.2992 | 0.2965 | 0.4979 | 0.5025 | 0.5940 | 0.3995 | 0.8947 | 0.8999 | 0.5107 |
| | (0.0328) | (0.0241) | (0.0357) | (0.0111) | (0.0673) | (0.0534) | (0.0097) | (0.0012) | |
| T | $\hat{\beta}_1^{[1]}$ | $\hat{\beta}_2^{[1]}$ | $\hat{\beta}_1^{[2]}$ | $\hat{\beta}_2^{[2]}$ | $\hat{\nu}_1 = 0.6$ | $\hat{\nu}_2 = 0.4$ | $\hat{\rho}_{11} = 0.3$ | $\hat{\rho}_{22} = 0.3$ | $\hat{\kappa}_{12,1}$ |
| 60 | 0.3184 | 0.3175 | 0.4894 | 0.4810 | 0.5886 | 0.3852 | 0.2898 | 0.2831 | 0.4812 |
| | (0.1057) | (0.0848) | (0.0901) | (0.0812) | (0.1280) | (0.1154) | (0.0758) | (0.0766) | |
| 100 | 0.3033 | 0.3068 | 0.5071 | 0.5017 | 0.5934 | 0.3991 | 0.2944 | 0.2916 | 0.4950 |
| | (0.0798) | (0.0672) | (0.0753) | (0.0616) | (0.1165) | (0.1015) | (0.0780) | (0.0667) | |
| 600 | 0.2977 | 0.2950 | 0.4963 | 0.5010 | 0.5966 | 0.3997 | 0.2959 | 0.2933 | 0.4968 |
| | (0.0362) | (0.0258) | (0.0304) | (0.0116) | (0.0675) | (0.0579) | (0.0013) | (0.0029) | |
| T | $\hat{\beta}_1^{[1]}$ | $\hat{\beta}_2^{[1]}$ | $\hat{\beta}_1^{[2]}$ | $\hat{\beta}_2^{[2]}$ | $\hat{\nu}_1 = 1.7$ | $\hat{\nu}_2 = 1.9$ | $\hat{\rho}_{11}$=0.9 | $\hat{\rho}_{22} = 0.9$ | $\hat{\kappa}_{12,1}$ |
| 60 | 0.3160 | 0.3117 | 0.4827 | 0.4895 | 1.6850 | 1.8817 | 0.8892 | 0.8851 | 0.4843 |
| | (0.1234) | (0.0869) | (0.0935) | (0.1030) | (0.1440) | (0.1411) | (0.0836) | (0.0810) | |
| 100 | 0.3079 | 0.3011 | 0.5056 | 0.5060 | 1.7063 | 1.9026 | 0.8911 | 0.8973 | 0.5092 |
| | (0.0716) | (0.0682) | (0.0621) | (0.0881) | (0.1201) | (0.1155) | (0.0677) | (0.0660) | |
| 600 | 0.2981 | 0.2995 | 0.4976 | 0.5015 | 1.6970 | 1.8983 | 0.8929 | 0.8988 | 0.4947 |
| | (0.0295) | (0.0170) | (0.0366) | (0.0318) | (0.0023) | (0.0086) | (0.0060) | (0.0059) | |
| T | $\hat{\beta}_1^{[1]}$ | $\hat{\beta}_2^{[1]}$ | $\hat{\beta}_1^{[2]}$ | $\hat{\beta}_2^{[2]}$ | $\hat{\nu}_1 = 1.7$ | $\hat{\nu}_2 = 1.9$ | $\hat{\rho}_{11} = 0.3$ | $\hat{\rho}_{21} = 0.3$ | $\hat{\kappa}_{12,1}$ |
| 60 | 0.3186 | 0.3124 | 0.4845 | 0.4897 | 1.6823 | 1.8819 | 0.2858 | 0.2898 | 0.4855 |
| | (0.1013) | (0.0931) | (0.0849) | (0.0816) | (0.1271) | (0.1122) | (0.0813) | (0.0741) | |
| 100 | 0.3052 | 0.3033 | 0.5087 | 0.5091 | 1.6978 | 1.8913 | 0.2927 | 0.2993 | 0.4946 |
| | (0.0785) | (0.0762) | (0.0684) | (0.0646) | (0.0951) | (0.0812) | (0.0617) | (0.0664) | |
| 600 | 0.2970 | 0.2985 | 0.4954 | 0.5040 | 1.6988 | 1.8927 | 0.2938 | 0.2997 | 0.4960 |
| | (0.0358) | (0.0219) | (0.0369) | (0.0146) | (0.0630) | (0.0557) | (0.0087) | (0.0080) | |

terms, $\kappa_{12,t}$, is reported only for $t = 1$ due to high dimension, but were reliable estimates between 0 and 1.

## 11.6 Summary

In this chapter, a constrained BINMA(1) model has been developed with COM-Poisson innovations under non-stationary moment conditions. It has been proved in trial experiments that this new BINMA(1) model is able to model bivariate series that may exhibit different levels and forms of dispersion. In particular, for the estimation of the unknown model parameters, a GQL approach has been developed and its asymptotic properties have also been established. Monte Carlo simulation experiments have been performed to assess the performance of the GQL method. The next chapter extends the constrained bivariate models to unconstrained bivariate models.

# 12

# *Unconstrained BINAR(1) Model with COM-Poisson Innovations*

## 12.1 Introduction

In Chapter 10, the constrained BINAR(1) model with COM-Poisson innovations has been developed based on the binomial thinning mechanism. This BINAR(1) model is significantly more meritorious than other BINAR(1) models with Poisson and NB innovations since it encompasses situations where the series exhibit under-dispersion or a mixture of over-, under- or equi-dispersion. However, one important limitation of this model is the current BINAR(1) does not consider the cross-correlation that may arise by relating current series observation with previous-lagged observation of the other series as in the unconstrained BINAR(1) model with Poisson and NB innovations introduced in Chapter 4 and 8, respectively. Thus, it is of interest to construct an unconstrained BINAR(1) model with COM-Poisson innovations taking into account the unconstrained nature of the cross-correlation structures. The main objectives of this chapter are to construct an unconstrained BINAR(1) model with COM-Poisson innovations and to develop a GQL approach to estimate the different sets of unknown parameters.

The organization of the chapter is as follows: In Section 12.2, an unconstrained BINAR(1) model with COM-Poisson innovations is constructed and its marginal and joint moments are derived. In Section 12.3, the GQL algorithm is implemented to estimate the regression, dispersion, serial- and cross-dependence parameters. In Section 12.4, we present some numerical experiments to illustrate the different scenarios of over-, under- and mixed-dispersion under various combinations of the serial- and cross-correlation parameters. The chapter ends with some concluding remarks.

 DOI: 10.1201/9781003600435-12

## 12.2 Model Development

Suppose that $\{[Y_t^{[1]}, Y_t^{[2]}]', t \in \mathbb{Z}\}$ is related by a random coefficient BINAR(1) model of the form

$$\begin{pmatrix} Y_t^{[1]} \\ Y_t^{[2]} \end{pmatrix} = \begin{pmatrix} \rho_{11} & \rho_{12} \\ \rho_{21} & \rho_{22} \end{pmatrix} \circ \begin{pmatrix} Y_{t-1}^{[1]} \\ Y_{t-1}^{[2]} \end{pmatrix} + \begin{pmatrix} R_t^{[1]} \\ R_t^{[2]} \end{pmatrix} \tag{12.1}$$

where $R_t^{[1]}$ and $R_t^{[2]}$ are the corresponding innovation terms for $Y_t^{[1]}$ and $Y_t^{[2]}$, respectively and $R_t^{[k]}$ is marginally COM-Poisson with probability function

$$P(R_t^{[k]} = r_k) = \frac{(\theta_t^{[k]})^{r_k}}{(r_k!)^{\nu_k}} \frac{1}{Z(\theta_t^{[k]}, \nu_k)}, \quad r_k = 0, 1, 2, 3, \ldots, \quad \text{for}$$
$$\theta_t^{[k]} > 0, \nu_k \geq 0, k \in \{1, 2\}. \tag{12.2}$$

Referring to the BINAR(1) model in Equation (12.1),

- '$\circ$' indicates the binomial thinning operator (Steutel & Van Harn (1979)) wherein $\rho_{jk} \circ Y_{t-1}^{[k]}$, for

$$\rho_{jk} \circ Y_{t-1}^{[k]} = \begin{cases} \sum_{l=1}^{Y_{t-1}^{[k]}} b_l(\rho_{jk}), & Y_{t-1}^{[k]} > 0, \\ 0 & Y_{t-1}^{[k]} = 0 \end{cases} \tag{12.3}$$

  where $\{b_l(\rho_{jk})\}$ is a sequence of independent and identically distributed Bernoulli random variable with $\mathrm{P}\{b_l(\rho_{jk})\} = 1 - \mathrm{P}\{b_l(\rho_{jk})\} = \rho_{jk}$, for $\rho_{jk} \in [0, 1]$, $j, k \in \{1, 2\}$.

- The innovation terms $R_t^{[1]}$ and $R_t^{[2]}$ follow marginally the COM-Poisson distribution, as follows: $R_t^{[k]} \sim \text{COM-Poisson}(\lambda_t^{[k]}, \nu_k)$ and $\text{Corr}(R_t^{[1]}, R_t^{[2]})$ is assumed to be zero. Note that $\lambda_t^{[k]} = (\theta_t^{[k]})^{1/\nu_k} - (\frac{\nu_k - 1}{2\nu_k})$, where $\theta_t^{[k]} = \exp(\boldsymbol{x}_t' \boldsymbol{\beta}^{[k]})$ with $\boldsymbol{x}_t = [x_{t1}, x_{t2}, \ldots, x_{tj}, \ldots, x_{tp}]'$ and $\boldsymbol{\beta}^{[k]} = [\beta_1^{[k]}, \beta_2^{[k]}, \ldots, \beta_j^{[k]}, \ldots, \beta_p^{[k]}]'$ for $k \in \{1, 2\}$.

- We assume that the counting series $Y_t^{[k]}$ has mean, $E(Y_t^{[k]}) = \mu_t^{[k]}$, that is yet to be determined.

Equation (12.1) can be written in vector form as

$$\boldsymbol{Y_t} = \boldsymbol{A} \circ \boldsymbol{Y_{t-1}} + \boldsymbol{R_t}, \tag{12.4}$$

where $\boldsymbol{Y_t} = [Y_t^{[1]}, Y_t^{[2]}]'$, $\boldsymbol{R_t} = [R_t^{[1]}, R_t^{[2]}]'$ and $\boldsymbol{A} = \begin{pmatrix} \rho_{11} & \rho_{12} \\ \rho_{21} & \rho_{22} \end{pmatrix}$.

From Lemma 4.1, we have

$$E(\boldsymbol{Y_t}) = \boldsymbol{A} E(\boldsymbol{Y_{t-1}}) + E(\boldsymbol{R_t}) \tag{12.5}$$

and

$$\boldsymbol{\Sigma}_{0,t} = \boldsymbol{A}\text{Var}(\boldsymbol{Y_{t-1}})\boldsymbol{A}^{'} + \text{diag}(\boldsymbol{B\mu_{t-1}}) + \text{Var}(\boldsymbol{R_t}) \tag{12.6}$$

where

$$\boldsymbol{B} = \begin{pmatrix} \rho_{11}(1-\rho_{11}) & \rho_{12}(1-\rho_{12}) \\ \rho_{21}(1-\rho_{21}) & \rho_{22}(1-\rho_{22}) \end{pmatrix}$$

and $\boldsymbol{\Sigma}_{0,t} = \begin{bmatrix} \text{Var}(Y_t^{[1]}) & \text{Cov}(Y_t^{[1]}, Y_t^{[2]}) \\ \text{Cov}(Y_t^{[2]}, Y_t^{[1]}) & \text{Var}(Y_t^{[2]}) \end{bmatrix}$.

Hence, expanding Equation (12.5), we obtain

$$\begin{aligned} E(Y_t^{[1]}) \equiv \mu_t^{[1]} &= \rho_{11}\mu_{t-1}^{[1]} + \rho_{12}\mu_{t-1}^{[2]} + \lambda_t^{[1]} \quad \text{and} \\ E(Y_t^{[2]}) \equiv \mu_t^{[2]} &= \rho_{21}\mu_{t-1}^{[1]} + \rho_{22}\mu_{t-1}^{[2]} + \lambda_t^{[2]} \end{aligned} \tag{12.7}$$

and expanding Equation (12.6), we have the variances and covariance as follows:

$$\begin{aligned} \text{Var}(Y_t^{[1]}) &= \rho_{11}(1-\rho_{11})\mu_{t-1}^{[1]} + \rho_{11}^2\text{Var}(Y_{t-1}^{[1]}) + \rho_{12}(1-\rho_{12})\mu_{t-1}^{[2]} \\ &\quad + \rho_{12}^2\text{Var}(Y_{t-1}^{[2]}) + 2\rho_{11}\rho_{12}\text{Cov}(Y_{t-1}^{[1]}, Y_{t-1}^{[2]}) + \frac{\lambda_t^{[1]}}{\nu_1} + \frac{\nu_1 - 1}{2{\nu_1}^2}, \end{aligned} \tag{12.8}$$

$$\begin{aligned} \text{Var}(Y_t^{[2]}) &= \rho_{21}(1-\rho_{21})\mu_{t-1}^{[1]} + \rho_{21}^2\text{Var}(Y_{t-1}^{[1]}) + \rho_{22}(1-\rho_{22})\mu_{t-1}^{[2]} \\ &\quad + \rho_{22}^2\text{Var}(Y_{t-1}^{[2]}) + 2\rho_{21}\rho_{22}\text{Cov}(Y_{t-1}^{[1]}, Y_{t-1}^{[2]}) + \frac{\lambda_t^{[2]}}{\nu_2} + \frac{\nu_2 - 1}{2{\nu_2}^2} \end{aligned} \tag{12.9}$$

and

$$\begin{aligned} \text{Cov}(Y_t^{[1]}, Y_t^{[2]}) &= (\rho_{11}\rho_{22} + \rho_{12}\rho_{21})\text{Cov}(Y_{t-1}^{[1]}, Y_{t-1}^{[2]}) + \rho_{11}\rho_{21}\text{Var}(Y_{t-1}^{[1]}) \\ &\quad + \rho_{22}\rho_{12}\text{Var}(Y_{t-1}^{[2]}) \end{aligned} \tag{12.10}$$

**Remark 12.1.** Under stationary moment conditions, we obtain

$$\mu_1^{[1]} = \frac{\rho_{12}\mu_1^{[2]} + \lambda_1^{[1]}}{1-\rho_{11}}, \quad \mu_1^{[2]} = \frac{\rho_{12}\mu_1^{[1]} + \lambda^{[2]}}{1-\rho_{22}}, \tag{12.11}$$

$$\begin{aligned} \text{Var}(Y_1^{[1]}) &= \left(\frac{1}{1-\rho_{11}^2}\right)\{\rho_{11}(1-\rho_{11})\mu_1^{[1]} + \rho_{12}(1-\rho_{12})\mu_1^{[2]} + \rho_{12}^2\text{Var}(Y_1^{[2]}) \\ &\quad + 2\rho_{11}\rho_{12}\text{Cov}(Y_1^{[1]}, Y_1^{[2]}) + \frac{\lambda_1^{[1]}}{\nu_1} + \frac{\nu_1 - 1}{2{\nu_1}^2}\}, \end{aligned} \tag{12.12}$$

$$\mathrm{Var}(Y_1^{[2]}) = \left(\frac{1}{1-\rho_{22}^2}\right)\{\rho_{21}(1-\rho_{21})\mu_1^{[1]} + \rho_{21}^2\mathrm{Var}(Y_1^{[1]}) + \rho_{22}(1-\rho_{22})\mu_1^{[2]}$$
$$+ 2\rho_{21}\rho_{22}\mathrm{Cov}(Y_1^{[1]}, Y_1^{[2]}) + \frac{\lambda_1^{[2]}}{\nu_2} + \frac{\nu_2 - 1}{2{\nu_2}^2}\}, \tag{12.13}$$

$$\mathrm{Cov}(Y_1^{[1]}, Y_1^{[2]}) = \frac{\rho_{11}\rho_{21}\mathrm{Var}(Y_1^{[1]}) + \rho_{22}\rho_{12}\mathrm{Var}(Y_1^{[2]})}{1-\rho_{11}\rho_{22}-\rho_{12}\rho_{21}} \tag{12.14}$$

and solving simultaneously, we obtain

$$\mu_1^{[1]} = \frac{(1-\rho_{22})\lambda_1^{[1]} + \rho_{12}\lambda_1^{[2]}}{(1-\rho_{11})(1-\rho_{22}) - \rho_{12}\rho_{21}} \quad , \quad \mu_1^{[2]} = \frac{(1-\rho_{11})\lambda_1^{[2]} + \rho_{21}\lambda_1^{[1]}}{(1-\rho_{11})(1-\rho_{22}) - \rho_{12}\rho_{21}}, \tag{12.15}$$

$$\mathrm{Var}(Y_1^{[1]}) = \{(\rho_{12}^2)\left[\rho_{21}(1-\rho_{21})\mu_1^{[1]} + \rho_{22}(1-\rho_{22})\mu_1^{[2]} + \frac{\lambda_1^{[2]}}{\nu_2} + \frac{\nu_2-1}{2\nu_2^2}\right]$$
$$\times (1-\rho_{11}\rho_{22}-\rho_{12}\rho_{21}) + (2\rho_{11}\rho_{12})$$
$$\times \left[\rho_{21}(1-\rho_{21})\mu_1^{[1]} + \rho_{22}(1-\rho_{22})\mu_1^{[2]} + \frac{\lambda_1^{[2]}}{\nu_2} + \frac{\nu_2-1}{2\nu_2^2}\right](\rho_{22}\rho_{12})$$
$$+ \left[\rho_{11}(1-\rho_{11})\mu_1^{[1]} + \rho_{12}(1-\rho_{12})\mu_1^{[2]} + \frac{\lambda_1^{[1]}}{\nu_1} + \frac{\nu_1-1}{2\nu_1^2}\right]$$
$$\times (1-\rho_{22}^2)(1-\rho_{11}\rho_{22}-\rho_{12}\rho_{21})$$
$$- \left[\rho_{11}(1-\rho_{11})\mu_1^{[1]} + \rho_{12}(1-\rho_{12})\mu_1^{[2]} + \frac{\lambda_1^{[1]}}{\nu_1} + \frac{\nu_1-1}{2\nu_1^2}\right]$$
$$\times (2\rho_{21}\rho_{22})(\rho_{22}\rho_{12})\}/(\delta), \tag{12.16}$$

$$\mathrm{Var}(Y_1^{[2]}) = \{(1-\rho_{22}^2)\left[\rho_{21}(1-\rho_{21})\mu_1^{[1]} + \rho_{22}(1-\rho_{22})\mu_1^{[2]} + \frac{\lambda_1^{[2]}}{\nu_2} + \frac{\nu_2-1}{2\nu_2^2}\right]$$
$$\times (1-\rho_{11}\rho_{22}-\rho_{12}\rho_{21}) - (2\rho_{11}\rho_{12})$$
$$\times \left[\rho_{21}(1-\rho_{21})\mu_1^{[1]} + \rho_{22}(1-\rho_{22})\mu_1^{[2]} + \frac{\lambda_1^{[2]}}{\nu_2} + \frac{\nu_2-1}{2\nu_2^2}\right](\rho_{11}\rho_{21})$$
$$+ \left[\rho_{11}(1-\rho_{11})\mu_1^{[1]} + \rho_{12}(1-\rho_{12})\mu_1^{[2]} + \frac{\lambda_1^{[1]}}{\nu_1} + \frac{\nu_1-1}{2\nu_1^2}\right]$$
$$\times (\rho_{21})(1-\rho_{11}\rho_{22}-\rho_{12}\rho_{21})$$
$$+ \left[\rho_{11}(1-\rho_{11})\mu_1^{[1]} + \rho_{12}(1-\rho_{12})\mu_1^{[2]} + \frac{\lambda_1^{[1]}}{\nu_1} + \frac{\nu_1-1}{2\nu_1^2}\right]$$
$$\times (2\rho_{21}\rho_{22})(\rho_{11}\rho_{21})\}/(\delta), \tag{12.17}$$

and

$$\begin{aligned}
&\mathrm{Cov}(Y_1^{[1]}, Y_1^{[2]})\\
&= \{(1-\rho_{22}^2)\left[\rho_{21}(1-\rho_{21})\mu_1^{[1]} + \rho_{22}(1-\rho_{22})\mu_1^{[2]} + \frac{\lambda_1^{[2]}}{\nu_2} + \frac{\nu_2 - 1}{2\nu_2^2}\right](\rho_{22}\rho_{12})\\
&\quad + (\rho_{12}^2)\left[\rho_{21}(1-\rho_{21})\mu_1^{[1]} + \rho_{22}(1-\rho_{22})\mu_1^{[2]} + \frac{\lambda_1^{[2]}}{\nu_2} + \frac{\nu_2 - 1}{2\nu_2^2}\right](\rho_{11}\rho_{21})\\
&\quad + \left[\rho_{11}(1-\rho_{11})\mu_1^{[1]} + \rho_{12}(1-\rho_{12})\mu_1^{[2]} + \frac{\lambda_1^{[1]}}{\nu_1} + \frac{\nu_1 - 1}{2\nu_1^2}\right](1-\rho_{22}^2)(\rho_{11}\rho_{21})\\
&\quad + \left[\rho_{11}(1-\rho_{11})\mu_1^{[1]} + \rho_{12}(1-\rho_{12})\mu_1^{[2]} + \frac{\lambda_1^{[1]}}{\nu_1} + \frac{\nu_1 - 1}{2\nu_1^2}\right](\rho_{21}^2)(\rho_{22}\rho_{12})\}/(\delta)
\end{aligned} \tag{12.18}$$

where $\delta = \{(1-\rho_{11}^2)(1-\rho_{22}^2)(1-\rho_{11}\rho_{22}-\rho_{12}\rho_{21}) - (1-\rho_{11}^2)(2\rho_{21}\rho_{22})(\rho_{22}\rho_{12}) - (\rho_{12}^2)(\rho_{21}^2)(1 - \rho_{11}\rho_{22} - \rho_{12}\rho_{21}) - (\rho_{12}^2)(2\rho_{21}\rho_{22})(\rho_{11}\rho_{21}) - (2\rho_{11}\rho_{12})(1 - \rho_{22}^2)(\rho_{11}\rho_{21}) - (2\rho_{11}\rho_{12})(\rho_{21}^2)(\rho_{22}\rho_{12})\}$.

As for the derivation of lag-$h$ covariance expressions, we refer to Section 4.2, where based on

$$\boldsymbol{\Sigma}_{h,t} = \begin{bmatrix} \mathrm{Cov}(Y_t^{[1]}, Y_{t+h}^{[1]}) & \mathrm{Cov}(Y_{t+h}^{[1]}, Y_t^{[2]}) \\ \mathrm{Cov}(Y_t^{[1]}, Y_{t+h}^{[2]}) & \mathrm{Cov}(Y_t^{[2]}, Y_{t+h}^{[2]}) \end{bmatrix} = \begin{bmatrix} \Sigma_{11,h,t} & \Sigma_{12,h,t} \\ \Sigma_{21,h,t} & \Sigma_{22,h,t} \end{bmatrix}, \tag{12.19}$$

we proved that

$$\boldsymbol{\Sigma}_{h,t} = \boldsymbol{A}^h \boldsymbol{\Sigma}_{0,t}. \tag{12.20}$$

Using this general lag-$h$ auto-covariance formula, it is easy to derive the different diagonal and off-diagonal covariances with lag-$h$.

**Remark 12.2.** Under $\nu_k = 1$, Equations (12.7)–(12.9) reduce to

$$E(Y_t^{[1]}) = \rho_{11}\mu_{t-1}^{[1]} + \rho_{12}\mu_{t-1}^{[2]} + \lambda_t^{[1]} \quad , \quad E(Y_t^{[2]}) = \rho_{21}\mu_{t-1}^{[1]} + \rho_{22}\mu_{t-1}^{[2]} + \lambda_t^{[2]}, \tag{12.21}$$

$$\begin{aligned}
\mathrm{Var}(Y_t^{[1]}) &= \rho_{11}(1-\rho_{11})\mu_{t-1}^{[1]} + \rho_{11}^2 \mathrm{Var}(Y_{t-1}^{[1]}) + \rho_{12}(1-\rho_{12})\mu_{t-1}^{[2]}\\
&\quad + \rho_{12}^2 \mathrm{Var}(Y_{t-1}^{[2]}) + 2\rho_{11}\rho_{12}\mathrm{Cov}(Y_{t-1}^{[1]}, Y_{t-1}^{[2]}) + \lambda_t^{[1]}
\end{aligned} \tag{12.22}$$

and

$$\begin{aligned}
\mathrm{Var}(Y_t^{[2]}) &= \rho_{21}(1-\rho_{21})\mu_{t-1}^{[1]} + \rho_{21}^2 \mathrm{Var}(Y_{t-1}^{[1]}) + \rho_{22}(1-\rho_{22})\mu_{t-1}^{[2]}\\
&\quad + \rho_{22}^2 \mathrm{Var}(Y_{t-1}^{[2]}) + 2\rho_{21}\rho_{22}\mathrm{Cov}(Y_{t-1}^{[1]}, Y_{t-1}^{[2]}) + \lambda_t^{[2]}.
\end{aligned} \tag{12.23}$$

These moment expressions are similar as in Equations (4.5), (4.7) and (4.8). Thus, the unconstrained BINAR(1) model with Poisson innovations introduced in Chapter 4 is a special case of the unconstrained BINAR(1) model with COM-Poisson innovations.

## 12.3 Estimation Methodology: The GQL Approach

This section develops a GQL estimation approach to estimate the parameters $\boldsymbol{\Theta} = [\boldsymbol{\beta}^{[k]}, \boldsymbol{\rho}, \boldsymbol{\nu}]$ in the BINAR(1) model specified in Section 12.2. The GQL equation is expressed as

$$\frac{\partial \boldsymbol{\mu}}{\partial \boldsymbol{\Theta}}' \boldsymbol{\Sigma_{\Theta}}^{-1}(\boldsymbol{f} - \boldsymbol{\mu}) = 0. \tag{12.24}$$

Details of the components of Equation (12.24) are given in the following subsections.

### 12.3.1 The Derivative Matrix ($\frac{\partial \boldsymbol{\mu}}{\partial \boldsymbol{\Theta}}$)

The derivative matrix $\frac{\partial \boldsymbol{\mu}}{\partial \boldsymbol{\Theta}}$ is expressed as $\frac{\partial \boldsymbol{\mu}}{\partial \boldsymbol{\Theta}} = \left[\frac{\partial \boldsymbol{\mu}_1}{\partial \boldsymbol{\Theta}}, \frac{\partial \boldsymbol{\mu}_2}{\partial \boldsymbol{\Theta}}, \ldots, \frac{\partial \boldsymbol{\mu}_t}{\partial \boldsymbol{\Theta}}, \ldots, \frac{\partial \boldsymbol{\mu}_T}{\partial \boldsymbol{\Theta}}\right]'$, where $\frac{\partial \boldsymbol{\mu}_t}{\partial \boldsymbol{\Theta}} = \left[\frac{\partial \mu_t^{[1]}}{\partial \boldsymbol{\Theta}}, \frac{\partial \mu_t^{[2]}}{\partial \boldsymbol{\Theta}}\right]'$ with $\frac{\partial \mu_t^{[k]}}{\partial \boldsymbol{\Theta}} = \left[\frac{\partial \mu_t^{[k]}}{\partial \boldsymbol{\beta}^{[1]}}, \frac{\partial \mu_t^{[k]}}{\partial \boldsymbol{\beta}^{[2]}}, \frac{\partial \mu_t^{[k]}}{\partial \rho_{11}}, \frac{\partial \mu_t^{[k]}}{\partial \rho_{12}}, \frac{\partial \mu_t^{[k]}}{\partial \rho_{21}}, \frac{\partial \mu_t^{[k]}}{\partial \rho_{22}}, \frac{\partial \mu_t^{[k]}}{\partial \nu_1}, \frac{\partial \mu_t^{[k]}}{\partial \nu_2}\right]'$, where

$$\frac{\partial \mu_t^{[k]}}{\partial \boldsymbol{\beta}^{[l]}} = \left[\frac{\partial \mu_t^{[k]}}{\partial \beta_1^{[l]}}, \frac{\partial \mu_t^{[k]}}{\partial \beta_2^{[l]}}, \ldots, \frac{\partial \mu_t^{[k]}}{\partial \beta_j^{[l]}}, \ldots, \frac{\partial \mu_t^{[k]}}{\partial \beta_p^{[l]}}\right]', \quad k, l \in \{1, 2\}.$$

From Equation (12.15),

$$\mu_1^{[1]} = \frac{(1-\rho_{22})\left[(\theta_1^{[1]})^{\frac{1}{\nu_1}} - \frac{\nu_1 - 1}{2\nu_1}\right] + \rho_{12}\left[(\theta_1^{[2]})^{\frac{1}{\nu_2}} - \frac{\nu_2 - 1}{2\nu_2}\right]}{(1-\rho_{11})(1-\rho_{22}) - \rho_{12}\rho_{21}} \tag{12.25}$$

and

$$\mu_1^{[2]} = \frac{(1-\rho_{11})\left[(\theta_1^{[2]})^{\frac{1}{\nu_2}} - \frac{\nu_2 - 1}{2\nu_2}\right] + \rho_{21}\left[(\theta_1^{[1]})^{\frac{1}{\nu_1}} - \frac{\nu_1 - 1}{2\nu_1}\right]}{(1-\rho_{11})(1-\rho_{22}) - \rho_{12}\rho_{21}}, \tag{12.26}$$

where $\lambda_1^{[k]} = (\theta_1^{[k]})^{\frac{1}{\nu_k}} - \frac{\nu_k - 1}{2\nu_k}$ with $\theta_1^{[k]} = \exp(\boldsymbol{x}_1' \boldsymbol{\beta}^{[k]})$.

Hence, the derivative entries with respect to $\beta_j^{[k]}$ for the $j^{th}$ covariate are obtained iteratively as follows:

$$\frac{\partial \mu_t^{[k]}}{\partial \beta_j^{[k]}} = \begin{cases} \dfrac{(1-\rho_{ii})\frac{(\theta_t^{[k]})^{\frac{1}{\nu_k}} x_{tj}}{\nu_k}}{(1-\rho_{kk})(1-\rho_{ii})-\rho_{kj}\rho_{ik}} & t=1 \\ \rho_{kk}\dfrac{\partial \mu_{t-1}^{[k]}}{\partial \beta_j^{[k]}} + \rho_{ki}\dfrac{\partial \mu_{t-1}^{[i]}}{\partial \beta_j^{[k]}} + \dfrac{(\theta_t^{[k]})^{\frac{1}{\nu_k}} x_{tj}}{\nu_k} & t=2,\ldots,T, \end{cases} \tag{12.27}$$

$$\frac{\partial \mu_t^{[k]}}{\partial \beta_j^{[i]}} = \begin{cases} \dfrac{\rho_{ki}\frac{(\theta_t^{[i]})^{\frac{1}{\nu_i}} x_{tj}}{\nu_i}}{(1-\rho_{kk})(1-\rho_{ii})-\rho_{ki}\rho_{ik}} & t=1 \\ \rho_{kk}\dfrac{\partial \mu_{t-1}^{[k]}}{\partial \beta_j^{[i]}} + \rho_{kj}\dfrac{\partial \mu_{t-1}^{[i]}}{\partial \beta_j^{[i]}} & t=2,\ldots,T, \end{cases} \tag{12.28}$$

for $i,k \in \{1,2\}, i \neq k$.

Under similar assumptions, the derivatives with respect to $\boldsymbol{\rho}$ are obtained iteratively as follows:

$$\frac{\partial \mu_t^{[k]}}{\partial \rho_{kk}} = \begin{cases} \dfrac{[(1-\rho_{ii})\lambda_t^{[k]}+\rho_{ki}\lambda_t^{[i]}][1-\rho_{ii}]}{[(1-\rho_{kk})(1-\rho_{ii})-\rho_{ki}\rho_{ik}]^2} & t=1 \\ \rho_{kk}\dfrac{\partial \mu_{t-1}^{[k]}}{\partial \rho_{kk}} + \mu_{t-1}^{[k]} + \rho_{ki}\dfrac{\partial \mu_{t-1}^{[i]}}{\partial \rho_{kk}} & t=2,\ldots,T, \end{cases} \tag{12.29}$$

$$\frac{\partial \mu_t^{[k]}}{\partial \rho_{ki}} = \begin{cases} \dfrac{\lambda_t^{[i]}[(1-\rho_{kk})(1-\rho_{ii})-\rho_{ki}\rho_{ik}]+[(1-\rho_{ii})\lambda_t^{[k]}+\rho_{ki}\lambda_t^{[i]}][\rho_{ik}]}{[(1-\rho_{kk})(1-\rho_{ii})-\rho_{ki}\rho_{ik}]^2} & t=1 \\ \rho_{kk}\dfrac{\partial \mu_{t-1}^{[k]}}{\partial \rho_{ki}} + \rho_{ki}\dfrac{\partial \mu_{t-1}^{[i]}}{\partial \rho_{ki}} + \mu_{t-1}^{[i]} & t=2,\ldots,T, \end{cases} \tag{12.30}$$

$$\frac{\partial \mu_t^{[k]}}{\partial \rho_{ik}} = \begin{cases} \dfrac{[(1-\rho_{ii})\lambda_t^{[k]}+\rho_{ki}\lambda_t^{[i]}][\rho_{ki}]}{[(1-\rho_{kk})(1-\rho_{ii})-\rho_{ki}\rho_{ik}]^2} & t=1 \\ \rho_{kk}\dfrac{\partial \mu_{t-1}^{[k]}}{\partial \rho_{ik}} + \rho_{ki}\dfrac{\partial \mu_{t-1}^{[i]}}{\partial \rho_{ik}} & t=2,\ldots,T, \end{cases} \tag{12.31}$$

$$\frac{\partial \mu_t^{[k]}}{\partial \rho_{ii}} = \begin{cases} \dfrac{[(1-\rho_{ii})\lambda_t^{[k]}+\rho_{ki}\lambda_t^{[i]}][1-\rho_{kk}]-\lambda_t^{[k]}[(1-\rho_{kk})(1-\rho_{ii})-\rho_{ki}\rho_{ik}]}{[(1-\rho_{kk})(1-\rho_{ii})-\rho_{ki}\rho_{ik}]^2} & t=1 \\ \rho_{kk}\dfrac{\partial \mu_{t-1}^{[k]}}{\partial \rho_{ii}} + \rho_{ki}\dfrac{\partial \mu_{t-1}^{[i]}}{\partial \rho_{ii}} & t=2,\ldots,T, \end{cases} \tag{12.32}$$

for $i,k \in \{1,2\}, i \neq k$.

Similarly, the derivative expressions with respect to $\boldsymbol{\nu}$ are computed iteratively as follows:

$$\frac{\partial \mu_t^{[k]}}{\partial \nu_k} = \begin{cases} \dfrac{(1-\rho_{ii})\left[-\frac{(\theta_t^{[k]})^{\frac{1}{\nu_k}}\ln(\theta_t^{[k]})}{\nu_k^2} - \frac{1}{2\nu_k} + \frac{\nu_k-1}{2\nu_k^2}\right]}{(1-\rho_{kk})(1-\rho_{ii})-\rho_{ki}\rho_{ik}} & t=1 \\ \rho_{kk}\dfrac{\partial \mu_{t-1}^{[k]}}{\partial \nu_k} + \rho_{ki}\dfrac{\partial \mu_{t-1}^{[i]}}{\partial \nu_k} - \dfrac{(\theta_t^{[k]})^{\frac{1}{\nu_k}}\ln(\theta_t^{[k]})}{\nu_k^2} - \dfrac{1}{2\nu_k} + \dfrac{\nu_k-1}{2\nu_k^2} & t=2,\ldots,T, \end{cases} \tag{12.33}$$

$$\frac{\partial \mu_t^{[k]}}{\partial \nu_i} = \begin{cases} \dfrac{\rho_{ki}\left[-\frac{(\theta_t^{[i]})^{\frac{1}{\nu_i}} \ln(\theta_t^{[i]})}{\nu_i^2} - \frac{1}{2\nu_i} + \frac{\nu_i - 1}{2\nu_i^2}\right]}{(1-\rho_{kk})(1-\rho_{ii}) - \rho_{ki}\rho_{ik}} & t = 1 \\ \rho_{kk}\dfrac{\partial \mu_{t-1}^{[k]}}{\partial \nu_i} + \rho_{ki}\dfrac{\partial \mu_{t-1}^{[i]}}{\partial \nu_i} & t = 2, \ldots, T, \end{cases} \tag{12.34}$$

for $i, k \in \{1, 2\}, i \neq k$.

### 12.3.2 The Auto-Covariance Matrix ($\boldsymbol{\Sigma_\Theta}$).

$\boldsymbol{\Sigma_\Theta}$ represents the $(2T \times 2T)$ auto-covariance matrix given by

$$\begin{pmatrix} \boldsymbol{\Sigma}_{0,1} & \boldsymbol{\Sigma}'_{1,1} & \boldsymbol{\Sigma}'_{2,1} & \cdots & \boldsymbol{\Sigma}'_{t-1,1} & \cdots & \boldsymbol{\Sigma}'_{t+h-1,1} & \cdots & \boldsymbol{\Sigma}'_{T-1,1} \\ \boldsymbol{\Sigma}_{1,1} & \boldsymbol{\Sigma}_{0,2} & \boldsymbol{\Sigma}'_{1,2} & \cdots & \boldsymbol{\Sigma}'_{t-2,2} & \cdots & \boldsymbol{\Sigma}'_{t+h-2,2} & \cdots & \boldsymbol{\Sigma}'_{T-2,2} \\ \boldsymbol{\Sigma}_{2,1} & \boldsymbol{\Sigma}_{1,2} & \boldsymbol{\Sigma}_{0,3} & \cdots & \boldsymbol{\Sigma}'_{t-3,3} & \cdots & \boldsymbol{\Sigma}'_{t+h-3,3} & \cdots & \boldsymbol{\Sigma}'_{T-3,3} \\ \vdots & \vdots & \vdots & \ddots & \vdots & \vdots & \vdots & \vdots & \vdots \\ \boldsymbol{\Sigma}_{t-1,1} & \boldsymbol{\Sigma}_{t-2,2} & \boldsymbol{\Sigma}_{t-3,3} & \cdots & \boldsymbol{\Sigma}_{0,t} & \cdots & \boldsymbol{\Sigma}'_{t+h-t,t} & \cdots & \boldsymbol{\Sigma}'_{T-t,t} \\ \vdots & \vdots & \vdots & \vdots & \vdots & \ddots & \vdots & \vdots & \vdots \\ \boldsymbol{\Sigma}_{t+h-1,1} & \boldsymbol{\Sigma}_{t+h-2,2} & \boldsymbol{\Sigma}_{t+h-3,3} & \cdots & \boldsymbol{\Sigma}_{t+h-t,t} & \cdots & \boldsymbol{\Sigma}_{0,t+h} & \cdots & \boldsymbol{\Sigma}'_{T-t,t} \\ \vdots & \vdots & \vdots & \vdots & \vdots & \vdots & \vdots & \ddots & \vdots \\ \boldsymbol{\Sigma}_{T-1,1} & \boldsymbol{\Sigma}_{T-2,2} & \boldsymbol{\Sigma}_{T-3,3} & \cdots & \boldsymbol{\Sigma}_{T-t,t} & \cdots & \boldsymbol{\Sigma}_{T-t+h,t} & \cdots & \boldsymbol{\Sigma}_{0,T} \end{pmatrix} \tag{12.35}$$

where $\boldsymbol{\Sigma_{h,t}}$ is specified as in (12.20). Note that the initial variance and covariance expressions at $t = 1$ is given by Equations (12.16)–(12.18) and thereon for $t = 2, \ldots, T$, Equations (12.8)–(12.10) are computed iteratively.

### 12.3.3 Score Vector

The score vector $\boldsymbol{f} = [\boldsymbol{f}_1, \boldsymbol{f}_2, \ldots, \boldsymbol{f}_t, \ldots, \boldsymbol{f}_{t+h}, \ldots, \boldsymbol{f}_T]'$ with $\boldsymbol{f}_t = [Y_t^{[1]}, Y_t^{[2]}]'$ and corresponding $\boldsymbol{\mu} = [\boldsymbol{\mu}_1, \boldsymbol{\mu}_2, \ldots, \boldsymbol{\mu}_t, \ldots, \boldsymbol{\mu}_T]'$ with $\boldsymbol{\mu}_t = [\mu_t^{[1]}, \mu_t^{[2]}]'$ for $t = 1, 2, \ldots, T$.

The GQL in Equation (12.24) is solved iteratively using the Newton-Raphson as follows

$$(\hat{\boldsymbol{\Theta}}_{r+1}) = (\hat{\boldsymbol{\Theta}}_r) + \left[\frac{\partial \boldsymbol{\mu}}{\partial \boldsymbol{\Theta}}' \boldsymbol{\Sigma_\Theta}^{-1} \frac{\partial \boldsymbol{\mu}}{\partial \boldsymbol{\Theta}}\right]_r^{-1} \left[\frac{\partial \boldsymbol{\mu}}{\partial \boldsymbol{\Theta}}' \boldsymbol{\Sigma_\Theta}^{-1} (\boldsymbol{f} - \boldsymbol{\mu})\right]_r \tag{12.36}$$

where $\hat{\boldsymbol{\Theta}}_r$ are the estimates at the $r^{th}$ iteration and $[.]_r$ are the values of the expression at the $r^{th}$ iteration.

The working of this algorithm is as follows: For an initial $\hat{\boldsymbol{\Theta}}$, we compute the entries of the auto-covariance and Hessian structures and thereon solve the iterative Equation (12.36) to obtain an updated $\hat{\boldsymbol{\Theta}}$. These updated values are again used to compute the entries of the auto-covariance and Hessian structures and in turn Equation (12.36) is solved to obtain a new set of parameters. This cycle of iteration continues until convergence. Under mild regulatory condition, the asymptotic normality and consistency properties of $\hat{\boldsymbol{\Theta}}$ follows the normal distribution with mean $\mathbf{0}$ and covariance $\left[\frac{\partial \boldsymbol{\mu}}{\partial \boldsymbol{\Theta}}' \boldsymbol{\Sigma}_{\boldsymbol{\Theta}}{}^{-1} \frac{\partial \boldsymbol{\mu}}{\partial \boldsymbol{\Theta}}\right]^{-1}$ (Refer to Subsection 2.3.1.2).

Using the converged estimates, we derive the one-step ahead forecasting equations of the unconstrained BINAR(1) model with COM-Poisson innovations. The conditional expectation and variance of the one-step ahead forecast $Y_{t+1}^{[k]}$ given $y_t^{[k]}$ are expressed as follows:

$$\begin{aligned} E(Y_{t+1}^{[1]}|y_t^{[1]}, y_t^{[2]}) &= E(\rho_{11} \circ y_t^{[1]}|y_t^{[1]}) + E(\rho_{12} \circ y_t^{[2]}|y_t^{[2]}) + E(R_{t+1}^{[1]}) \\ &= \rho_{11} y_t^{[1]} + \rho_{12} y_t^{[2]} + \lambda_{t+1}^{[1]}, \end{aligned} \tag{12.37}$$

$$\begin{aligned} E(Y_{t+1}^{[2]}|y_t^{[1]}, y_t^{[2]}) &= E(\rho_{21} \circ y_t^{[1]}|y_t^{[1]}) + E(\rho_{22} \circ y_t^{[2]}|y_t^{[2]}) + E(R_{t+1}^{[2]}) \\ &= \rho_{21} y_t^{[1]} + \rho_{22} y_t^{[2]} + \lambda_{t+1}^{[1]}, \end{aligned} \tag{12.38}$$

with the corresponding variances

$$\begin{aligned} \text{Var}(Y_{t+1}^{[1]}|y_t^{[1]}, y_t^{[2]}) &= \text{Var}(\rho_{11} \circ y_t^{[1]}|y_t^{[1]}) + \text{Var}(\rho_{12} \circ y_t^{[2]}|y_t^{[2]}) + \text{Var}(R_{t+1}^{[1]}) \\ &= \rho_{11}(1-\rho_{11}) y_t^{[1]} + \rho_{12}(1-\rho_{12}) y_t^{[2]} + \frac{\lambda_{t+1}^{[1]}}{\nu_1} + \frac{\nu_1 - 1}{2\nu_1^2} \end{aligned} \tag{12.39}$$

and

$$\begin{aligned} \text{Var}(Y_{t+1}^{[2]}|y_t^{[1]}, y_t^{[2]}) &= \text{Var}(\rho_{21} \circ y_t^{[1]}|y_t^{[1]}) + \text{Var}(\rho_{22} \circ y_t^{[2]}|y_t^{[2]}) + \text{Var}(R_{t+1}^{[2]}) \\ &= \rho_{21}(1-\rho_{21}) y_t^{[1]} + \rho_{22}(1-\rho_{22}) y_t^{[2]} + \frac{\lambda_{t+1}^{[2]}}{\nu_2} + \frac{\nu_2 - 1}{2\nu_2^2}. \end{aligned} \tag{12.40}$$

## 12.4 Numerical Evaluation

Using Equations (12.15)–(12.17) we illustrate (Tables 12.1–12.4) under the stationary moment condition, the possible scenarios where $\text{Var}(Y_1^{[k]})$ exceeds or is lower than $\mu_1^{[k]}$ for the different values of $\{\rho_{11}, \rho_{12}, \rho_{21}\rho_{22}\}$, $\{\nu_1, \nu_2\}$ and $\{\theta_1^{[1]}, \theta_1^{[2]}\}$.

**TABLE 12.1**
Experimental Results under Different Serial and Dispersion Combinations $[\nu_1 = 1, \nu_2 = 1]$

| $\rho_{11}$ | $\rho_{12}$ | $\rho_{21}$ | $\rho_{22}$ | $\nu_1$ | $\nu_2$ | $\theta_1^{[1]}$ | $\theta_1^{[2]}$ | $\mu_1^{[1]}$ | $\text{Var}(Y_1^{[1]})$ | $\mu_1^{[2]}$ | $\text{Var}(Y_1^{[2]})$ |
|---|---|---|---|---|---|---|---|---|---|---|---|
| 0.1 | 0.15 | 0.2 | 0.25 | 1 | 1 | 0.1 | 0.2 | 0.1627 | 0.1633 | 0.31 | 0.296 |
| | | | | | | 0.9 | 0.95 | 1.267 | 1.270 | 1.60 | 1.53 |
| | | | | | | 1.1 | 1.2 | 1.558 | 1.562 | 2.016 | 1.59 |
| | | | | | | 2 | 2.1 | 2.814 | 2.821 | 3.55 | 3.40 |
| 0.5 | 0.55 | 0.6 | 0.65 | 1 | 1 | 0.1 | 0.2 | 0.57 | 0.68 | 0.93 | 1 |
| | | | | | | 0.9 | 0.95 | 3.93 | 4.63 | 5.32 | 5.88 |
| | | | | | | 1.1 | 1.2 | 4.86 | 5.73 | 6.64 | 7.33 |
| | | | | | | 2 | 2.1 | 8.7 | 10.3 | 11.8 | 13 |
| 0.9 | 0.1 | 0.11 | 0.8 | 1 | 1 | 0.1 | 0.2 | 4.4 | 9.8 | 3.4 | 12.4 |
| | | | | | | 0.9 | 0.95 | 30.6 | 66 | 21.6 | 78.4 |
| | | | | | | 1.1 | 1.2 | 37.8 | 81.6 | 26.8 | 97.3 |
| | | | | | | 2 | 2.1 | 67.8 | 146 | 48 | 174 |

**TABLE 12.2**
Experimental Results under Different Serial and Dispersion Combinations for $[\nu_1 < 1, \nu_2 < 1]$

| $\rho_{11}$ | $\rho_{12}$ | $\rho_{21}$ | $\rho_{22}$ | $\nu_1$ | $\nu_2$ | $\theta_1^{[1]}$ | $\theta_1^{[2]}$ | $\mu_1^{[1]}$ | $\text{Var}(Y_1^{[1]})$ | $\mu_1^{[2]}$ | $\text{Var}(Y_1^{[2]})$ |
|---|---|---|---|---|---|---|---|---|---|---|---|
| 0.1 | 0.15 | 0.2 | 0.25 | 0.1 | 0.15 | 0.1 | 0.2 | 5.9 | 1.3 | 5.3 | 2 |
| | | | | | | 0.9 | 0.95 | 6.5 | 5.1 | 6.4 | 5.1 |
| | | | | | | 1.1 | 1.2 | 9.7 | 29.2 | 10.9 | 27.8 |
| | | | | | | 2 | 2.1 | 1978 | 17186 | 719 | 2184 |
| 0.5 | 0.55 | 0.6 | 0.65 | 0.1 | 0.15 | 0.1 | 0.2 | 16.9 | 12.5 | 19.7 | 16 |
| | | | | | | 0.9 | 0.95 | 18.9 | 20.4 | 23 | 27 |
| | | | | | | 1.1 | 1.2 | 29.1 | 65.2 | 37.4 | 74.8 |
| | | | | | | 2 | 2.1 | 3143 | 17314 | 2716 | 8567 |
| 0.9 | 0.1 | 0.11 | 0.8 | 0.1 | 0.15 | 0.1 | 0.2 | 131.5 | 233 | 86.5 | 281.8 |
| | | | | | | 0.9 | 0.95 | 147 | 304 | 98 | 368 |
| | | | | | | 1.1 | 1.2 | 227 | 692 | 156 | 755 |
| | | | | | | 2 | 2.1 | 24450 | 137849 | 14165 | 81309 |
| 0.1 | 0.15 | 0.2 | 0.25 | 0.8 | 0.9 | 0.1 | 0.2 | 0.26 | 0.15 | 0.37 | 0.31 |
| | | | | | | 0.9 | 0.95 | 1.4 | 1.5 | 1.7 | 1.691 |
| | | | | | | 1.1 | 1.2 | 1.8 | 1.9 | 2.17 | 2.19 |
| | | | | | | 2 | 2.1 | 3.45 | 3.94 | 4.03 | 4.08 |
| 0.5 | 0.55 | 0.6 | 0.65 | 0.8 | 0.9 | 0.1 | 0.2 | 0.84 | 0.82 | 1.18 | 1.19 |
| | | | | | | 0.9 | 0.95 | 4.29 | 5.2 | 5.7 | 6.4 |
| | | | | | | 1.1 | 1.2 | 5.4 | 6.6 | 7.3 | 8.2 |
| | | | | | | 2 | 2.1 | 10.5 | 13 | 13.7 | 15.7 |
| 0.9 | 0.1 | 0.11 | 0.8 | 0.8 | 0.9 | 0.1 | 0.2 | 6.5 | 13 | 5 | 16.4 |
| | | | | | | 0.9 | 0.95 | 33.4 | 72.9 | 23.4 | 85.6 |
| | | | | | | 1.1 | 1.2 | 42 | 92 | 30 | 108 |
| | | | | | | 2 | 2.1 | 82 | 180 | 57 | 209 |

**TABLE 12.3**
Experimental Results under Different Serial and Dispersion Combinations $[\nu_1 > 1, \nu_2 > 1]$

| $\rho_{11}$ | $\rho_{12}$ | $\rho_{21}$ | $\rho_{22}$ | $\nu_1$ | $\nu_2$ | $\theta_1^{[1]}$ | $\theta_1^{[2]}$ | $\mu_1^{[1]}$ | $\text{Var}(Y_1^{[1]})$ | $\mu_1^{[2]}$ | $\text{Var}(Y_1^{[2]})$ |
|---|---|---|---|---|---|---|---|---|---|---|---|
| 0.1 | 0.15 | 0.2 | 0.25 | 1.1 | 1.2 | 0.1 | 0.2 | 0.13 | 0.17 | 0.27 | 0.30 |
| | | | | | | 0.9 | 0.95 | 1.21 | 1.17 | 1.49 | 1.34 |
| | | | | | | 1.1 | 1.2 | 1.47 | 1.41 | 1.83 | 1.63 |
| | | | | | | 2 | 2.1 | 2.54 | 2.42 | 3.04 | 2.67 |
| 0.5 | 0.55 | 0.6 | 0.65 | 1.1 | 1.2 | 0.1 | 0.2 | 0.48 | 0.63 | 0.80 | 0.94 |
| | | | | | | 0.9 | 0.95 | 3.72 | 4.31 | 4.97 | 5.37 |
| | | | | | | 1.1 | 1.2 | 4.53 | 5.23 | 6.1 | 6.6 |
| | | | | | | 2 | 2.1 | 7.77 | 8.91 | 10.26 | 10.97 |
| 0.9 | 0.1 | 0.11 | 0.8 | 1.1 | 1.2 | 0.1 | 0.2 | 3.71 | 8.59 | 2.93 | 10.94 |
| | | | | | | 0.9 | 0.95 | 28.9 | 61.8 | 20.3 | 73 |
| | | | | | | 1.1 | 1.2 | 35.2 | 75.3 | 24.8 | 89 |
| | | | | | | 2 | 2.1 | 60 | 128 | 42 | 151 |
| 0.1 | 0.15 | 0.2 | 0.25 | 2 | 2.5 | 0.1 | 0.2 | 0.129 | 0.228 | 0.335 | 0.308 |
| | | | | | | 0.9 | 0.95 | 0.97 | 0.74 | 1.17 | 0.81 |
| | | | | | | 1.1 | 1.2 | 1.11 | 0.83 | 1.33 | 0.91 |
| | | | | | | 2 | 2.1 | 1.6 | 1.12 | 1.82 | 1.20 |
| 0.5 | 0.55 | 0.6 | 0.65 | 2 | 2.5 | 0.1 | 0.2 | 0.511 | 0.752 | 0.947 | 1.03 |
| | | | | | | 0.9 | 0.95 | 2.97 | 3.1 | 3.92 | 3.81 |
| | | | | | | 1.1 | 1.2 | 3.39 | 3.52 | 4.48 | 4.32 |
| | | | | | | 2 | 2.1 | 4.82 | 4.89 | 6.23 | 5.95 |
| 0.9 | 0.1 | 0.11 | 0.8 | 2 | 2.5 | 0.1 | 0.2 | 3.98 | 9.71 | 3.31 | 11.9 |
| | | | | | | 0.9 | 0.95 | 23.1 | 47.1 | 16.1 | 55.5 |
| | | | | | | 1.1 | 1.2 | 26.37 | 53.6 | 18.4 | 63.2 |
| | | | | | | 2 | 2.1 | 37.5 | 75.4 | 25.8 | 88.7 |

**TABLE 12.4**
Experimental Results under Different Serial and Dispersion Combinations for $[\nu_1 < 1, \nu_2 > 1]$

| $\rho_{11}$ | $\rho_{12}$ | $\rho_{21}$ | $\rho_{22}$ | $\nu_1$ | $\nu_2$ | $\theta_1^{[1]}$ | $\theta_1^{[2]}$ | $\mu_1^{[1]}$ | $\text{Var}(Y_1^{[1]})$ | $\mu_1^{[2]}$ | $\text{Var}(Y_1^{[2]})$ |
|---|---|---|---|---|---|---|---|---|---|---|---|
| 0.1 | 0.15 | 0.2 | 0.25 | 0.8 | 1.5 | 0.1 | 0.2 | 0.25 | 0.14 | 0.30 | 0.33 |
| | | | | | | 0.9 | 0.95 | 1.35 | 1.44 | 1.42 | 1.21 |
| | | | | | | 1.1 | 1.2 | 1.68 | 1.84 | 1.73 | 1.45 |
| | | | | | | 2 | 2.1 | 3.25 | 3.7 | 2.83 | 2.34 |
| 0.5 | 0.55 | 0.6 | 0.65 | 0.8 | 1.5 | 0.1 | 0.2 | 0.768 | 0.751 | 1.014 | 1.155 |
| | | | | | | 0.9 | 0.95 | 4 | 4.8 | 5 | 5.4 |
| | | | | | | 1.1 | 1.2 | 4.95 | 5.98 | 6.12 | 6.6 |
| | | | | | | 2 | 2.1 | 9.26 | 11.4 | 10.6 | 11.7 |
| 0.9 | 0.1 | 0.11 | 0.8 | 0.8 | 1.5 | 0.1 | 0.2 | 5.98 | 11.94 | 4.16 | 15.37 |
| | | | | | | 0.9 | 0.95 | 31 | 67 | 21 | 76 |
| | | | | | | 1.1 | 1.2 | 39 | 83 | 26 | 94 |
| | | | | | | 2 | 2.1 | 72 | 157 | 47 | 170 |

These above results could only demonstrate that for different parameter inputs, there exists the possibility of the counting series to be mutually or mixed over- and under-dispersed with different levels of dispersion.
Some common remarks are:

(a) For $\{0.1 \leq \rho_{jk} \leq 0.25\}$ and $\{0.5 \leq \rho_{jk} \leq 0.65\}$, $\text{Var}(Y_1^{[k]})$ is lesser than $\mu_1^{[k]}$ for small $\theta_1^{[1]}$ and $\theta_1^{[2]}$.

(b) For high serial-correlation and low pair of cross-correlations, as in $\{0.9, 0.1, 0.11, 0.8\}$, the counting series become hugely over-dispersed.

(c) For relatively lower serial/cross values, the counting series are subject to under-dispersion/over-dispersion for both series or to some combination of over- and under-dispersion, but as the serial-correlations increase, both counting series express huge over-dispersion.

(d) For mixed dispersion, that is $\nu_1 < 1, \nu_2 > 1$, we note that for low serial/cross-correlations, there is a mixed dispersed series and as serial-correlation increase, the data tends to over-dispersion under both series.

In general, irrespective of any serial/cross parameters and values of $\nu_1, \nu_2$, as $\theta_1^{[k]}$ increases, $\mu_1^{[k]}$ also increases. Hence, the proposed model is suitable for analyzing dispersed series of high and low level of dispersion.

## 12.5 Summary

In this chapter, a novel unconstrained BINAR(1) model with COM-Poisson innovations under time-dependent moments has been developed and has shown numerically to accommodate all forms of mutual and mixed levels of dispersion. The GQL estimation approach has also been developed to estimate the unknown model parameters. A trial experiment has been conducted to show that this BINAR(1) model can analyze data which exhibit equi-, over- or under-dispersion or mixed dispersion. The next chapter provides a similar investigation for the unconstrained BINMA(1) model.

# 13

# *Unconstrained BINMA(1) Model with COM-Poisson Innovations*

## 13.1 Introduction

In Chapter 11, we introduced the constrained BINMA(1) model with COM-Poisson innovations based on a diagonal correlation structure. In this chapter we extend this model to an unconstrained BINMA(1) with COM-Poisson innovations by allowing for a full correlation structure. As for the inferential analysis, we develop a GQL approach, as it has been able to generate reliable estimates in the earlier chapters.

The chapter is laid out as follows: In Section 13.2, an unconstrained BINMA(1) model with COM-Poisson innovations is introduced and its marginal and joint moments are derived. In Section 13.3, a trial experiment specifying the different scenarios is conducted where the series are mutually or mixed equi-, over- and under-dispersed. In Section 13.4, a GQL equation is formulated to estimate the regression, dispersion and dependence parameters and the forecasting equations of the BINMA(1) model are derived. A Monte-Carlo experiment is presented in Section 13.5 to assess the performance of GQL approach. The conclusion is presented in Section 13.7.

## 13.2 Model Development

The unconstrained BINMA(1) model is given by

$$\begin{pmatrix} Y_t^{[1]} \\ Y_t^{[2]} \end{pmatrix} = \begin{pmatrix} \rho_{11} & \rho_{12} \\ \rho_{21} & \rho_{22} \end{pmatrix} \circ \begin{pmatrix} R_{t-1}^{[1]} \\ R_{t-1}^{[2]} \end{pmatrix} + \begin{pmatrix} R_t^{[1]} \\ R_t^{[2]} \end{pmatrix}. \tag{13.1}$$

In the above model, the sequence $R_t^{[k]}$ is identically and independently COM-Poisson distributed such that $\text{Corr}(R_i^{[k]}, R_j^{[k]}) = 0$ for $i, j \in \mathbb{Z}$, with $R_t^{[k]} \sim \text{COM-Poisson}(\lambda_t^{[k]}, \nu_k)$. Hence, it follows that $E(Y_t^{[k]}) = \lambda_t^{[k]}$ and $\text{Var}(Y_t^{[k]}) = \frac{\lambda_t^{[k]}}{\nu_k} + \frac{\nu_k - 1}{2\nu_k^2}$ for $k \in \{1, 2\}$.

 DOI: 10.1201/9781003600435-13

As for $\{\rho_{k'k} \circ R_{t-1}^{[k]}\}$, the '$\circ$' indicates the binomial thinning operator (Steutel & Van Harn (1979)) such that

$$\rho_{k'k} \circ R_{t-1}^{[k]} = \begin{cases} \sum_{l=1}^{R_{t-1}^{[k]}} b_l(\rho_{k'k}), & R_{t-1}^{[k]} > 0, \\ 0 & R_{t-1}^{[k]} = 0 \end{cases} \tag{13.2}$$

or $\{\rho_{k'k} \circ R_{t-1}^{k} | R_{t-1}^{[k]}\} \sim \text{Binomial}(R_{t-1}^{[k]}, \rho_{k'k})$.

In the context that there exists some $p$ covariates at the $t^{th}$ time point, that is, $\boldsymbol{x}_t = [x_{t1}, x_{t2}, \ldots, x_{tj}, \ldots, x_{tp}]'$, we assume $\lambda_t^{[k]} = {\theta_t^{[k]}}^{1/\nu_k} - \frac{\nu_k - 1}{2\nu_k}$ where $\theta_t^{[k]} = \exp(\boldsymbol{x}_t' \boldsymbol{\beta}^{[k]})$ with $\boldsymbol{\beta}^{[k]} = [\beta_1^{[k]}, \beta_2^{[k]}, \ldots, \beta_j^{[k]}, \ldots, \beta_p^{[k]}]'$ for $k \in \{1, 2\}$.

To simplify the derivation of the marginal and joint moments, the BINMA(1) given by Equation (13.1) is rewritten in vector form as

$$\boldsymbol{Y_t} = \boldsymbol{A} \circ \boldsymbol{R_{t-1}} + \boldsymbol{R_t} \tag{13.3}$$

where $\boldsymbol{Y_t} = [Y_t^{[1]}, Y_t^{[2]}]'$, $\boldsymbol{R_t} = [R_t^{[1]}, R_t^{[2]}]'$ and $\boldsymbol{A} = \begin{pmatrix} \rho_{11} & \rho_{12} \\ \rho_{21} & \rho_{22} \end{pmatrix}$.

From Lemma 5.1,

$$E(\boldsymbol{Y_t}) = \boldsymbol{A} E(\boldsymbol{R_{t-1}}) + E(\boldsymbol{R_t}) \tag{13.4}$$

and

$$\boldsymbol{\Sigma}_{0,t} = \boldsymbol{A} \text{Var}(\boldsymbol{R_{t-1}}) \boldsymbol{A}' + \text{diag}(\boldsymbol{B} \boldsymbol{\lambda_{t-1}}) + \text{Var}(\boldsymbol{R_t}) \tag{13.5}$$

where

$$\boldsymbol{B} = \begin{pmatrix} \rho_{11}(1-\rho_{11}) & \rho_{12}(1-\rho_{12}) \\ \rho_{21}(1-\rho_{21}) & \rho_{22}(1-\rho_{22}) \end{pmatrix}$$

and $\boldsymbol{\Sigma}_{0,t} = \text{Var}(\boldsymbol{Y_t}) = \begin{bmatrix} \text{Var}(Y_t^{[1]}) & \text{Cov}(Y_t^{[1]}, Y_t^{[2]}) \\ \text{Cov}(Y_t^{[2]}, Y_t^{[1]}) & \text{Var}(Y_t^{[2]}) \end{bmatrix}$.

Based on the above conditions and expanding Equation (13.4), we obtain

$$\begin{aligned} E(Y_t^{[1]}) \equiv \mu_t^{[1]} &= \rho_{11} \lambda_{t-1}^{[1]} + \rho_{12} \lambda_{t-1}^{[2]} + \lambda_t^{[1]} \quad \text{and} \\ E(Y_t^{[2]}) \equiv \mu_t^{[2]} &= \rho_{21} \lambda_{t-1}^{[1]} + \rho_{22} \lambda_{t-1}^{[2]} + \lambda_t^{[2]}. \end{aligned} \tag{13.6}$$

As for the variance-covariance components, by expanding Equation (13.5), we have

$$\begin{aligned} \text{Var}(Y_t^{[1]}) = \sigma_t^{2[1]} &= \rho_{11}^2 \left[ \frac{\lambda_{t-1}^{[1]}}{\nu_1} + \frac{\nu_1 - 1}{2\nu_1^2} \right] + \rho_{11}(1-\rho_{11}) \lambda_{t-1}^{[1]} \\ &+ \rho_{12}^2 \left[ \frac{\lambda_{t-1}^{[2]}}{\nu_2} + \frac{\nu_2 - 1}{2\nu_2^2} \right] + \rho_{12}(1-\rho_{12}) \lambda_{t-1}^{[2]} + \frac{\lambda_t^{[1]}}{\nu_1} + \frac{\nu_1 - 1}{2\nu_1^2}, \end{aligned} \tag{13.7}$$

$$\begin{aligned}\mathrm{Var}(Y_t^{[2]}) = \sigma_t^{2\,[2]} &= \rho_{21}^2\left[\frac{\lambda_{t-1}^{[1]}}{\nu_1} + \frac{\nu_1 - 1}{2\nu_1^2}\right] + \rho_{21}(1-\rho_{21})\lambda_{t-1}^{[1]} \\ &+ \rho_{22}^2\left[\frac{\lambda_{t-1}^{[2]}}{\nu_2} + \frac{\nu_2 - 1}{2\nu_2^2}\right] + \rho_{22}(1-\rho_{22})\lambda_{t-1}^{[2]} + \frac{\lambda_t^{[2]}}{\nu_2} + \frac{\nu_2-1}{2\nu_2^2}\end{aligned} \tag{13.8}$$

and

$$\mathrm{Cov}(Y_t^{[1]}, Y_t^{[2]}) = \rho_{11}\rho_{21}\left[\frac{\lambda_{t-1}^{[1]}}{\nu_1} + \frac{\nu_1 - 1}{2\nu_1^2}\right] + \rho_{12}\rho_{22}\left[\frac{\lambda_{t-1}^{[2]}}{\nu_2} + \frac{\nu_2 - 1}{2\nu_2^2}\right]. \tag{13.9}$$

**Remark 13.1.** Under stationary moment conditions, we obtain

$$\mu_t^{[1]} = (1+\rho_{11})\lambda_1^{[1]} + \rho_{12}\lambda_1^{[2]} \quad , \quad \mu_t^{[2]} = \rho_{21}\lambda_1^{[1]} + (1+\rho_{22})\lambda_1^{[2]}, \tag{13.10}$$

$$\begin{aligned}\mathrm{Var}(Y_t^{[1]}) &= \frac{\lambda_1^{[1]}}{\nu_1}[1 + \rho_{11}\nu_1 + \rho_{11}^2(1-\nu_1)] + \frac{\lambda_1^{[2]}}{\nu_2}[\rho_{12}\nu_2 + \rho_{12}^2(1-\nu_2)] \\ &+ (1+\rho_{11}^2)\left(\frac{\nu_1 - 1}{2\nu_1^2}\right) + \rho_{12}^2\left(\frac{\nu_2-1}{2\nu_2^2}\right),\end{aligned} \tag{13.11}$$

$$\begin{aligned}\mathrm{Var}(Y_t^{[2]}) &= \frac{\lambda_1^{[1]}}{\nu_1}[\rho_{21}\nu_1 + \rho_{21}^2(1-\nu_1)] + \frac{\lambda_1^{[2]}}{\nu_2}[1 + \rho_{22}\nu_2 + \rho_{22}^2(1-\nu_2)] \\ &+ (1+\rho_{22}^2)\left(\frac{\nu_2 - 1}{2\nu_2^2}\right) + \rho_{21}^2\left(\frac{\nu_1-1}{2\nu_1^2}\right)\end{aligned} \tag{13.12}$$

and

$$\mathrm{Cov}(Y_t^{[1]}, Y_t^{[2]}) = \rho_{11}\rho_{21}\left[\frac{\lambda_1^{[1]}}{\nu_1} + \frac{\nu_1 - 1}{2\nu_1^2}\right] + \rho_{12}\rho_{22}\left[\frac{\lambda_1^{[2]}}{\nu_2} + \frac{\nu_2 - 1}{2\nu_2^2}\right]. \tag{13.13}$$

By denoting covariances at lag-$h$ as $\boldsymbol{\Sigma}_{h,t}$, that is,

$$\boldsymbol{\Sigma}_{h,t} = \begin{bmatrix} \mathrm{Cov}(Y_t^{[1]}, Y_{t+h}^{[1]}) & \mathrm{Cov}(Y_{t+h}^{[1]}, Y_t^{[2]}) \\ \mathrm{Cov}(Y_t^{[1]}, Y_{t+h}^{[2]}) & \mathrm{Cov}(Y_t^{[2]}, Y_{t+h}^{[2]}) \end{bmatrix} = \begin{bmatrix} \Sigma_{11,h,t} & \Sigma_{12,h,t} \\ \Sigma_{21,h,t} & \Sigma_{22,h,t} \end{bmatrix}, \tag{13.14}$$

it is easily proved that

$$\boldsymbol{\Sigma}_{h,t} = \begin{cases} \begin{bmatrix} \rho_{11} & \rho_{12} \\ \rho_{21} & \rho_{22} \end{bmatrix}^h \begin{bmatrix} \mathrm{Var}(R_t^{[1]}) & 0 \\ 0 & \mathrm{Var}(R_t^{[2]}) \end{bmatrix}, & \text{for} \quad h = 1 \\ 0 & \quad\;\; h > 1. \end{cases} \tag{13.15}$$

(Refer to Section 5.2).

**Remark 13.2.** Under $\rho_{12} = \rho_{21} = 0$, the moments are

$$E(Y_t^{[1]}) = \rho_{11}\lambda_{t-1}^{[1]} + \lambda_t^{[1]}, \qquad E(Y_t^{[2]}) = \rho_{22}\lambda_{t-1}^{[2]} + \lambda_t^{[2]}, \tag{13.16}$$

$$\begin{aligned}\mathrm{Var}(Y_t^{[1]}) &= \rho_{11}^2\left[\frac{\lambda_{t-1}^{[1]}}{\nu_1} + \frac{\nu_1 - 1}{2\nu_1^2}\right] + \rho_{11}(1-\rho_{11})\lambda_{t-1}^{[1]} \\ &\quad + \frac{\lambda_t^{[1]}}{\nu_1} + \frac{\nu_1 - 1}{2\nu_1^2}\end{aligned} \tag{13.17}$$

and

$$\begin{aligned}\mathrm{Var}(Y_t^{[2]}) &= \rho_{21}^2\left[\frac{\lambda_{t-1}^{[1]}}{\nu_1} + \frac{\nu_1 - 1}{2\nu_1^2}\right] + \rho_{21}(1-\rho_{21})\lambda_{t-1}^{[1]} \\ &\quad + \frac{\lambda_t^{[2]}}{\nu_2} + \frac{\nu_2 - 1}{2\nu_2^2}.\end{aligned} \tag{13.18}$$

The above moments are same as the moments in Equations (11.4) and (11.5) and hence, the constrained BINMA(1) model with COM-Poisson innovations is a special case of this unconstrained BINMA(1) model with COM-Poisson innovations.

**Remark 13.3.** For $\nu_1 = 1$, $\nu_2 = 1$, Equations (13.6)-(13.8) reduce to

$$E(Y_t^{[1]}) = \mathrm{Var}(Y_t^{[1]}) = \rho_{11}\lambda_{t-1}^{[1]} + \rho_{12}\lambda_{t-1}^{[2]} + \lambda_t^{[1]} \tag{13.19}$$

$$E(Y_t^{[2]}) = \mathrm{Var}(Y_t^{[2]}) = \rho_{21}\lambda_{t-1}^{[1]} + \rho_{22}\lambda_{t-1}^{[2]} + \lambda_t^{[2]} \tag{13.20}$$

which are same as Equations (5.5) and (5.6). Hence, the unconstrained BINMA(1) model with Poisson innovations as well is a special case of the unconstrained BINMA(1) model with COM-Poisson innovations.

## 13.3 Trial Experiment

Since the expectations and variances presented in Section 13.2 under stationary and non-stationary conditions are complicated, it is worth investigating under which parameter values, if possible, $Y_t^{[k]}$ is over-, equi- or under-dispersed or more interestingly mixed-dispersed. This can only be achieved via a trial experiment under different set of assumed values for $\{\rho_{11}, \rho_{12}, \rho_{21}, \rho_{22}\}$, $\{\nu_1, \nu_2\}$, $\{\theta_1^{[1]}, \theta_1^{[2]}\}$. We consider 4 cases: $\nu_1 = 1, \nu_2 = 1$, $\nu_1 < 1, \nu_2 < 1$, $\nu_1 > 1, \nu_2 > 1$, $\nu_1 < 1, \nu_2 > 1$ with some values of $0 < \rho_{jk} < 1$ and $\theta_1^{[k]} > 0$, for $k, j \in \{1, 2\}$. The trial experiment is run under simple case of stationarity (Tables 13.1–13.4).

**TABLE 13.1**

Experimental Results under Different Serial and Dispersion Combinations for $[\nu_1 = 1, \nu_2 = 1]$

| $\rho_{11}$ | $\rho_{12}$ | $\rho_{21}$ | $\rho_{22}$ | $\nu_1$ | $\nu_2$ | $\theta_1^{[1]}$ | $\theta_1^{[2]}$ | $\mu_t^{[1]}$ | $\mathrm{Var}(Y_t^{[1]})$ | $\mu_t^{[2]}$ | $\mathrm{Var}(Y_t^{[2]})$ |
|---|---|---|---|---|---|---|---|---|---|---|---|
| 0.1 | 0.15 | 0.2 | 0.25 | 1 | 1 | 0.1 | 0.2 | 0.14 | 0.14 | 0.27 | 0.27 |
| | | | | | | 0.9 | 0.95 | 1.1325 | 1.1325 | 1.3675 | 1.3675 |
| | | | | | | 1.1 | 1.2 | 1.39 | 1.39 | 1.72 | 1.72 |
| | | | | | | 2 | 2.1 | 2.515 | 2.515 | 3.025 | 3.025 |
| 0.5 | 0.55 | 0.6 | 0.65 | 1 | 1 | 0.1 | 0.2 | 0.26 | 0.26 | 0.39 | 0.39 |
| | | | | | | 0.9 | 0.95 | 1.8725 | 1.8725 | 2.1075 | 2.1075 |
| | | | | | | 1.1 | 1.2 | 2.31 | 2.31 | 2.64 | 2.64 |
| | | | | | | 2 | 2.1 | 14.155 | 14.155 | 4.665 | 4.665 |
| 0.9 | 0.1 | 0.11 | 0.8 | 1 | 1 | 0.1 | 0.2 | 0.21 | 0.21 | 0.371 | 0.371 |
| | | | | | | 0.9 | 0.95 | 1.805 | 1.805 | 1.809 | 1.809 |
| | | | | | | 1.1 | 1.2 | 2.21 | 2.21 | 2.281 | 2.281 |
| | | | | | | 2 | 2.1 | 4.01 | 4.01 | 4 | 4 |

**TABLE 13.2**

Experimental Results under Different Serial and Dispersion Combinations for $[\nu_1 < 1, \nu_2 < 1]$

| $\rho_{11}$ | $\rho_{12}$ | $\rho_{21}$ | $\rho_{22}$ | $\nu_1$ | $\nu_2$ | $\theta_1^{[1]}$ | $\theta_1^{[2]}$ | $\mu_t^{[1]}$ | $\mathrm{Var}(Y_t^{[1]})$ | $\mu_t^{[2]}$ | $\mathrm{Var}(Y_t^{[2]})$ |
|---|---|---|---|---|---|---|---|---|---|---|---|
| 0.1 | 0.15 | 0.2 | 0.25 | 0.1 | 0.15 | 0.1 | 0.2 | 5.375 | 0.766 | 4.44 | 1.25 |
| | | | | | | 0.9 | 0.95 | 5.865 | 4.516 | 5.399 | 6.612 |
| | | | | | | 1.1 | 1.2 | 8.734 | 28.132 | 9.175 | 27.22 |
| | | | | | | 2 | 2.1 | 1153 | 10474 | 385 | 1597 |
| 0.5 | 0.55 | 0.6 | 0.65 | 0.1 | 0.15 | 0.1 | 0.2 | 8.308 | 1.826 | 7.375 | 1.725 |
| | | | | | | 0.9 | 0.95 | 9.222 | 7.88 | 8.756 | 9.962 |
| | | | | | | 1.1 | 1.2 | 14.05 | 42.53 | 14.49 | 44.43 |
| | | | | | | 2 | 2.1 | 1622 | 13376 | 854 | 6000 |
| 0.9 | 0.1 | 0.11 | 0.8 | 0.1 | 0.15 | 0.1 | 0.2 | 8.833 | 0.66 | 5.595 | 0.894 |
| | | | | | | 0.9 | 0.95 | 9.567 | 7.114 | 6.912 | 8.851 |
| | | | | | | 1.1 | 1.2 | 14.1 | 48.37 | 11.95 | 38.87 |
| | | | | | | 2 | 2.1 | 1968 | 18649 | 371 | 1785 |
| 0.1 | 0.15 | 0.2 | 0.25 | 0.8 | 0.9 | 0.1 | 0.2 | 0.2328 | 0.1199 | 0.3147 | 0.2710 |
| | | | | | | 0.9 | 0.95 | 1.2518 | 1.348 | 1.451 | 1.506 |
| | | | | | | 1.1 | 1.2 | 1.5687 | 1.729 | 1.8504 | 1.942 |
| | | | | | | 2 | 2.1 | 3.104 | 3.583 | 3.421 | 3.65 |
| 0.5 | 0.55 | 0.6 | 0.65 | 0.8 | 0.9 | 0.1 | 0.2 | 0.3944 | 0.2445 | 0.4764 | 0.3838 |
| | | | | | | 0.9 | 0.95 | 2.053 | 2.185 | 2.251 | 2.355 |
| | | | | | | 1.1 | 1.2 | 2.581 | 2.801 | 2.863 | 3.034 |
| | | | | | | 2 | 2.1 | 5.04 | 5.687 | 5.356 | 5.807 |
| 0.9 | 0.1 | 0.11 | 0.8 | 0.8 | 0.9 | 0.1 | 0.2 | 0.3666 | 0.1655 | 0.421 | 0.359 |
| | | | | | | 0.9 | 0.95 | 2.003 | 2.174 | 1.910 | 1.993 |
| | | | | | | 1.1 | 1.2 | 2.506 | 2.790 | 2.442 | 2.576 |
| | | | | | | 2 | 2.1 | 4.99 | 5.842 | 4.48 | 4.81 |

**TABLE 13.3**
Experimental Results under Different Serial and Dispersion Combinations for $[\nu_1 > 1, \nu_2 > 1]$

| $\rho_{11}$ | $\rho_{12}$ | $\rho_{21}$ | $\rho_{22}$ | $\nu_1$ | $\nu_2$ | $\theta_1^{[1]}$ | $\theta_1^{[2]}$ | $\mu_t^{[1]}$ | $\text{Var}(Y_t^{[1]})$ | $\mu_t^{[2]}$ | $\text{Var}(Y_t^{[2]})$ |
|---|---|---|---|---|---|---|---|---|---|---|---|
| 0.1 | 0.15 | 0.2 | 0.25 | 1.1 | 1.2 | 0.1 | 0.2 | 0.1123 | 0.1478 | 0.2383 | 0.2819 |
| | | | | | | 0.9 | 0.95 | 1.0808 | 1.0415 | 1.2662 | 1.1836 |
| | | | | | | 1.1 | 1.2 | 1.3117 | 1.255 | 1.56 | 1.44 |
| | | | | | | 2 | 2.1 | 2.2815 | 2.15 | 2.582 | 2.3369 |
| 0.5 | 0.55 | 0.6 | 0.65 | 1.1 | 1.2 | 0.1 | 0.2 | 0.2148 | 0.2696 | 0.3407 | 0.4096 |
| | | | | | | 0.9 | 0.95 | 1.776 | 1.706 | 1.961 | 1.8394 |
| | | | | | | 1.1 | 1.2 | 2.162 | 2.061 | 2.41 | 2.234 |
| | | | | | | 2 | 2.1 | 3.723 | 3.499 | 4.024 | 3.657 |
| 0.9 | 0.1 | 0.11 | 0.8 | 1.1 | 1.2 | 0.1 | 0.2 | 0.1657 | 0.2281 | 0.3293 | 0.3949 |
| | | | | | | 0.9 | 0.95 | 1.728 | 1.66 | 1.67 | 1.544 |
| | | | | | | 1.1 | 1.2 | 2.0937 | 1.995 | 2.060 | 1.878 |
| | | | | | | 2 | 2.1 | 3.659 | 3.430 | 3.392 | 3.020 |
| 0.1 | 0.15 | 0.2 | 0.25 | 2 | 2.5 | 0.1 | 0.2 | 0.1066 | 0.1991 | 0.295 | 0.282 |
| | | | | | | 0.9 | 0.95 | 0.8705 | 0.6374 | 0.989 | 0.6746 |
| | | | | | | 1.1 | 1.2 | 0.995 | 0.71 | 1.129 | 0.7514 |
| | | | | | | 2 | 2.1 | 1.4375 | 0.964 | 1.5397 | 0.9824 |
| 0.5 | 0.55 | 0.6 | 0.65 | 2 | 2.5 | 0.1 | 0.2 | 1.1331 | 0.2477 | 0.4115 | 0.4230 |
| | | | | | | 0.9 | 0.95 | 1.15 | 0.863 | 1.541 | 1.051 |
| | | | | | | 1.1 | 1.2 | 1.3146 | 0.9638 | 1.1759 | 1.169 |
| | | | | | | 2 | 2.1 | 1.903 | 1.32 | 2.423 | 1.537 |
| 0.9 | 0.1 | 0.11 | 0.8 | 2 | 2.5 | 0.1 | 0.2 | 0.1484 | 0.3145 | 0.4128 | 0.389 |
| | | | | | | 0.9 | 0.95 | 1.395 | 0.986 | 1.3003 | 0.8256 |
| | | | | | | 1.1 | 1.2 | 1.595 | 1.095 | 1.484 | 0.9143 |
| | | | | | | 2 | 2.1 | 2.317 | 1.484 | 2.010 | 1.172 |

**TABLE 13.4**
Experimental Results under Different Serial and Dispersion Combinations for $[\nu_1 < 1, \nu_2 > 1]$

| $\rho_{11}$ | $\rho_{12}$ | $\rho_{21}$ | $\rho_{22}$ | $\nu_1$ | $\nu_2$ | $\theta_1^{[1]}$ | $\theta_1^{[2]}$ | $\mu_t^{[1]}$ | $\text{Var}(Y_t^{[1]})$ | $\mu_t^{[2]}$ | $\text{Var}(Y_t^{[2]})$ |
|---|---|---|---|---|---|---|---|---|---|---|---|
| 0.1 | 0.15 | 0.2 | 0.25 | 0.8 | 1.5 | 0.1 | 0.2 | 0.2257 | 0.1148 | 0.2554 | 0.3069 |
| | | | | | | 0.9 | 0.95 | 1.2217 | 1.3133 | 1.20 | 1.039 |
| | | | | | | 1.1 | 1.2 | 1.521 | 1.675 | 1.454 | 1.237 |
| | | | | | | 2 | 2.1 | 2.975 | 3.44 | 2.342 | 1.957 |
| 0.5 | 0.55 | 0.6 | 0.65 | 0.8 | 1.5 | 0.1 | 0.2 | 0.314 | 0.201 | 0.235 | 0.317 |
| | | | | | | 0.9 | 0.95 | 1.942 | 2.013 | 1.92 | 1.733 |
| | | | | | | 1.1 | 1.2 | 2.407 | 2.539 | 2.339 | 2.097 |
| | | | | | | 2 | 2.1 | 4.565 | 5.037 | 3.933 | 3.561 |
| 0.9 | 0.1 | 0.11 | 0.8 | 0.8 | 1.5 | 0.1 | 0.2 | 0.3619 | 0.1616 | 0.3355 | 0.4206 |
| | | | | | | 0.9 | 0.95 | 1.983 | 2.152 | 1.55 | 1.296 |
| | | | | | | 1.1 | 1.2 | 2.474 | 2.756 | 1.87 | 1.528 |
| | | | | | | 2 | 2.1 | 4.903 | 5.75 | 2.927 | 2.31 |

From Tables 13.1–13.4, the following highlights are made:

(a) As shown in Equations (13.19) and (13.20) and from Table 13.1, both counting series are equi-dispersed but at different levels.

(b) From Table 13.2, we conclude that for large $\theta_1^{[k]}$ values, $\theta_1^{[k]} > 1$ and irrespective of $\{\rho_{jk}\}$, both counting series exhibit over-dispersion but at different levels, for $j, k \in \{1, 2\}$.

(c) From Table 13.3, we remark that for $\theta_1^{[k]} > 0.9$ and different combinations of $\{\rho_{jk}\}$, both counting series exhibit different levels of under-dispersion, for $j, k \in \{1, 2\}$.

(d) From Table 13.4, we note that for $\theta_1^{[k]} > 0.9$ and any combinations of $\{\rho_{jk}\}$, the first series is over-dispersed while the second series is under-dispersed for $j, k \in \{1, 2\}$.

## 13.4 Estimation of Parameters

The GQL equation is specified as

$$\boldsymbol{D_\psi}'\boldsymbol{\Sigma_\psi}^{-1}(\boldsymbol{f} - \boldsymbol{\mu}) = 0, \tag{13.21}$$

where in this case $\hat{\boldsymbol{\psi}} = [\hat{\beta}_1^{[1]}, \hat{\beta}_2^{[1]}, \hat{\beta}_1^{[2]}, \hat{\beta}_2^{[2]}, \hat{\rho}_{11}, \hat{\rho}_{12}, \hat{\rho}_{21}, \hat{\rho}_{22}, \hat{\nu}_1, \hat{\nu}_2]$, the score vector $\boldsymbol{f} = [\boldsymbol{f}_1, \boldsymbol{f}_2, \ldots, \boldsymbol{f}_t, \ldots, \boldsymbol{f}_{t+h}, \ldots, \boldsymbol{f}_T]$, where $\boldsymbol{f}_t = [Y_t^{[1]}, Y_t^{[2]}]'$ and $\boldsymbol{\mu} = [\boldsymbol{\mu}_1, \boldsymbol{\mu}_2, \ldots, \boldsymbol{\mu}_t, \ldots, \boldsymbol{\mu}_T]$, where $\boldsymbol{\mu}_t = [\mu_t^{[1]}, \mu_t^{[2]}]'$ for $t = 1, 2, \ldots, T$.

From Equation (13.15), the covariance matrix $\boldsymbol{\Sigma_\psi}$ is given by

$$\boldsymbol{\Sigma_\psi} = \begin{pmatrix} \boldsymbol{\Sigma}_{0,1} & \boldsymbol{\Sigma}'_{1,1} & 0 & 0 & \ldots & 0 & 0 \\ \boldsymbol{\Sigma}_{1,1} & \boldsymbol{\Sigma}_{0,2} & \boldsymbol{\Sigma}'_{1,2} & 0 & \ldots & 0 & 0 \\ 0 & \boldsymbol{\Sigma}_{1,2} & \boldsymbol{\Sigma}_{0,3} & \boldsymbol{\Sigma}'_{1,3} & \ldots & 0 & 0 \\ \vdots & \vdots & \vdots & \vdots & \ddots & \vdots & \vdots \\ 0 & 0 & 0 & 0 & \ldots & \boldsymbol{\Sigma}_{0,T-1} & \boldsymbol{\Sigma}'_{1,T-1} \\ 0 & 0 & 0 & 0 & \ldots & \boldsymbol{\Sigma}_{1,T-1} & \boldsymbol{\Sigma}_{0,T} \end{pmatrix}_{2T \times 2T} \tag{13.22}$$

where the entries are computed using Equations (13.7)–(13.9).

As for the derivative matrix, $\boldsymbol{D}_\psi = [\boldsymbol{D}_1, \boldsymbol{D}_2, \ldots, \boldsymbol{D}_t, \ldots, \boldsymbol{D}_T]'$ with

$$\boldsymbol{D}_t = \begin{pmatrix} \frac{\partial\mu_t^{[1]}}{\partial\boldsymbol{\beta}^{[1]}} & \frac{\partial\mu_t^{[2]}}{\partial\boldsymbol{\beta}^{[1]}} \\ \frac{\partial\mu_t^{[1]}}{\partial\boldsymbol{\beta}^{[2]}} & \frac{\partial\mu_t^{[2]}}{\partial\boldsymbol{\beta}^{[2]}} \\ \frac{\partial\mu_t^{[1]}}{\partial\rho_{11}} & 0 \\ \frac{\partial\mu_t^{[1]}}{\partial\rho_{12}} & 0 \\ 0 & \frac{\partial\mu_t^{[2]}}{\partial\rho_{21}} \\ 0 & \frac{\partial\mu_t^{[2]}}{\partial\rho_{22}} \\ \frac{\partial\mu_t^{[1]}}{\partial\nu_1} & \frac{\partial\mu_t^{[2]}}{\partial\nu_1} \\ \frac{\partial\mu_t^{[1]}}{\partial\nu_2} & \frac{\partial\mu_t^{[2]}}{\partial\nu_2} \end{pmatrix}_{(2p+6)\times 2} \tag{13.23}$$

where $\mu_t^{[k]} = \rho_{kk}\left[(\theta_{t-1}^{[k]})^{1/\nu_k} - (\frac{\nu_k-1}{2\nu_k})\right] + \rho_{kk'}\left[(\theta_{t-1}^{[k']})^{1/\nu_{k'}} - (\frac{\nu_{k'}-1}{2\nu_{k'}})\right] + (\theta_t^{[k]})^{1/\nu_k} - (\frac{\nu_k-1}{2\nu_k})$ and

$$\frac{\partial\mu_t^{[k]}}{\partial\boldsymbol{\beta}^{[l]}} = \left[\frac{\partial\mu_t^{[k]}}{\partial\beta_1^{[l]}}, \frac{\partial\mu_t^{[k]}}{\partial\beta_2^{[l]}}, \ldots, \frac{\partial\mu_t^{[k]}}{\partial\beta_j^{[l]}}, \ldots, \frac{\partial\mu_t^{[k]}}{\partial\beta_p^{[l]}}\right]', \quad k, k', l \in \{1, 2\}, k \neq k'.$$

The entries of the derivative matrix are:

1. $\frac{\partial\mu_t^{[k]}}{\partial\boldsymbol{\beta}_j^{[k]}} = \rho_{kk}\frac{(\theta_{t-1}^{[k]})^{\frac{1}{\nu_k}}x_{t-1,j}}{\nu_k} + \frac{(\theta_t^{[k]})^{\frac{1}{\nu_k}}x_{tj}}{\nu_k}$

2. $\frac{\partial\mu_t^{[k]}}{\partial\boldsymbol{\beta}_j^{[k']}} = \rho_{kk'}\frac{(\theta_{t-1}^{[k']})^{\frac{1}{\nu_{k'}}}x_{t-1,j}}{\nu_{k'}}$

3. $\frac{\partial\mu_t^{[1]}}{\partial\rho_{11}} = \frac{\partial\mu_t^{[2]}}{\partial\rho_{21}} = \lambda_{t-1}^{[1]}$

4. $\frac{\partial\mu_t^{[1]}}{\partial\rho_{12}} = \frac{\partial\mu_t^{[2]}}{\partial\rho_{22}} = \lambda_{t-1}^{[2]}$

5. $\frac{\partial\mu_t^{[k]}}{\partial\nu_k} = \rho_{kk}\left[-\frac{(\theta_{t-1}^{[k]})^{\frac{1}{\nu_k}}ln(\theta_{t-1}^{[k]})}{\nu_k^2} - \frac{1}{2\nu_k} + \frac{\nu_k-1}{2\nu_k^2}\right] - \frac{(\theta_t^{[k]})^{\frac{1}{\nu_k}}ln(\theta_t^{[k]})}{\nu_k^2} - \frac{1}{2\nu_k} + \frac{\nu_k-1}{2\nu_k^2}$

6. $\frac{\partial\mu_t^{[k]}}{\partial\nu_{k'}} = \rho_{kk'}\left[-\frac{(\theta_{t-1}^{[k']})^{\frac{1}{\nu_{k'}}}ln(\theta_{t-1}^{[k']})}{\nu_{k'}^2} - \frac{1}{2\nu_{k'}} + \frac{\nu_{k'}-1}{2\nu_{k'}^2}\right]$

for $k, k' \in \{1, 2\}, k \neq k'$.

Equation (13.21) is solved using the Newton-Raphson iteration procedure yielding an iterative formula of the form

$$(\hat{\boldsymbol{\psi}}_{r+1}) = (\hat{\boldsymbol{\psi}}_r) + [\boldsymbol{D_\psi}'\boldsymbol{\Sigma_\psi}^{-1}\boldsymbol{D_\psi}]_r^{-1}[\boldsymbol{D_\psi}'\boldsymbol{\Sigma_\psi}^{-1}(\boldsymbol{f}-\boldsymbol{\mu})]_r, \qquad (13.24)$$

where $\hat{\boldsymbol{\psi}}_r$ are the estimates at the $r^{th}$ iteration and $[.]_r$ are the values of the expression at the $r^{th}$ iteration.

The algorithm works as follows: For an initial $\hat{\boldsymbol{\psi}}$, we compute the score and the Hessian structures and solve the iterative Equation (13.24) to obtain an updated $\hat{\boldsymbol{\psi}}$. These updated values are again used to re-calculate the score and the Hessian structure which are in turn used in the iterative Equation (13.24) to get another set of values. This cycle of iteration continues until convergence. Under mild regulatory condition, it is shown that $(\hat{\boldsymbol{\psi}}-\boldsymbol{\psi})'$ is asymptotic normal with mean $\boldsymbol{0}$ and covariance matrix $[\boldsymbol{D}_\psi'\boldsymbol{\Sigma_\psi}^{-1}\boldsymbol{D}_\psi]^{-1}$ (Refer to Subsection 2.3.1.2).

Using the converged estimates from the GQL Equation (13.24), we derive the one-step ahead forecasting equations of the unconstrained BINMA(1) model with COM-Poisson innovations. The conditional expectation and variance of the one-step ahead forecast $Y_{t+1}^{[k]}$ given $r_t^{[k]}$ are as follows:

$$\begin{aligned} E(Y_{t+1}^{[1]}|r_t^{[1]},r_t^{[2]}) &= E(\rho_{11}\circ r_t^{[1]}|r_t^{[1]}) + E(\rho_{12}\circ r_t^{[2]}|r_t^{[2]}) + E(R_{t+1}^{[1]}) \\ &= \rho_{11}r_t^{[1]} + \rho_{12}r_t^{[2]} + \lambda_{t+1}^{[1]}, \end{aligned} \qquad (13.25)$$

$$\begin{aligned} E(Y_{t+1}^{[2]}|r_t^{[1]},r_t^{[2]}) &= E(\rho_{21}\circ r_t^{[1]}|r_t^{[1]}) + E(\rho_{22}\circ r_t^{[2]}|r_t^{[2]}) + E(R_{t+1}^{[2]}) \\ &= \rho_{21}r_t^{[1]} + \rho_{22}r_t^{[2]} + \lambda_{t+1}^{[2]}, \end{aligned} \qquad (13.26)$$

with corresponding variances

$$\begin{aligned} \mathrm{Var}(Y_{t+1}^{[1]}|r_t^{[1]},r_t^{[2]}) &= \mathrm{Var}(\rho_{11}\circ r_t^{[1]}|r_t^{[1]}) + \mathrm{Var}(\rho_{12}\circ r_t^{[2]}|r_t^{[2]}) + \mathrm{Var}(R_{t+1}^{[1]}) \\ &= \rho_{11}(1-\rho_{11})r_t^{[1]} + \rho_{12}(1-\rho_{12})r_t^{[2]} + \frac{\lambda_{t+1}^{[1]}}{\nu_1} + \frac{\nu_1-1}{2\nu_1^2} \end{aligned} \qquad (13.27)$$

and

$$\begin{aligned} \mathrm{Var}(Y_{t+1}^{[2]}|r_t^{[1]},r_t^{[2]}) &= \mathrm{Var}(\rho_{21}\circ r_t^{[1]}|r_t^{[1]}) + \mathrm{Var}(\rho_{22}\circ r_t^{[2]}|r_t^{[2]}) + \mathrm{Var}(R_{t+1}^{[2]}) \\ &= \rho_{21}(1-\rho_{21})r_t^{[1]} + \rho_{22}(1-\rho_{22})r_t^{[2]} + \frac{\lambda_{t+1}^{[2]}}{\nu_2} + \frac{\nu_2-1}{2\nu_2^2} \end{aligned} \qquad (13.28)$$

where $r_t^{[k]}$ are approximated by their respective means.

## 13.5 Monte Carlo Experiment

In this section, we establish a simulation experiment to generate non-stationary BINMA(1) series of counts using Equation (13.1) based on $\beta_1^{[1]} = 0.3, \beta_2^{[1]} = 0.3, \beta_1^{[2]} = 0.5, \beta_2^{[2]} = 0.5$ with different combinations of correlation coefficients displayed as $[\rho_{11}, \rho_{22}, \rho_{12}, \rho_{21}] = [0.9, 0.9, 0.3, 0.3]$ and $[0.3, 0.3, 0.9, 0.9]$. The estimation algorithm is applied to the simulated BINMA(1) series for different time sizes T=60,100,600. The link predictor $\theta_t^{[k]}$ is computed using

$$\theta_t^{[k]} = \exp(\beta_1^{[k]'} x_{t1}' + \beta_2^{[k]'} x_{t2}') \tag{13.29}$$

where the covariate structure is of the form

$$x_{t1} = \begin{cases} -t/T & (t = 1, \ldots, T/4), \\ \text{rnorm}(1, 0, 1) & (t = (T/4) + 1, \ldots, 3T/4), \\ t/T & (t = (3T/4) + 1, \ldots, T), \end{cases}$$

and $x_{t2} = runif(1, 0, 1)$ is a randomly generated number following the uniform (0,1) distribution in R for each $t = 1, \ldots, T$. As for the innovation terms, $R_t^{[k]} \sim$ COM-Poisson$(\lambda_t^{[k]}, \nu_k)$, with the pair of $[\nu_1, \nu_2]$ assuming values $[\nu_1, \nu_2] = [1, 1], [0.5, 0.5], [2.0, 2.0], [0.5, 2.0]$ and $[2.0, 0.5]$. Five thousand Monte Carlo replications for each combination are made and the simulated mean estimates for the different parameters are displayed in Tables 13.5–13.6.

**TABLE 13.5**
Estimates of the Parameters and Standard Errors under Non-Stationary BINMA(1) Process for $[\nu_1, \nu_2] = [1, 1], [0.5, 0.5], [2.0, 2.0]$

| T | $\hat{\beta}_1^{[1]}$ | $\hat{\beta}_2^{[1]}$ | $\hat{\beta}_1^{[2]}$ | $\hat{\beta}_2^{[2]}$ | $\hat{\nu}_1 = 1$ | $\hat{\nu}_2 = 1$ | $\hat{\rho}_{11} = 0.9$ | $\hat{\rho}_{22} = 0.9$ | $\hat{\rho}_{12} = 0.3$ | $\hat{\rho}_{21} = 0.3$ |
|---|---|---|---|---|---|---|---|---|---|---|
| 60 | 0.3139 | 0.2806 | 0.4810 | 0.4841 | 0.9851 | 0.9866 | 0.9180 | 0.8821 | 0.3196 | 0.2888 |
| | (0.0918) | (0.0880) | (0.1083) | (0.0811) | (0.1446) | (0.1377) | (0.0716) | (0.0814) | (0.0765) | (0.0718) |
| 100 | 0.3065 | 0.3015 | 0.5026 | 0.5013 | 0.9944 | 0.9953 | 0.8981 | 0.8927 | 0.2917 | 0.3062 |
| | (0.0892) | (0.0729) | (0.0662) | (0.0690) | (0.1333) | (0.1194) | (0.0532) | (0.0616) | (0.0558) | (0.0523) |
| 600 | 0.2980 | 0.2990 | 0.4985 | 0.5008 | 0.9963 | 0.9978 | 0.8993 | 0.8974 | 0.2924 | 0.3015 |
| | (0.0218) | (0.0379) | (0.0140) | (0.0387) | (0.0740) | (0.0693) | (0.0096) | (0.0061) | (0.0120) | (0.0183) |
| T | $\hat{\beta}_1^{[1]}$ | $\hat{\beta}_2^{[1]}$ | $\hat{\beta}_1^{[2]}$ | $\hat{\beta}_2^{[2]}$ | $\hat{\nu}_1 = 1$ | $\hat{\nu}_2 = 1$ | $\hat{\rho}_{11} = 0.3$ | $\hat{\rho}_{22} = 0.3$ | $\hat{\rho}_{12} = 0.9$ | $\hat{\rho}_{21} = 0.9$ |
| 60 | 0.3107 | 0.2838 | 0.4824 | 0.4843 | 0.9818 | 0.9876 | 0.2848 | 0.2812 | 0.8882 | 0.8892 |
| | (0.1156) | (0.0887) | (0.0919) | (0.0816) | (0.1241) | (0.1376) | (0.0748) | (0.0885) | (0.0710) | (0.0713) |
| 100 | 0.3004 | 0.3031 | 0.5039 | 0.5091 | 0.9949 | 0.9983 | 0.2928 | 0.2932 | 0.9070 | 0.9068 |
| | (0.0839) | (0.0663) | (0.0747) | (0.0615) | (0.1086) | (0.1198) | (0.0648) | (0.0564) | (0.0565) | (0.0550) |
| 600 | 0.2998 | 0.2981 | 0.4967 | 0.5054 | 0.9967 | 0.9996 | 0.2955 | 0.2947 | 0.9009 | 0.9011 |
| | (0.0304) | (0.0279) | (0.0310) | (0.0192) | (0.0637) | (0.0507) | (0.0067) | (0.0015) | (0.0154) | (0.0191) |
| T | $\hat{\beta}_1^{[1]}$ | $\hat{\beta}_2^{[1]}$ | $\hat{\beta}_1^{[2]}$ | $\hat{\beta}_2^{[2]}$ | $\hat{\nu}_1 = 0.5$ | $\hat{\nu}_2 = 0.5$ | $\hat{\rho}_{11} = 0.9$ | $\hat{\rho}_{22} = 0.9$ | $\hat{\rho}_{12} = 0.3$ | $\hat{\rho}_{21} = 0.3$ |
| 60 | 0.3142 | 0.2881 | 0.4887 | 0.4823 | 0.4871 | 0.4825 | 0.8896 | 0.9120 | 0.2896 | 0.2898 |
| | (0.1268) | (0.0832) | (0.0999) | (0.0816) | (0.1410) | (0.1390) | (0.0896) | (0.0851) | (0.0719) | (0.0842) |
| 100 | 0.3074 | 0.3065 | 0.5014 | 0.5070 | 0.4959 | 0.4929 | 0.8983 | 0.8962 | 0.3064 | 0.2907 |
| | (0.0897) | (0.0630) | (0.0769) | (0.0682) | (0.1243) | (0.1222) | (0.0551) | (0.0679) | (0.0512) | (0.0579) |
| 600 | 0.2981 | 0.2988 | 0.4990 | 0.5019 | 0.4973 | 0.4956 | 0.8991 | 0.8977 | 0.2960 | 0.3007 |
| | (0.0237) | (0.0378) | (0.0124) | (0.0345) | (0.0780) | (0.0690) | (0.0084) | (0.0058) | (0.0147) | (0.0151) |

*(Continued)*

**TABLE 13.5**
*(Continued)*

| T | $\hat{\beta}_1^{[1]}$ | $\hat{\beta}_2^{[1]}$ | $\hat{\beta}_1^{[2]}$ | $\hat{\beta}_2^{[2]}$ | $\hat{\nu}_1=0.5$ | $\hat{\nu}_2=0.5$ | $\hat{\rho}_{11}=0.3$ | $\hat{\rho}_{22}=0.3$ | $\hat{\rho}_{12}=0.9$ | $\hat{\rho}_{21}0.9$ |
|---|---|---|---|---|---|---|---|---|---|---|
| 60 | 0.3149 | 0.2825 | 0.4865 | 0.4887 | 0.4818 | 0.4892 | 0.2812 | 0.2882 | 0.8875 | 0.8896 |
| | (0.1110) | (0.0821) | (0.0954) | (0.0881) | (0.1331) | (0.1288) | (0.0871) | (0.0758) | (0.0775) | (0.0752) |
| 100 | 0.3032 | 0.3069 | 0.5083 | 0.5030 | 0.4926 | 0.4983 | 0.2993 | 0.2945 | 0.8962 | 0.8958 |
| | (0.0725) | (0.0675) | (0.0761) | (0.0657) | (0.1060) | (0.1041) | (0.0653) | (0.0695) | (0.0569) | (0.0520) |
| 600 | 0.2978 | 0.2959 | 0.4953 | 0.5015 | 0.4945 | 0.4990 | 0.2996 | 0.2966 | 0.8991 | 0.8980 |
| | (0.0352) | (0.0245) | (0.0376) | (0.0119) | (0.0691) | (0.0545) | (0.0028) | (0.0066) | (0.0109) | (0.0155) |
| T | $\hat{\beta}_1^{[1]}$ | $\hat{\beta}_2^{[1]}$ | $\hat{\beta}_1^{[2]}$ | $\hat{\beta}_2^{[2]}$ | $\hat{\nu}_1=2.0$ | $\hat{\nu}_2=2.0$ | $\hat{\rho}_{11}=0.9$ | $\hat{\rho}_{22}=0.9$ | $\hat{\rho}_{12}=0.3$ | $\hat{\rho}_{21}=0.3$ |
| 60 | 0.3132 | 0.2886 | 0.4894 | 0.4883 | 1.9845 | 1.9892 | 0.8838 | 0.8852 | 0.2892 | 0.2889 |
| | (0.1017) | (0.0865) | (0.0910) | (0.0869) | (0.1415) | (0.1312) | (0.0823) | (0.0781) | (0.0711) | (0.0778) |
| 100 | 0.3075 | 0.3087 | 0.5040 | 0.5013 | 1.9928 | 1.9940 | 0.8931 | 0.8965 | 0.3063 | 0.3070 |
| | (0.0758) | (0.0640) | (0.0771) | (0.0606) | (0.1295) | (0.1249) | (0.0607) | (0.0680) | (0.0538) | (0.0517) |
| 600 | 0.2978 | 0.2960 | 0.4986 | 0.5008 | 1.9989 | 1.9978 | 0.9017 | 0.8975 | 0.3014 | 0.3011 |
| | (0.0338) | (0.0210) | (0.0326) | (0.0118) | (0.0668) | (0.0632) | (0.0076) | (0.0083) | (0.0135) | (0.0171) |
| T | $\hat{\beta}_1^{[1]}$ | $\hat{\beta}_2^{[1]}$ | $\hat{\beta}_1^{[2]}$ | $\hat{\beta}_2^{[2]}$ | $\hat{\nu}_1=2.0$ | $\hat{\nu}_2=2.0$ | $\hat{\rho}_{11}=0.3$ | $\hat{\rho}_{22}=0.3$ | $\hat{\rho}_{12}=0.9$ | $\hat{\rho}_{21}=0.9$ |
| 60 | 0.3121 | 0.2818 | 0.4890 | 0.4835 | 1.9875 | 1.9820 | 0.2895 | 0.2831 | 0.8897 | 0.8888 |
| | (0.1122) | (0.0894) | (0.0970) | (0.0899) | (0.1262) | (0.1216) | (0.0712) | (0.0645) | (0.0764) | (0.0767) |
| 100 | 0.3011 | 0.3021 | 0.5023 | 0.5006 | 1.9946 | 1.9970 | 0.2950 | 0.2958 | 0.9065 | 0.9074 |
| | (0.0835) | (0.0662) | (0.0773) | (0.0652) | (0.0942) | (0.0935) | (0.0669) | (0.0549) | (0.0537) | (0.0568) |
| 600 | 0.2995 | 0.2987 | 0.4989 | 0.5003 | 1.9977 | 1.9984 | 0.2963 | 0.2977 | 0.9016 | 0.9010 |
| | (0.0315) | (0.0292) | (0.0305) | (0.0138) | (0.0549) | (0.0511) | (0.0098) | (0.0005) | (0.0106) | (0.0130) |

**TABLE 13.6**
Estimates of the Parameters and Standard Errors under Non-Stationary BINMA(1) Process $[\nu_1, \nu_2] = [2.0, 0.5], [0.5, 2.0]$

| T | $\hat{\beta}_1^{[1]}$ | $\hat{\beta}_2^{[1]}$ | $\hat{\beta}_1^{[2]}$ | $\hat{\beta}_2^{[2]}$ | $\hat{\nu}_1=2.0$ | $\hat{\nu}_2=0.5$ | $\hat{\rho}_{11}=0.9$ | $\hat{\rho}_{22}=0.9$ | $\hat{\rho}_{12}=0.3$ | $\hat{\rho}_{21}=0.3$ |
|---|---|---|---|---|---|---|---|---|---|---|
| 60 | 0.3187 | 0.2836 | 0.4820 | 0.4831 | 1.9807 | 0.4814 | 0.8880 | 0.8886 | 0.3190 | 0.3186 |
| | (0.1053) | (0.0890) | (0.0938) | (0.0834) | (0.1380) | (0.1214) | (0.0722) | (0.0865) | (0.0718) | (0.0742) |
| 100 | 0.3010 | 0.3043 | 0.5020 | 0.5050 | 1.9932 | 0.4915 | 0.8951 | 0.8990 | 0.3067 | 0.3063 |
| | (0.0762) | (0.0638) | (0.0783) | (0.0627) | (0.1193) | (0.1027) | (0.0613) | (0.0671) | (0.0591) | (0.0560) |
| 600 | 0.2997 | 0.2979 | 0.4986 | 0.5022 | 1.9954 | 0.4930 | 0.8970 | 0.8995 | 0.3004 | 0.3017 |
| | (0.0327) | (0.0293) | (0.0371) | (0.0137) | (0.0615) | (0.0661) | (0.0065) | (0.0077) | (0.0142) | (0.0054) |
| T | $\hat{\beta}_1^{[1]}$ | $\hat{\beta}_2^{[1]}$ | $\hat{\beta}_1^{[2]}$ | $\hat{\beta}_2^{[2]}$ | $\hat{\nu}_1=2.0$ | $\hat{\nu}_2=0.5$ | $\hat{\rho}_{11}=0.3$ | $\hat{\rho}_{22}=0.3$ | $\hat{\rho}_{12}=0.9$ | $\hat{\rho}_{21}=0.9$ |
| 60 | 0.3195 | 0.2894 | 0.4850 | 0.4816 | 1.9852 | 0.5197 | 0.2882 | 0.2821 | 0.9181 | 0.8885 |
| | (0.1108) | (0.0805) | (0.0925) | (0.0822) | (0.1135) | (0.1269) | (0.0852) | (0.0747) | (0.0725) | (0.0701) |
| 100 | 0.3091 | 0.3018 | 0.5067 | 0.5092 | 1.9944 | 0.4972 | 0.2994 | 0.2902 | 0.9063 | 0.9060 |
| | (0.0722) | (0.0614) | (0.0758) | (0.0633) | (0.0960) | (0.1013) | (0.0669) | (0.0618) | (0.0592) | (0.0573) |
| 600 | 0.2933 | 0.2990 | 0.4966 | 0.5013 | 1.9959 | 0.4985 | 0.2996 | 0.2977 | 0.9008 | 0.9012 |
| | (0.0364) | (0.0271) | (0.0362) | (0.0135) | (0.0532) | (0.0586) | (0.0086) | (0.0050) | (0.0194) | (0.0086) |
| T | $\hat{\beta}_1^{[1]}$ | $\hat{\beta}_2^{[1]}$ | $\hat{\beta}_1^{[2]}$ | $\hat{\beta}_2^{[2]}$ | $\hat{\nu}_1=0.5$ | $\hat{\nu}_2=2.0$ | $\hat{\rho}_{11}=0.9$ | $\hat{\rho}_{22}=0.9$ | $\hat{\rho}_{12}=0.3$ | $\hat{\rho}_{21}=0.3$ |
| 60 | 0.3136 | 0.2829 | 0.4847 | 0.4839 | 0.4831 | 2.0127 | 0.8832 | 0.8892 | 0.2889 | 0.2895 |
| | (0.1293) | (0.0841) | (0.0955) | (0.0862) | (0.1407) | (0.1303) | (0.0712) | (0.0783) | (0.0799) | (0.0716) |
| 100 | 0.3094 | 0.3041 | 0.5093 | 0.5016 | 0.4931 | 1.9993 | 0.8975 | 0.8934 | 0.3066 | 0.3059 |
| | (0.0790) | (0.0614) | (0.0791) | (0.0609) | (0.1078) | (0.0921) | (0.0585) | (0.0637) | (0.0584) | (0.0553) |
| 600 | 0.2935 | 0.2972 | 0.4939 | 0.5010 | 0.4960 | 1.9998 | 0.8985 | 0.8955 | 0.3012 | 0.3001 |
| | (0.0254) | (0.0313) | (0.0138) | (0.0384) | (0.0553) | (0.0652) | (0.0067) | (0.0039) | (0.0125) | (0.0087) |
| T | $\hat{\beta}_1^{[1]}$ | $\hat{\beta}_2^{[1]}$ | $\hat{\beta}_1^{[2]}$ | $\hat{\beta}_2^{[2]}$ | $\hat{\nu}_1=0.5$ | $\hat{\nu}_2=2.0$ | $\hat{\rho}_{11}=0.3$ | $\hat{\rho}_{22}=0.3$ | $\hat{\rho}_{12}=0.9$ | $\hat{\rho}_{21}=0.9$ |
| 60 | 0.3151 | 0.2875 | 0.4812 | 0.4826 | 0.4830 | 1.9852 | 0.2887 | 0.2833 | 0.8885 | 0.8895 |
| | (0.1167) | (0.1052) | (0.0808) | (0.0841) | (0.1319) | (0.1246) | (0.0773) | (0.0787) | (0.0769) | (0.0760) |
| 100 | 0.3029 | 0.3011 | 0.5020 | 0.5067 | 0.4972 | 1.9911 | 0.2915 | 0.2978 | 0.9072 | 0.9067 |
| | (0.0718) | (0.0761) | (0.0639) | (0.0647) | (0.1018) | (0.0961) | (0.0636) | (0.0606) | (0.0582) | (0.0512) |
| 600 | 0.2986 | 0.2993 | 0.4990 | 0.5031 | 0.4988 | 1.9927 | 0.2936 | 0.2990 | 0.9013 | 0.9008 |
| | (0.0310) | (0.0337) | (0.0207) | (0.0241) | (0.0634) | (0.0624) | (0.0069) | (0.0017) | (0.0043) | (0.0122) |

From the above results, the estimates of the different parameters are consistent with the parameter values and with increasing time points, the standard errors of the different estimators decrease, thus satisfying the asymptotic properties. However, a slightly higher standard errors are noted for the over-dispersion coefficients, as compared to the standard errors of the other parameters. In terms of convergence failures, for $\rho_{11} = \rho_{22} = 0.9$, $\rho_{11} = \rho_{22} = 0.3$,

$\nu_1 = \nu_2 = 1$, the number of non-convergent simulations is 325 for T=60, 275 for T=100 and 200 for T=600. For $\nu_1, \nu_2 > 1$, the number of non-convergent simulations slightly increases. For $\rho_{11} = \rho_{22} = 0.9$, $\rho_{11} = \rho_{22} = 0.3$, $\nu_1 = \nu_2 = 2.0$, the number of non-convergent simulations is 350 for T=60, 320 for T=100 and 215 for T=600.

## 13.6 Summary

In this chapter, we have developed a flexible BINMA(1) model with COM-Poisson innovations that enable handling different levels of over-, equi- and under-dispersed data or combination of various types of dispersed data. The different scenarios of dispersion have been illustrated through some trial experiments. The GQL estimation procedure has been constructed to estimate the regression, dispersion and dependence effects. The GQL estimators are consistent and asymptotically normal and these findings have also been supported by the Monte Carlo simulation results.

# 14

# *Conclusion and Future Directions*

## 14.1 Conclusion

This book has dealt with two important issues: Firstly, the development of the constrained and unconstrained bivariate integer-valued time series models of autoregressive and moving average nature of order 1, namely BINAR(1) and BINMA(1), for non-stationary counts under different innovation distributions that incorporate different types of dispersion levels and secondly, the development of the GQL estimation methodology to estimate the unknown parameters of each model.

With respect to the development of models, the constrained BINAR(1) and BINMA(1) models have been constructed under the correlated Poisson, NB and Com-Poisson innovations to cater for different types of dispersion levels of the bivariate counts. In particular, the constrained models with NB and Com-Poisson innovations are appropriate for analyzing bivariate series with different levels of over-dispersion. The constrained BINAR(1) and BINMA(1) models with COM-Poisson innovations can accommodate different types of dispersion: Equi-, over- and under-dispersion and including mixed levels of dispersion where one series may be over-dispersed while the other under-dispersed or vice versa. However, it was remarked that the simulation of the bivariate COM-Poisson innovations could not be easily performed under the bivariate COM-Poisson model of Sellers et al. (2016) and this motivated us to develop an alternative OD algorithm to generate bivariate COM-Poisson innovations under different types of dispersion, which is more versatile and straight forward approach.

As for the development of the unconstrained BINAR(1) and BINMA(1) models with Poisson, NB and COM-Poisson innovation distributions, it was remarked that by assuming the innovation series to follow marginally the above distributions, the marginal distribution of the counting series was quite difficult to specify given the forms of the means and variances of the different counting series, except for the unconstrained BINMA(1) model with Poisson innovations. Interestingly, the derivation of the serial and cross autocovariances under these unconstrained structures could be performed suitably under the multivariate binomial thinning properties.

Due to the complexity of the marginal moments of the counting series under the unconstrained BINAR(1) and BINMA(1) models and the difficulty


DOI: 10.1201/9781003600435-14

in deriving the joint likelihood function for the bivariate models under the non-stationary set-up (Nastic et al. (2016a)), we extended the univariate GQL estimation method to a bivariate GQL estimation method by constructing novel robust auto-covariance structures. Under the constrained BINAR(1) and BINMA(1) models, the development of the GQL equation has been rather easy as we focused on the use of the robust moment autocorrelation estimators from Sutradhar & Das (1999), Jowaheer & Sutradhar (2005), Mallick & Sutradhar (2008) and Mamode Khan (2015) to build the covariance structure. However, under the two parameter NB and COM-Poisson models wherein the GQL was implemented into two phases, the second GQL required some lemmas to facilitate the derivation of some higher-order moments in the covariance expressions.

Attention is drawn here toward the constrained BINAR(1) case, where the performance of GQL with CMLE and FGLS under stationary moment conditions was compared through the Monte Carlo simulation experiments: Firstly, the CMLE implementation under the stationary conditions was computationally very intensive and yielded higher non-convergence failures than the other estimation methods though it provided lesser biased estimates. Secondly, the GQL yielded asymptotically equally efficient estimates as CMLE and with considerably lesser number of failing simulations with respect to convergence. As for the BINMA(1) case, a similar investigation was made by comparing the GQL and FGLS estimation approaches under the stationary moment conditions, where GQL yielded far better estimates than FGLS.

Under the unconstrained BINAR(1) and BINMA(1) models, the auto-covariance structure of the GQL approach was re-constructed based on the multivariate thinning properties. Due to large number of unknown parameters that include the regression and dispersion effects, the serial- and cross-dependence parameters, the re-formulated GQL was also implemented into two phases. Remarkably here, under the unconstrained BINAR(1) structure, the score function under the second GQL was conditioned on the previous observations since it was utmost difficult to compute directly some of the higher-order moments, especially in the case of the unconstrained BINAR(1) model with NB innovations. In fact, these higher-order moments required the formulation and use of some lemmas to facilitate the derivation of the covariance expressions. However, the score function for the unconstrained BINMA(1) model did not require any conditional assumption, but some lemmas were used to derive some higher-order moments in the covariance expressions. Under all set-ups, the simulation results have shown that the GQL method of estimation yielded consistent parameter estimates with inevitably some non-convergent simulations. As expected, the standard errors of the estimates decreased as sample size increased and under the unconstrained situation, the number of non-convergent simulations was slightly higher than under the constrained set-ups, unconstrained model being much more complex as compared to the corresponding constrained models. We also demonstrated that in some cases the constrained models could be derived as a special case of the unconstrained models. However, no simulation study was conducted for the unconstrained BINAR(1) models given the difficulty in identifying the marginal distributions of the counting series.

## 14.2 Future Work

The work covered in this book provides several opportunities for future research. More specifically, the BINAR and BINMA models were developed only for order 1 and hence extensions to higher orders can be done, for example BINAR(p) and BINMA(p). However, given the complications involved in modeling BINAR(1) and BINMA(1) models in this book, one would expect huge amount of complications with the extended models, such that representing the models in an appropriate form itself may be cumbersome. Also, the derivation of moments and estimation of parameters for such extended models may be quite complex and may be computationally very intensive. To illustrate some of the possible models, one can consider the following constrained and unconstrained BINAR(p) and BINMA(p) models:

**Constrained BINAR(p) Model**

The constrained BINAR(p) model can be specified as:

$$\begin{aligned} Y_t^{[1]} &= \rho_{11}^{(1)} \circ Y_{t-1}^{[1]} + \rho_{11}^{(2)} \circ Y_{t-2}^{[1]} + \ldots + \rho_{11}^{(p)} \circ Y_{t-p}^{[1]} + R_t^{[1]} \\ &= \sum_{i=1}^{p} \rho_{11}^{(i)} \circ Y_{t-i}^{[1]} + R_t^{[1]} \end{aligned} \tag{14.1}$$

$$\begin{aligned} Y_t^{[2]} &= \rho_{22}^{(1)} \circ Y_{t-1}^{[2]} + \rho_{22}^{(2)} \circ Y_{t-2}^{[2]} + \ldots + \rho_{22}^{(p)} \circ Y_{t-p}^{[2]} + R_t^{[2]} \\ &= \sum_{i=1}^{p} \rho_{22}^{(i)} \circ Y_{t-i}^{[2]} + R_t^{[2]}, \quad \text{for} \quad t \in \mathbb{Z}. \end{aligned} \tag{14.2}$$

Equations (14.1) and (14.2) may be combined in vector form as follows:

$$\boldsymbol{Y_t} = \boldsymbol{A}^{(i)} \circ \boldsymbol{Y}_{t-1} + \boldsymbol{R}_t$$

$$\begin{bmatrix} Y_t^{[1]} \\ Y_t^{[2]} \end{bmatrix} = \sum_{i=1}^{p} \begin{bmatrix} \rho_{11}^{(i)} & 0 \\ 0 & \rho_{22}^{(i)} \end{bmatrix} \circ \begin{bmatrix} Y_{t-i}^{[1]} \\ Y_{t-i}^{[2]} \end{bmatrix} + \begin{bmatrix} R_t^{[1]} \\ R_t^{[2]} \end{bmatrix} \tag{14.3}$$

where $\boldsymbol{A}^{(i)}$ represent a $(2 \times 2)$ matrix of autocorrelation coefficients with $\{\rho_{kk}^{(i)}\}_{i=1}^{p}$ for $k \in \{1, 2\}$ and the superscript $(i)$ indicates the autocorrelation coefficients at lag-$i$.

**Constrained BINMA(p) Model**

The constrained BINMA(p) model can be specified as:

$$\begin{aligned} Y_t^{[1]} &= \rho_{11}^{(1)} \circ R_{t-1}^{[1]} + \rho_{11}^{(2)} \circ R_{t-2}^{[1]} + \ldots + \rho_{11}^{(p)} \circ R_{t-p}^{[1]} + R_t^{[1]} \\ &= \sum_{i=1}^{p} \rho_{11}^{(i)} \circ R_{t-i}^{[1]} + R_t^{[1]} \end{aligned} \tag{14.4}$$

$$\begin{aligned} Y_t^{[2]} &= \rho_{22}^{(1)} \circ R_{t-1}^{[2]} + \rho_{22}^{(2)} \circ R_{t-2}^{[2]} + \ldots + \rho_{22}^{(p)} \circ R_{t-p}^{[2]} + R_t^{[2]} \\ &= \sum_{i=1}^{p} \rho_{22}^{(i)} \circ R_{t-i}^{[2]} + R_t^{[2]}, \quad \text{for} \quad t \in \mathbb{Z}. \end{aligned} \tag{14.5}$$

The above BINMA(1) model can be re-written in vector form as follows:

$$\begin{aligned} \boldsymbol{Y_t} &= \boldsymbol{A}^{(i)} \circ \boldsymbol{R}_{t-1} + \boldsymbol{R}_t \\ \begin{bmatrix} Y_t^{[1]} \\ Y_t^{[2]} \end{bmatrix} &= \sum_{i=1}^{p} \begin{bmatrix} \rho_{11}^{(i)} & 0 \\ 0 & \rho_{22}^{(i)} \end{bmatrix} \circ \begin{bmatrix} R_{t-i}^{[1]} \\ R_{t-i}^{[2]} \end{bmatrix} + \begin{bmatrix} R_t^{[1]} \\ R_t^{[2]} \end{bmatrix}. \end{aligned} \tag{14.6}$$

**Unconstrained BINAR(p) Model**

The unconstrained BINAR(p) model can be specified as:

$$\begin{aligned} Y_t^{[1]} &= \rho_{11}^{(1)} \circ Y_{t-1}^{[1]} + \rho_{12}^{(1)} \circ Y_{t-1}^{[2]} + \rho_{11}^{(2)} \circ Y_{t-2}^{[1]} \\ &\quad + \rho_{12}^{(2)} \circ Y_{t-2}^{[2]} + \ldots + \rho_{11}^{(p)} \circ Y_{t-p}^{[1]} + \rho_{12}^{(p)} \circ Y_{t-p}^{[2]} + R_t^{[1]} \\ &= \sum_{i=1}^{p} \rho_{11}^{(i)} \circ Y_{t-i}^{[1]} + \rho_{12}^{(i)} \circ Y_{t-i}^{[2]} + R_t^{[1]} \end{aligned} \tag{14.7}$$

$$\begin{aligned} Y_t^{[2]} &= \rho_{21}^{(1)} \circ Y_{t-1}^{[1]} + \rho_{22}^{(1)} \circ Y_{t-1}^{[2]} + \rho_{21}^{(2)} \circ Y_{t-2}^{[1]} \\ &\quad + \rho_{22}^{(2)} \circ Y_{t-2}^{[2]} + \ldots + \rho_{21}^{(p)} \circ Y_{t-p}^{[1]} + \rho_{22}^{(p)} \circ Y_{t-p}^{[2]} + R_t^{[2]} \\ &= \sum_{i=1}^{p} \rho_{21}^{(i)} \circ Y_{t-i}^{[1]} + \rho_{22}^{(i)} \circ Y_{t-i}^{[2]} + R_t^{[2]}, \quad \text{for} \quad t \in \mathbb{Z} \end{aligned} \tag{14.8}$$

or in vector form, one can obtain

$$\begin{aligned} \boldsymbol{Y_t} &= \sum_{i=1}^{p} \boldsymbol{A}^{(i)} \circ \boldsymbol{Y}_{t-1} + \boldsymbol{R}_t \\ \begin{bmatrix} Y_t^{[1]} \\ Y_t^{[2]} \end{bmatrix} &= \sum_{i=1}^{p} \begin{bmatrix} \rho_{11}^{(i)} & \rho_{12}^{(i)} \\ \rho_{21}^{(i)} & \rho_{22}^{(i)} \end{bmatrix} \circ \begin{bmatrix} Y_{t-i}^{[1]} \\ Y_{t-i}^{[2]} \end{bmatrix} + \begin{bmatrix} R_t^{[1]} \\ R_t^{[2]} \end{bmatrix}. \end{aligned} \tag{14.9}$$

**Unconstrained BINMA(p) Model**

The unconstrained BINMA(p) model can be specified as:

$$\begin{aligned} Y_t^{[1]} &= \rho_{11}^{(1)} \circ R_{t-1}^{[1]} + \rho_{12}^{(1)} \circ R_{t-1}^{[2]} + \rho_{11}^{(2)} \circ R_{t-2}^{[1]} \\ &\quad + \rho_{12}^{(2)} \circ R_{t-2}^{[2]} + \ldots + \rho_{11}^{(p)} \circ R_{t-p}^{[1]} + \rho_{12}^{(p)} \circ R_{t-p}^{[2]} + R_t^{[1]} \\ &= \sum_{i=1}^{p} \rho_{11}^{(i)} \circ R_{t-i}^{[1]} + \rho_{12}^{(i)} \circ R_{t-i}^{[2]} + R_t^{[1]} \end{aligned} \tag{14.10}$$

$$\begin{aligned} Y_t^{[2]} &= \rho_{21}^{(1)} \circ R_{t-1}^{[1]} + \rho_{22}^{(1)} \circ R_{t-1}^{[2]} + \rho_{21}^{(2)} \circ R_{t-2}^{[1]} \\ &\quad + \rho_{22}^{(2)} \circ R_{t-2}^{[2]} + \ldots + \rho_{21}^{(p)} \circ R_{t-p}^{[1]} + \rho_{22}^{(p)} \circ R_{t-p}^{[2]} + R_t^{[2]} \\ &= \sum_{i=1}^{p} \rho_{21}^{(i)} \circ R_{t-i}^{[1]} + \rho_{22}^{(i)} \circ R_{t-i}^{[2]} + R_t^{[2]}, \quad \text{for} \quad t \in \mathbb{Z} \end{aligned} \tag{14.11}$$

or in matrix form, it can be expressed as

$$\boldsymbol{Y_t} = \sum_{i=1}^{p} \boldsymbol{A}^{(i)} \circ \boldsymbol{R}_{t-1} + \boldsymbol{R}_t$$

$$\begin{bmatrix} Y_t^{[1]} \\ Y_t^{[2]} \end{bmatrix} = \sum_{i=1}^{p} \begin{bmatrix} \rho_{11}^{(i)} & \rho_{12}^{(i)} \\ \rho_{21}^{(i)} & \rho_{22}^{(i)} \end{bmatrix} \circ \begin{bmatrix} R_{t-i}^{[1]} \\ R_{t-i}^{[2]} \end{bmatrix} + \begin{bmatrix} R_t^{[1]} \\ R_t^{[2]} \end{bmatrix}. \tag{14.12}$$

In addition, other thinning operators such as the NB thinning or signed binomial thinning can be used to develop new models. GQL equations for such models can be formulated as well.

# *Bibliography*


Al-Osh, M. and Alzaid, A. (1987), 'First-order integer-valued autoregressive process', *Journal of Time Series Analysis* **8**, 261–275.

Al-Osh, M. and Alzaid, A. (1988a), 'Integer-valued moving average (INMA) process', *Statistical Papers* **29**, 281–300.

Aly, E. and Bouzar, N. (2005), 'Stationary solutions for integer-valued autoregressive processes', *International Journal of Mathematics and Mathematical Sciences* **2005(1)**, 1–18.

Alzaid, A. and Al-Osh, M. (1990), 'An integer-valued pth order autoregressive structure (INAR(p)) process', *Journal of Applied Probability* **27**, 314–324.

Amemiya, T. (1985), 'Advanced econometrics', *Cambridge: Harvard University Press.*

Best, K. (2001), 'Probability distributions of language entities', *Journal of Quantitative Linguistics* **8**, 1–11.

Boatwright, P., Borle, S., and Kadane, J. (2003), 'A model of the joint distribution of purchase quantity and timing', *Journal of the American Statistical Association* **98**, 564–572.

Borges, P., Bourguignon, M., and Molinares, F. (2017), 'A generalised NG-INAR(1) process with inflated-parameter geometric counting series', *Australian and New Zealand Journal of Statistics* **59(1)**, 137–150.

Borle, S., Boatwright, P., and Kadane, J. (2006), 'The timing of bid placement and extent of multiple bidding: an empirical investigation using eBay online auctions', *Statistical Science* **21**, 194–205.

Borle, S., Boatwright, P., Kadane, J., Nunes, J., and Shmueli, G. (2005), 'Effect of product assortment changes on consumer retention', *Marketing Science* **24**, 616–622.

Borle, S., Dholakia, U., Singh, S., and Westbrook, R. (2007), 'The impact of survey participation on subsequent behavior: an empirical investigation', *Marketing Science* **26**, 711–726.

Boudreault, M. and Charpentier, A. (2011), 'Multivariate integer-valued autoregressive models applied to earthquake counts'. HAL Id: hal-00646848. https://hal.archives-ouvertes.fr/hal-00646848.

Bourguignon, M. and Vasconcellos, K. (2015), 'Improved estimation for Poisson INAR(1) models', *Journal of Statistical Computation and Simulation* **85(12)**, 2425–2441.

Brannas, K. (1995), 'Explanatory variable in the AR(1) count data model', *Department of Economics, Umea University* **381**, 1–21.

Brannas, K. and Hall, A. (2001), 'Estimation integer-valued moving average models', *Applied Stochastic models in Business and Industry* **17**, 277–291.

Brannas, K. and Lonnbark, C. (2006), 'Effects of explanatory variables in count data moving average models', *Department of Economics, Umea University* **679**, 1–6.

Brannas, K. and Quoreshi, A. (2010), 'Integer-valued moving average modelling of the number of transactions in stocks', *Applied Financial Economics* **20(18)**, 1429–1440.

Brijs, T., Karlis, D., and Wets, G. (2008), 'Studying the effect of weather conditions on daily crash counts using a discrete time series model', *Accident Analysis and Prevention* **40(3)**, 1180–1190.

Chan, K. and Ledolter, J. (1995), 'Monte Carlo EM estimation for time series models involving counts', *Journal of the American Statistical Association* **90(429)**, 242–252.

Comte, F. and Lieberman, O. (2003), 'Asymptotic theory for multivariate GARCH processes', *Journal of Time Series Analysis* **84**, 61–84.

Cox, D. (1981), 'Statistical analysis of time series: Some recent developments', *Scandinavian Journal of Statistics* **8(2)**, 93–115.

Crowder, M. (1995), 'On the use of a working correlation matrix in using generalized linear models for repeated measures', *Biometrika* **82**, 407–410.

Durbin, J. and Koopman, S. (2000), 'Time series analysis of non-Gaussian observations based on state space models from both classical and Bayesian perspectives', *Journal of the Royal Statistical Society* **62**, 3–56.

Enciso-Mora, V., Neal, P., and Rao, T. (2009), 'Integer-valued AR processes with explanatory variables', *Sankhya: The Indian Journal of Statistics* **71(2)**, 248–263.

Engle, R. and Bollerslev, T. (1986), 'Modelling the persistence of conditional variances', *Econometric Reviews* **5(1)**, 1–50.

Franke, J. and Rao, T. (1995), Multivariate first-order integer-valued autoregressions, Technical report, University of Kaiserslautern.

FruHwirth-Schnatter, S. and Wagner, H. (2006), 'Auxiliary mixture sampling for parameter-driven models of time series of counts with application to state space modeling', *Biometrica* **93**, 827–841.

Guikema, S. and Coffelt, J. (2009), 'A flexible count data regression model for risk analysis', *Risk Analysis* **79**, 1–9.

Hansen, L. (1982), 'Large sample properties of generalized method of moments estimators', *Econometrika* **50**, 1029–1054.

Jazi, M., Jones, G., and Lai, C. (2012), 'First order integer-valued AR processes with zero inflated Poisson innovations', *Journal of Time Series Analysis* **33**, 954–963.

Joe, H. (1996), 'Time series models with univariate margins in the convolution-closed infinitely divisible class', *Journal of Applied Probability* **33**, 664–677.

Jowaheer, V. and Sutradhar, B. (2002), 'Analysing longitudinal count data with overdipsersion', *Biometrika* **89**, 389–399.

Jowaheer, V. and Sutradhar, B. (2005), 'Fitting lower order non-stationary autocorrelation models to the time series of Poisson counts', *Transactions on Mathematics* **4**, 427–434.

Jowaheer, V. and Sutradhar, B. (2009), 'GMM versus GQL for panel count data analysis', *Statistics and Probability Letters* **79(18)**, 1928–1934.

Jung, R. and Tremayne, A. (2003), 'Testing for serial dependence in time series models of counts', *Journal of Time Series Analysis* **24**(1), 65–84.

Kalyanam, K., Borle, S., and Boatwright, P. (2007), 'Deconstructing each item's category contribution', *Marketing Science* **26**, 327–341.

Karlis, D. and Pedeli, X. (2013), 'Flexible bivariate INAR(1) processes using copulas', *Communications in Statistics-Theory and Methods* **42**, 723–740.

Kim, H. and Lee, S. (2017), 'On first order integer-valued autoregressive process with Katz family innovations', *Statistical Computation and Simulation* **87**, 546–562.

Kim, H. and Park, Y. (2008), 'A non-stationary integer-valued autoregressive model', *Statistical Papers* **49**, 485–502.

Kocherlakota, S. and Kocherlakota, K. (2001), 'Regression in the bivariate Poisson distribution', *Communications in Statistics-Theory and Methods* **30(5)**, 815–825.

Latour, A. (1997), 'The multivariate GINAR(p) process', *Advances in Applied Probability* **29**, 228–248.

Ling, S. and McAleer, M. (2003), 'Asymptotic theory for a new vector ARMA GARCH model', *Economic Theory* **19**, 280–310.

Lord, D., Geedipally, S., and Guikema, S. (2010), 'Extension of the application of Conway-Maxwell-Poisson models: analysing traffic crash data exhibiting underdispersion', *Risk Analysis* **30**, 1268–1276.

Lord, D., Guikema, S., and Geedipally, S. (2008), 'Application of Conway-Maxwell-Poisson for analyzing motor vehicle crashes', *Accident Analysis and Prevention* **40**, 1123–1134.

Lutkepohl, H. (2005), *New introduction to multiple time series analysis*, First edn, Springer-Verlag Berlin Heidelberg.

Mallick, T. and Sutradhar, B. (2008), 'GQL versus conditional GQL inferences for non-stationary time series of counts with overdispersion', *Journal of Time Series Analysis* **29(2)**, 402–420.

Mamode Khan, N. (2015), 'Simulation of MA(1) longitudinal negative binomial counts and using generalized methods of moment equation approach', *Chaotic Modeling and Simulation (CMSIM)* **3**, 235–241.

Mamode Khan, N. (2016), 'A robust algorithm for estimating regression and dispersion parameters in non-stationary longitudinally correlated COM-Poisson', *Journal of Computational Mathematics* **19(1)**, 25–36.

Mamode Khan, N. and Jowaheer, V. (2013), 'Comparing joint GQL estimation and GMM adaptive estimation in COM-Poisson longitudinal regression model', *Communication in Statistics-Simulation and Computation* **42(4)**, 755–770.

Mamodekhan, N. and Jowaheer, V. (2010), 'Analyzing the factors effecting the passenger car breakdowns using COM-Poisson GLM', *International Journal of Mathematical, Computational, Physical, Electrical and Computer Engineering* **4(1)**, 52–54.

Marshall, A. and Olkin, I. (1990), 'Multivariate distribution generated from mixtures of convolution and product families', *Topics in Statistical Dependence, Block, Sampson y Sanits (Eds), Institute of Mathematical Statistics.*

McDonald, D. (2005), 'The local limit theorem: A historical perspective', *Journal of Iranian Statistical Society* **4**, 73–86.

McDonald, L. and Zucchini, W. (1997), *Hidden Markov and other models for discrete-valued time series*, Chapman and Hall, pp. 1–256.

McKenzie, E. (1985), 'Some simple models for discrete variate time series', *Water Resources Bulletin* **21(4)**, 645–650.

McKenzie, E. (1986), 'Autoregressive moving average processes with Negative Binomial and geometric marginal distributions', *Advanced Applied Probability* **18**, 679–705.

McKenzie, E. (1988), 'Some ARMA models for dependent sequences of Poisson counts', *Advances in Applied Probability* **20**, 822–835.

McKenzie, E. (1991), 'Linear characterizations of the Poisson distribution', *Statistics and Probability Letters* **11**, 459–461.

Minka, T., Shmueli, G., Kadane, J., Borle, S., and Boatwright, P. (2003), Computing with the COM-Poisson distribution, Technical report, Carnegie Mellon University.

Nadarajah, S. (2009), 'Useful moment and CDF formulations for the COM-Poisson distribution', *Statistical papers* **50**, 617–622.

Nastic, A., Laketa, P., and Ristic, M. (2016a), 'Random environment integer-valued autoregressive process', *Journal of Time Series Analysis* **37(2)**, 267–287.

Nastic, A., Ristic, M., and Popovic, P. (2016b), 'Estimation in a bivariate integer-valued autoregressive process', *Communication in Statistics-Theory and Methods* **45(19)**, 5660–5678.

Ng, C., Ong, S., and Srivastava, H. (2010), 'A class of bivariate negative binomial distributions with different index parameters in the marginals', *Applied Mathematics and Computation* **217**, 3069–3087.

Pedeli, X. and Karlis, D. (2009), Bivariate INAR(1) models, Technical report, Athens University of Economics.

Pedeli, X. and Karlis, D. (2011), 'A bivariate INAR(1) process with application', *Statistical Modelling: An International Journal* **11**, 325–349.

Pedeli, X. and Karlis, D. (2013a), 'Some properties of multivariate INAR(1) processes', *Computational Statistics and Data Analysis* **67**, 213–225.

Pedeli, X. and Karlis, D. (2013b), 'On estimation of the bivariate Poisson INAR process', *Communications in Statistics-Theory and Methods* **35**, 514–533.

Pedeli, X. and Karlis, D. (2013c), 'On composite likelihood estimation of a multivariate INAR(1) model', *Journal of Time Series Analysis* **34**, 206–220.

Popovic, P. (2016), 'Random coefficient bivariate INAR(1) process', *Facta Universitatis, Series: Mathematics and Informatics* **57**, 731–753.

Popovic, P., Ristic, M., and Nastic, A. (2016), 'A geometric bivariate time series with different marginal parameters', *Statistical Papers* **57**, 731–753.

Qu, A. and Lindsay, B. (2003), 'Building adaptive estimating equations when inverse of covariance estimation is difficult', *Journal of Royal Statistical Society* **65**, 127–142.

Qu, A., Lindsay, B., and Li, B. (2000), 'Improving generalised estimating equations using quadratic inference functions', *Biometrika* **87**, 823–836.

Quoreshi, A. (2006a), 'Bivariate time series modeling of financial count data', *Communication in Statistics-Theory and Methods* **35**, 1343–1358.

Quoreshi, A. (2008), 'A vector integer-valued moving average model for high frequency financial count data', *Economics Letters* **101**, 258–261.

Ravishanker, N., Serhiyenko, V., and Willig, M. (2014), 'Hierarchical dynamic models for multivariate time series of counts', *Statistics and Its Interface* **7**, 559–570.

Ristic, M., Bakouch, H., and Nastic, A. (2009), 'A new geometric first-order integer-valued autoregressive (NGINAR(1)) process', *Journal of Statistical Planning and Inference* **136**, 2218–2226.

Ristic, M., Nastic, A., Jayakumar, K., and Bakouch, H. (2012), 'A bivariate INAR(1) time series model with geometric marginals', *Applied Mathematical Letters* **25(3)**, 481–485.

Schweer, S. and Wei$\beta$, C. (2014), 'Compound Poisson INAR(1) processes: stochastic properties and testing for overdispersion', *Computational Statistics and Data Analysis* **77**, 267–284.

Scotto, M., Wei$\beta$, C., Silva, M., and Pereira, I. (2014), 'Bivariate binomial autoregressive models', *Journal of Multivariate Time Series* **125**, 233–251.

Sellers, K., Morris, D., and Balakrishnan, N. (2016), 'Bivariate Conway-Maxwell-Poisson distribution: Formulation, properties and inference', *Journal of Multivariate Analysis* **150**, 152–168.

Sellers, K., Borle, S., and Shmueli, G. (2012), 'The COM-Poisson model for count data: A survey of methods and applications', *Applied Stochastic Models in Business and Industry,* **22**, 104–116.

Shephard, N. (1995), 'Generalized linear autoregressions', *Economics Papers 8., Economics Group, Nuffield College, University of Oxford.*

Shmueli, G., Minka, T., Borle, J., and Boatwright, P. (2005), 'A useful distribution for fitting discrete data: revival of the Conway-Maxwell-Poisson distribution', *Journal of Royal Statistical Society* **54(1)**, 127–142.

Silva, M. and Oliveira, V. (2004), 'Difference equations for the higher order moments and cumulants of the INAR(1) model', *Journal of Time Series Analysis* **25**, 317–333.

Singh, S., Borle, S., and Jain, D. (2009), 'A generalized framework for estimating customer lifetime value when customer lifetimes are not observed', *Quantitative Marketing and Economics* **7**, 181–205.

Steutel, F. and Van Harn, K. (1979), 'Discrete analogues of self-decomposability and stability', *The Annals of Probability* **7**, 3893–3899.

Steutel, F. and Van Harn, K. (1986), 'Discrete operator self-decomposability and queing networks', *Stochastic Models* **2**, 161–169.

Sutradhar, B. (2003), 'An overview on regression models for discrete longitudinal responses', *Statistical Science* **18**(3), 377–393.

Sutradhar, B. (2008), Best practice recommendation for forecasting counts, Technical report, Department of Mathematics and Statistics, Memorial University of Newfoundland, Canada.

Sutradhar, B. and Das, K. (1999), 'On the efficiency of regression estimators in generalised linear models for longitudinal data', *Biometrika* **86**, 459–465.

Sutradhar, B., Jowaheer, V., and Rao, P. (2014), 'Remarks on asymptotic efficient estimation for regression effects in stationary and non-stationary models for panel count data', *Brazilian Journal of Probability and Statistics* **28(2)**, 241–254.

Wedderburn, R. (1974), 'Quasi-likelihood functions, generalized linear models, and the Gauss-Newton method', *Biometrika* **61**(3), 439–447.

Weiβ, C. (2008b), 'Thinning operations for modelling time series of counts-a survey', *A StA Advances in Statistical Analysis* **92(3)**, 319–341.

Wimmer, G., Kohler, R., Grotjahn, R., and Altmann, G. (1994), 'Toward a theory of word length distribution', *Journal of Quantitative Linguistics* **1**, 98–106.

Yahav, I. and Shmueli, G. (2011), 'On generating multivariate Poisson data in management science applications', *Applied Stochastic Models in Business and Industry* **28(1)**, 91–102.

Zeger, S. (1988), 'A regression model for time series of counts', *Biometrika* **75**, 621–629.

Zeger, S. and Qaqish, B. (1988), 'Markov regression models for time series: a quasi-likelihood approach', *Biometrics* **44(4)**, 1019–1031.

Zhang, H., Wang, D., and Zhu, F. (2009), 'Inference for INAR(p) processes with signed generalized power series thinning operator', *Journal of Statistical Planning and Inference* **140(3)**, 667–683.

Zheng, H. and Basawa, I. (2008), 'First-order observation-driven integer-valued autoregressive processes', *Statistics and Probability Letters* **33(3)**, 335–356.

Zheng, H., Basawa, I., and Datta, S. (2007), 'First-order random coefficient integer-valued autoregressive processes', *Journal of Statistical Planning and Inference* **173(1)**, 212–229.

Zhu, F. (2012), 'Modelling time series of counts with COM-Poisson INGARCH models', *Mathematical and Computer Modelling* **56**, 191–203.

# Index

**Note:** Page numbers in **bold** indicate tables.